Peptide Nano-Chemistry and Nanotechnology

Gang Wei
Editor

Peptide Nano-Chemistry and Nanotechnology

From Molecular Design, Self-Assembly, Biomimetic Synthesis to Applications

Springer

Editor
Gang Wei
Qingdao University
Qingdao, Shandong, China

ISBN 978-981-96-2386-0 ISBN 978-981-96-2387-7 (eBook)
https://doi.org/10.1007/978-981-96-2387-7

This Springer imprint is published by the registered company Springer Nature Singapore Pte Ltd.
The registered company address is: 152 Beach Road, #21-01/04 Gateway East, Singapore 189721, Singapore

About This Book

Peptide with desired structure and functions has been widely used for materials science and nanotechnology due to their tailored sequence, high activity, self-assembly ability, biomimetic activity, and others. To present the progress of peptides in nano-chemistry and nanotechnology, this bok presents 14 chapters on the aspects of (i) synthesis and properties of peptides, (ii) functionalization and hybridization of peptides, and (iii) applications of peptide-based nanomaterials. This book demonstrates firstly the self-assembly of designed peptides to 1D, 2D, and 3D nanostructures. Secondly, this book introduces the hybridization of peptides with polymers, nanoparticles, carbon materials, and 2D materials through specific binding and biomimetic synthesis to create bioactive nanomaterials. Finally, the applications of peptide-based nanomaterials in materials science, nanotechnology, and biomedicine are demonstrated. This book will be helpful for readers to understand the chemical, physical, and biological properties of peptides, and further inspire the design and synthesis of functional peptide nanomaterials for advanced applications.

Contents

1 **Design and Self-Assembly of Peptide Amphiphiles** 1
Xin Luan and Gang Wei

2 **Self-Assembly of Peptide to One-Dimensional Nanomaterials** 21
Danzhu Zhu and Gang Wei

3 **Synthesis and Properties of Peptide Hydrogels** 35
Guozheng Yang and Gang Wei

4 **Design and Self-Assembly of Stimuli-Responsive Peptides** 51
Peng He and Gang Wei

5 **Hybridization of Peptides and Metallic Nanoparticles** 67
Youyin Xu and Gang Wei

6 **Hybridization and Applications of Peptides and Two-Dimensional Materials** 85
Lei Guo

7 **Peptide Nanomaterials for Cancer Therapy** 101
Yi Cao, Pin Wang, Xiaojiao Ge, Xiaolin Duan, Yuanyuan Wei, Li Sun, Aiguo Wu, and Juan Li

8 **Peptide Assembling Nanomaterials for Cancer Therapy** 133
Yi Cao, Xiaojiao Ge, Xiaolin Duan, Li Sun, Aiguo Wu, and Juan Li

9 **Peptide Hydrogels for Drug and Gene Delivery** 155
Zixuan Xiao and Zhiqiang Su

10 **Peptide Nanomaterials for Bioimaging Applications** 169
Ziyu Zhou and Xin Zhou

11 **Peptide Nanomaterials for Energy and Environmental Science Applications** 189
Yan Wang, Keming Wan, Caiyu Liu, and Wenliang Zhang

12 Preparation and Applications of Antimicrobial Peptides 205
Han Wu, Bing Yu, and Hailin Cong

13 Peptide Nanomaterials for Tissue Engineering and Regenerative Medicine 227
Huiqiang Hu, Yanni Sun, Zhengang Sun, and Yongming Xi

14 Peptide-Based Nanomaterials for Sensors and Biosensors 247
Hani Nasser Abdelhamid

Chapter 1
Design and Self-Assembly of Peptide Amphiphiles

Xin Luan and Gang Wei

Abstract Peptide amphiphile (PA) is a small molecule containing both hydrophilic and hydrophobic motifs in the peptide sequence. As it is designed from natural amino acids, it exhibits good biocompatibility, high bioactivity, and high biodegradability. Therefore, PA can be used as a flexible building block for the molecular design and utilization in various ways and has been widely used in various scientific and technical fields. In addition, PA shows the ability to self-assemble into various amyloid aggregates and can self-assemble into a series of functional nanostructures, such as nanofibers, nanotubes, spherical micelles, vesicles, nanosheets, and hydrogels. In this chapter, we focus on the design, synthesis, and self-assembly mechanisms of PAs. The effects of environmental adjusting, such as the interactions, pH, ionic strength, and temperature, on the self-assembly of PAs and the formation of supramolecular nanostructures are presented and discussed. This chapter will provide valuable information and inspirations for the rational design and synthesis of PA-based functional nanomaterials for biomedical and nanotechnological applications.

Keywords Peptide amphiphiles · Self-assembly · Nanomaterials · Biomedicine · Nanotechnology

1.1 Introduction

The self-assembly of biomolecules is a spontaneous behavior [1], in which biomolecules undergo a bottom-up way in various noncovalent (such as hydrogen bonds, electrostatic interactions, hydrophobic interactions, aromatic interactions, and van der Waals forces) and covalent interactions to form ordered nanostructures [2, 3]. Although the energy of the noncovalent driving forces in the self-assembly process is much lower than that of covalent bonds, it can still drive biomolecules to self-assemble into robust nanostructures [4]. Meanwhile, the formation of molecular

X. Luan · G. Wei (✉)
College of Chemistry and Chemical Engineering, Qingdao University, Qingdao 266071, China
e-mail: weigroup@qdu.edu.cn

G. Wei (ed.), *Peptide Nano-Chemistry and Nanotechnology*,
https://doi.org/10.1007/978-981-96-2387-7_1

aggregates that driven by noncovalent interactions was influenced by external conditions, such as the pH, temperature, and ionic strength [5]. When triggered by external or internal factors, the structure of molecular aggregates may collapse or transform [6].

Peptides are short oligomers of natural amino acids that linked by amide bonds [7]. Peptide-based biomolecules have attracted much attention due to their high biocompatibility, biodegradability, adjustable functionality, and design flexibility [8]. Previously, cyclic peptides, amphiphilic peptides, and ion-complementary peptides have been used as bioactive building blocks to construct functional nanostructures with different shapes via peptide self-assembly [9, 10]. Peptide amphiphiles (PAs) are small molecules containing both hydrophilic and hydrophobic moieties that coupled to a peptide bone sequence [11]. Among peptides for constructing functional nanostructures, PAs are becoming important due to their high biological activity and structural advantages [12]. The structural diversity of PAs determines the modes of their self-assembly processes and leads to the bottom-up self-assembly into ordered nanostructures, such as micelles, microemulsions, vesicles, liquid crystals, fibers, nanosheets, and hydrogels, showing potential applications in drug delivery, wound healing, tissue engineering, material science, enzymatic protection, and many others [13–15].

Although PAs showed great potential in the development of materials science and nanotechnology, the aggregation behavior of PAs was significantly affected by the structure of PA molecule [16]. In addition, the self-assembly pathways affected the shape, internal structure, and size of PA-based nanostructures, thereby affecting their biological activity [17]. Therefore, the exploration of the self-assembly mechanisms of PAs and the rational design of PAs with good effectiveness and stability are of great significance for the synthesis of functional peptide nanomaterials [18].

In this chapter, we discuss the knowledges of PAs and bring together recent progress on the self-assembly mechanisms and the design strategies of PAs. In addition, the effects of environmental factors such as the molecular interactions, pH, ionic strength, and temperature on the formation of PA nanostructures are introduced and discussed.

1.2 Classification of PAs

PAs can be divided into two categories, including (i) amphiphilic peptides and (ii) lipidized PAs. In this section, we focus on the structure and self-assembly behavior of two different types of PAs.

1.2.1 Amphiphilic Peptides

Amphiphilic peptides can be divided into two distinct sub-classes, including those containing alternating hydrophilic or hydrophobic amino acid residues, and those consisting of long stretches of hydrophobic amino acids that linked to hydrophilic sequences. The properties and aggregation behavior of these modified amphiphilic peptides show great difference.

Liang and co-workers designed and prepared a pH-responsive amphiphilic peptide (Fig. 1.1a) with the sequence of VVVVVVKKGRGDS [19], in which the hydrophilic head is the KKGRGDS sequence, and the hydrophobic tail composes of the VVVVVV sequence. The V (valine) residue in the hydrophobic tail contained a hydrophobic side chain, which can provide the hydrophobic forces. Therefore, the designed PA was driven by the hydrogen bonds and hydrophobic forces to self-assemble into spherical micelles in neutral or alkaline media and then transformed into nanofibers as the PA concentration increased. Subsequently, the anti-tumor drug, doxorubicin (DOX), was successfully loaded into the hydrophobic core of the self-assembled PA micelles. The lysine (K) residues in the hydrophilic head provided pH-responsiveness, and their ionization in the acidic tumor microenvironment (TME) provided electrostatic repulsion. Therefore, the self-assembled PA nanostructures were destroyed under this force, allowing the dissolution of PA micelles to achieve controlled release of DOX. In another case, Cheng et al. synthesized peptide nanoparticles for dual-targeted cancer therapy by tailoring the sequences of both hydrophilic and hydrophobic domains in a PA [20]. The hydrophilic domain composes of ^{D}PPA-1 (NYSJPTDRQYHF, all $_{D}$-type amino acids), and the hydrophobic domain consists of one lysine, two leucines, and two glycines as linkers to couple functional 3-diethylaminopropyl isothiocyanate (DEAP) molecules and matrix metalloproteinase-2 (MMP-2). The resulting PA and indoleamine 2,3-dioxygenase (NLG919) co-assembled into micellar nanoparticles at pH 7.4. At lower pH, the formed PA nanoparticles collapsed, and NLG919 and ^{D}PPA-1 were precisely released and targeted to tumor cells, inhibiting the growth of melanoma effectively.

Kim et al. used PA to construct a supramolecular nanoplatform for cancer theranostics [21] (Fig. 1.1b). In their study, hydrophobic peptide blocks with a tendency to form β-sheets were first selected, which were then driven to self-assemble into proto-nanofibers. The formed nanofibers were successfully loaded with anticancer drug, DOX. Then the targeted octreotide-labeled PA molecule was selected, and the paramagnetic metallic ion (Gd^{3+}) chelating ligand, 1, 4, 7, 10-tetraazacyclododecane-1, 4, 7, 10 -tetraacetic acid (DOTA), was connected for cancer imaging. The co-assembly of PAs with biomedical functions and drug-carrying fibrils realized the transformation from proto-fibrils to helical fibrils, forming peptide nanofibers (PNFs) that could be targeted to human breast cancer cells (MCF-7) for magnetic resonance (MR) imaging.

Chen et al. designed a short PA molecule with high selectivity and antibacterial properties (Fig. 1.1c) [22]. The general chemical structure of the designed PA was

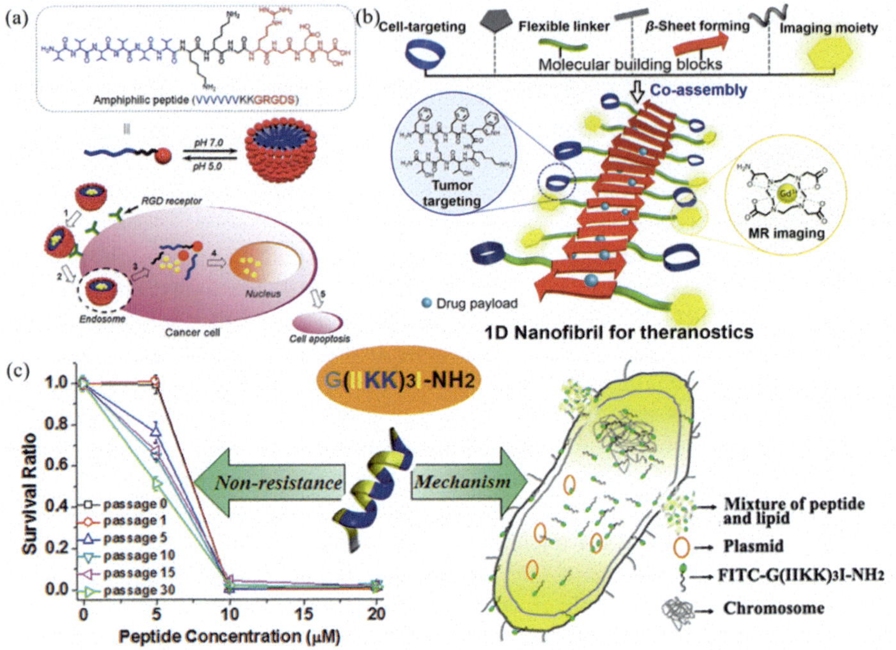

Fig. 1.1 Design and self-assembly of PAs: **a** pH-responsive self-assembly of the VVVVVVKKGRGDS PA. Reproduced with the permission from Ref. [19]. Copyright 2014, Elsevier. **b** 1D supramolecular nanoplatform fabricated by co-assembly of PA cancer therapy. Reproduced with the permission from Ref. [21]. Copyright 2016, American Chemical Society. **c** Antibacterial mechanism of G(IIKK)$_3$I-NH$_2$ PA. Reproduced with the permission from Ref. [22]. Copyright 2014, American Chemical Society

G(IIKK)$_n$I-NH$_2$ ($n = 2 - 4$). By regulating the number of the motif of IIKK, the designed PA revealed the highest cell selectivity and antibacterial toxicity. The short PA was positively charged under physiological conditions and could interact with the negatively charged cell membranes. Therefore, this PA transited from a random coil to an α-helical conformation, killing bacteria by selectively disrupting cell membranes with low toxicity to normal cells. The isoleucine in the peptide sequence provided the driving force for the formation of α-helical conformation, and the N-terminal-linked glycine and C-terminal-linked NH$_2$ improved the stability of the α-helical structure and the resistance to peptidases. The above factors all improved the selectivity of this short peptide with simple structure to the cell membranes of bacteria.

The preparation of functional PA co-assemblies for theragnostic purposes can be achieved by adding peptide building blocks with β-sheet propensity. The nanoscale morphology of PA aggregates in solution can determine the microstructure and properties of hierarchical self-assembled PA membranes. In a typical study, Zha et al. used a cytotoxic peptide to fabricate supramolecular membranes that could be used for cancer therapy [23]. A cytotoxic PA sequence of KLAKLAKKLAKLAK (KLAK PA) was selected, which specifically destroyed the cell membrane of cancer cells at a low concentration and leaded to the death of cells. The KLAK PA contains two types of α-helical sequences that can self-assemble into small spherical aggregates in solution. It was found that the diameter of the created spherical aggregates was adjustable by changing the PA concentration. In addition, the change of PA nanostructures from cylindrical to spherical aggregates was further discussed. Hydrophobic hyaluronic acid (HA) could be combined with cancer cells to accelerate the degradation of PA nanomaterials, and therefore, HA was co-assembled with the cytotoxic PA to form functional PA-HA membrane for drug delivery and cancer therapy.

1.2.2 Lipid PAs

In addition to PAs consisting of only natural amino acids, peptides bound to other components were also investigated. One of the most studied examples is that of PAs using an alkyl group as a hydrophobic tail. In the work of Xiao et al., a mesoporous silica nanoparticle (MSNPs) coated with a self-assembling PA was designed [24]. The hydrophilic amino acid sequence DSDDSDSDSRGDS was coupled with a hydrophobic alkyl side chain to form the lipid PA, C_{18}-DSDSDSDSRGDS. The alkyl chain stearic acid (C_{18}) was linked to the surface of MSNPs through disulfide bonds, as shown in Fig. 1.2a. The side chains formed a dense and compact network structure on the surface of the MSNPs under hydrophobic interaction, which firmly sealed the loaded DOX in the core of the MSNPs. In the presence of dithiothreitol (DTT) or glutathione (GSH), the disulfide bond was broken, and the drug was released successfully to the tumor sites. This study provided a new idea for the design and synthesis of self-assembled PAs as smart drug nano-platforms for cancer therapy.

Similarly, Ruan et al. mixed cell suspensions of bone marrow mesenchymal stem cells (BMSCs) with a PA containing IKVAV and fabricated peptide hydrogels with 3D nanofibrous structures via the peptide self-assembly [25]. The hydrophobic tail of the PA ($C_{16}H_{31}O$-$A_4G_3D_2$IKVAV) is a long alkyl chain composed of 16 carbon atoms, which provides hydrophobic interaction to drive molecular self-assembly. Four alanines were combined with three consecutive glycines to form A_4G_3, and the alkyl group acted synergistically, creating increasingly more hydrophobic residues away from the epitope. Aspartic acid carried negative charge at $pH < 8.5$ and electrostatically interacted with cations in solution. Driven by multiple forces, the designed PA self-assembled into PA hydrogels. The active sites of the IKVAV on the surface of the formed PNFs promoted the cell adhesion and axon outgrowth, thereby promoting successful differentiation of BMSCs into neural cells.

To deliver sufficient drugs into tumor cells, Du and co-workers designed a PA nanoparticle targeting the epidermal growth factor receptor (EGFR) [26]. The hydrophilic head of the PA composed of the EGFR targeting peptide (GE11) and the amino acid sequence of YHWYGYTPQNVI, and an additional Gly and two Glu-containing carboxyl groups were added to the hydrophilic sequence to increase the hydrophilicity of PA, as shown in Fig. 1.2b. C_{18} was attached to the hydrophilic sequence as a hydrophobic tail, which driven by the hydrophobic interaction to

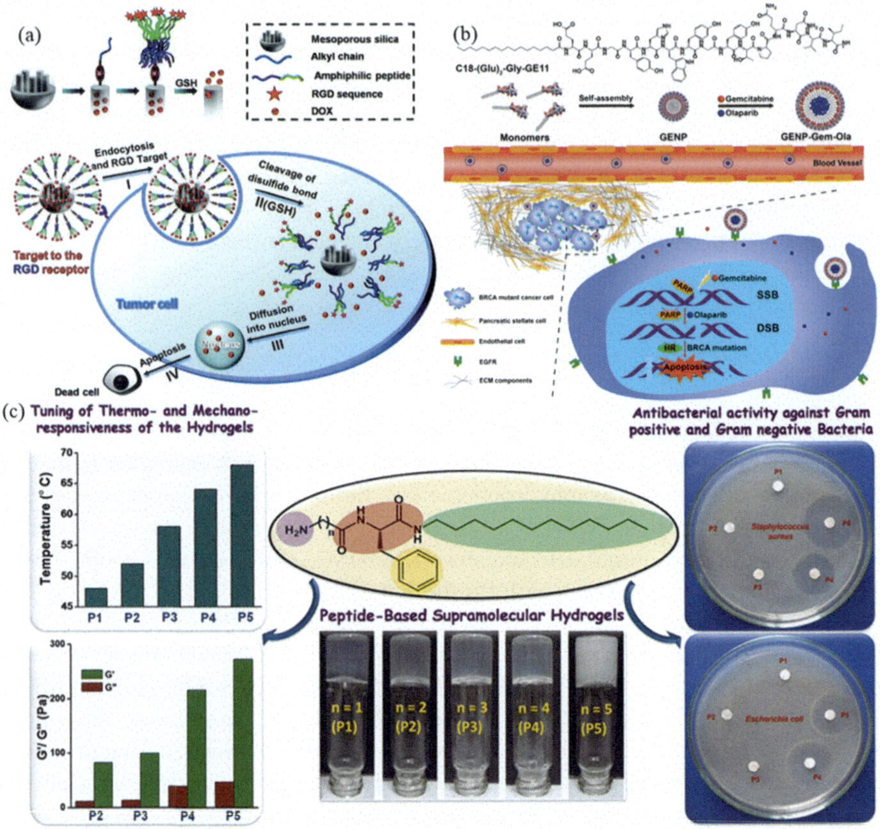

Fig. 1.2 Design and self-assembly of lipid PAs: **a** Structural schematic of RRMSN/DOX and redox-responsive cancer-targeted drug delivery. Reproduced with the permission from Ref. [24]. Copyright 2015, Royal Society of Chemistry. **b** Mechanism of EGFR-targeted self-assembly of PA nanoparticles for cancer therapy. Reproduced with the permission from Ref. [26]. Copyright 2018, American Chemical Society. **c** Molecular structures and self-assembly of amphiphilic peptides to hydrogels. Reproduced with the permission from Ref. [28]. Copyright 2017, American Chemical Society

form peptide nanoparticles. The synthesized PA nanoparticles could co-deliver two antitumor drugs, gemcitabine (Gem) and olaparib (Ola), to achieve high antitumor efficacy with precise targeting the overexpressed EGFR onto pancreatic cancer cells.

Furthermore, PAs with longer alkyl chains were more thermally stable than PAs with shorter alkyl chains [27]. Nandi et al. achieved the regulation of peptide self-assembly and the gel behavior by adjusting the length of the alkyl chain of amino acid residues at the terminal of the PA sequence [28]. In their work, they developed a series of lipid PAs with general chemical formula of [$H_2N–(CH_2)_nCONH–Phe–CONHC_{12}$ ($n = 1 - 5$, C_{12} = dodecylamine)], as shown in Fig. 1.2c. Among them, the C-terminal of the PAs was covalently linked to a long fatty acyl chain, and the N-terminal glycine/ω-amino acid residue was linked to a free amino group. The phenylalanine residues in the side chains provided the $\pi - \pi$ and hydrophobic interactions to drive the PA self-assembly into a thermo-responsive hydrogel with nanofibrous network structure. By changing the number of "*n*" in the general formula of PAs, the thermal stability, recovery time, and regulation of the affinity of the hydrogel to bacterial membrane were realized. When the alkyl chain at the end of the amino acid residue was longer, the self-assembled PA hydrogel exhibited clear antibacterial activity against *S. aureus*, *B. subtilis*, and *E. coli*.

Moyer and colleagues investigated the effect of pH and the sequence of PA on the self-assembled structures and corresponding drug loading efficiency [29]. First, a PA molecule consisting of an oligohistidine (H_6), an aliphatic hydrophobic sequence, and a solubilizing oligo(ethylene glycol) (OEG) group was designed. Tuning the nanostructure was achieved by attaching aliphatic hydrophobic tails to different peptide sequences. In one case, the aliphatic tail palmitic acid was attached to the N-terminus of the PA sequence, the OEG group was attached to the C-terminus, and the amphiphilic molecules of this sequence self-assembled into PNFs at pH 7.5. In another case, the C-terminus of the amino acid was connected to the aliphatic tail lauric acid, the N-terminus was connected to the OEG sequence, and the PA molecules self-assembled into spherical micelles at pH 7.5. In addition, the reversible self-assembly was achieved through adjusting the pH of system. Both self-assembled PA nanostructures were selectively dis-assembled under the influence of the histidine protonation at pH 6.0 to 6.5. Finally, compared with spherical micelles, PNFs exhibited superior antitumor drug encapsulation ability and tumor accumulation. This study demonstrated that the biomedical efficiency of peptide-based nanomaterials can be optimized by regulating the structure and sequence of peptide molecules.

1.3 Strategies for Regulating the Amphiphilicity

The number of amino acids can change the length of amphiphilic peptides, which directly affects the balance between hydrophilic and hydrophobic domains. Ashwanikumar and colleagues investigated the effect of the number of added amino acids on the self-assembled morphology of PA [30]. A single phenylalanine was added consecutively at the N-terminus of the PA (RADARADA). Increasing the number of phenylalanine changed the length of the peptide chain, increased the hydrophobicity, and broken the balance between the hydrophilic and the hydrophobic domains, as shown in Fig. 1.3a. The secondary structure of the cell-penetrating self-assembling peptide nanomaterials (CSPNs) transformed from random coils to twisted α-helices and β-sheets, eventually forming a stable α-helix. Driven by the $\pi - \pi$ interactions, PA molecules containing two, three, or four phenylalanines self-assembled into "nano-drill-like structures", which could effectively encapsulate hydrophobic molecules with a better internalization effect. Based on this study, it is concluded that functional PA nanomaterials with different structures can be designed by adjusting the number of amino acids.

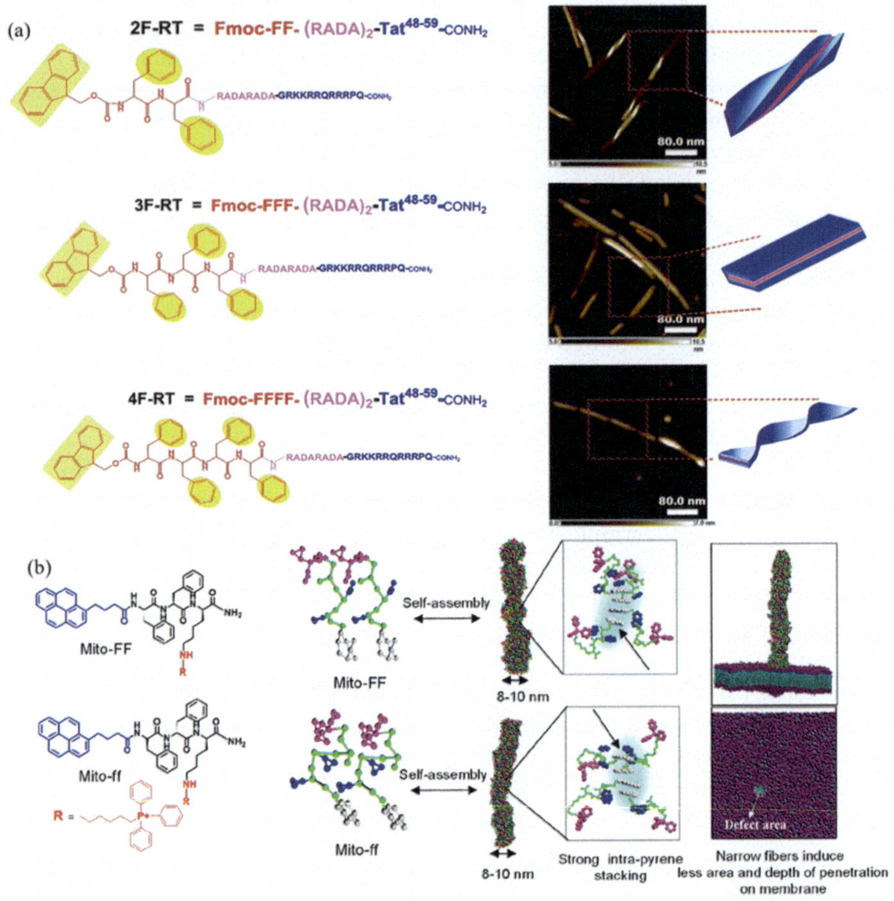

Fig. 1.3 **a** Structure and schematic diagram of CSPNs. Reproduced with the permission from Ref. [30]. Copyright 2018, Elsevier. **b** Chemical structures and self-assembled structure of two mitochondrial penetrating peptides. Reproduced with the permission from Ref. [35]. Copyright 2019, American Chemical Society

In a similar study, Lin et al. designed self-assembled monolayer Janus nanosheets from amphiphilic peptides [31]. The amphiphilic peptide was designed from the hexaphenylalanine, a hydrophobic tail, and glutamic acid and could self-assemble into Janus nanosheets with a monolayer thickness of about 5 nm. The driving forces for the formation of two-dimensional (2D) nanosheets were mainly from the configuration of aromatic amino acids and hydrocarbon chains. Among them, the alkyl chains promoted hydrophobic interactions and enhanced peptide association in the x- and y-axis directions. Meanwhile, phenylalanine promoted the β-sheet stacking in the y-axis direction. When the number of phenylalanines in the peptide chain was reduced, the self-assembled PA nanostructure also changed from 2D nanosheets to distorted 1D fibrils.

In addition, changing the relative positions of amino acids can also affect the self-assembly of PAs. For instance, Zhou et al. introduced a negatively charged glutamic acid residue and asparagine residue into the N-terminal or C-terminal of the peptide PTP-7 (FLGALFKALSKLL), designing two anticancer peptides, EN (ENFLGALFKALSKLL) and NE (FLGALFKALSKLLNE) [32]. The obtained results have shown that the introduction of amino acids at different positions in the peptide chain changed the amphipathic and secondary structure content of PA. The peptide EN with glutamate and aspartate added at the N-terminus tended to form nanofibers. In contrast, the distance between the glutamic acid residue and the two lysine residues in the peptide NE was relatively short, which was conducive to stable intramolecular interactions and thus tended to form micelles. The obtained two PA nanostructures changed the time-action curve, improved the stability of serum, and revealed great clinical application prospects. Therefore, we suggest that the formation of PA nanostructures can also be regulated effectively by changing the position of adding amino acids when designing amphiphilic peptides.

Simultaneously changing the position and number of amino acids can also affect the self-assembly behavior of PAs. For instance, Pashuck et al. explored the effect of the PA sequence on the stiffness of formed PNFs [33]. The peptide composed of three parts, namely hydrophobic tail, ionizable side chains, and fragment of β-sheet, which promoted the formation of nanofibers. By changing the position and quantity of both lysine and alanine in the sequence, the mechanical properties of self-assembled PA structures were analyzed. It was found that the mechanical stiffness of the formed PA hydrogel could be increased by increasing the amount of lysine, while increasing the amount of alanine conversely decreased the mechanical stiffness of the hydrogel. It is ascribed to that lysine has the highest tendency to form the β-sheets, whereas alanine has a weaker tendency to form the β-sheets. In addition, it was found that the position of lysine and alanine in the peptide sequence also had a certain influence on the mechanical properties of the PA hydrogel. The closer the valine to the hydrophobic alkyl tail, the stiffer the resulting hydrogel was.

Changing the type of amino acids has also been considered as a way to control the amphiphilicity of peptides. Xu and co-workers showed that the short peptides of I_3K and L_3K have the ability to self-assemble into stable nanostructures [34]. Due to the slight difference in the structure of leucine and isoleucine, I_3K and L_3K self-assembled into different nanostructures under various driving forces in aqueous solution. Since isoleucine is a β-branched hydrophobic amino acid, it has a strong driving force to form β-sheets, so I_3K mainly exhibited a β-sheet structure in aqueous solution to form nanotubes with a diameter of 10 nm. While L_3K has no obvious β-sheet structure. It was driven by hydrophobic interactions to self-assemble into nanospheres with a diameter of about 10–15 nm in aqueous solution.

Finally, the rational exchange of D, L-chirality has been often used to enhance the stability of self-assembled peptides. Heterochiral co-assembly systems (racemic mixtures) containing both D- and L-isomers generally showed faster assembly kinetics and higher aggregation ability than enantiomers. Jeena et al. developed a heterochiral co-assembling amphiphilic peptide in mitochondria [35], as shown in Fig. 1.3b. In their study, a PA with mitochondrial targeting ability that coupled with triphenylphosphine (TPP) was designed, which contained two phenylalanines, named Mito-FF. Enantiomers could only assemble into nanofibers with a diameter of 9–11 nm. However, Mito-FF and its mirror-image structure (Mito-ff) could co-assemble in mitochondria into dispersed nanofibrils with a diameter of 100 nm and a length of 0.5–1 μm, ascribing to the higher binding energy between phenylalanine groups of opposite chirality.

1.4 Factors for Affecting PA Self-Assembly

1.4.1 Internal Interactions

Sun et al. explored the oligomeric structure and self-assembly mechanism of two short PAs, A_6K and V_6K [36]. The self-assembly behavior of both short peptides was affected by hydrogen bonds, hydrophobic interactions, and electrostatic interactions. The increased hydrophobicity made V_6K to show stronger self-assembly ability than A_6K. In addition, V_6K had a tendency to form compact β-sheet that can self-assemble into plate-like structure. However, due to the lack of hydrophobic interaction, A_6K presented a collapsed random coil structure under hydrogen bond interaction and only self-assembled into a monolayered sheet structure. In another case, Dasgupta et al. found that the hydrophobic interactions induced by the hydrophilic headgroup of PA and the number of hydrophobic groups affected the self-assembly of PA [37]. They explored the self-assembled nanostructure of simple double-chain PA, and the results proved that, in the self-assembly process of double-chain PA, although the hydrophilic head of PA tended to control the nanostructure of PA, the hydrophobic interaction still dominated. It tended to form β-sheet nanofibers firstly and then hydrogel, and the double-chain PA also exhibited cooperative self-assembly and improved aggregation behavior. This interesting self-assembly phenomenon provided useful information and inspiration for the development of functional biomaterials based on PAs.

Intrinsically disordered peptide amphiphiles (IDPAs) consist of an intrinsically ordered peptide bound to a hydrocarbon chain, resulting in an amphiphile with a polymeric head group and a hydrophobic anchor. Ehm et al. designed IDPA systems with four different amino acid sequences and observed the self-assembled nanostructures [38]. It was found that although the chain conformation of the peptide was disordered, the hydrophobic interactions between the head groups of the peptide affected the morphology of the self-assembled nanostructure. Structural phase transitions occurred in the number and length of PAs, which indicated that the mesoscopic structures (micelles, micellar tubes, or condensed phases) formed by the self-assembly of the sequence-designed IPDA and their structural transitions could be applied for the drug delivery systems. Alternatively, a functional peptide could be designed to render IDPA sensitive to enzymatic cleavage and induce an enzymatically activated phase transition, which enabled IDPA to serve as a nanocarrier that responsive to the metabolism of target sites. Similarly, Fu et al. explored the effect of hydrophobic interactions on the self-assembly behavior of PAs [39]. The effect of hydrophobic interactions was mainly adjusted by modifying the alkyl tail or peptide sequence. It was found that enhanced hydrophobic interactions led to the self-assembly of IDPA from open networks of secondary structures to closed cylindrical nanostructures containing β-sheets or random coils. In addition, since the hydrophobicity was affected by the system temperature, further studies revealed that the temperature also played an important role in regulating the formation of secondary structure within these self-assembled nanostructures.

In the process of molecular self-assembly, the anion-π interaction between the electron-deficient aromatic moieties of negatively charged groups is also important for final PA superstructures. For example, Yu et al. designed and synthesized three types of PAs [40], which contained a pentafluoro-phenylalanine at different positions in the main chain. The self-assembly of PAs and fatty acids that driven by the anion-π interaction was explored. When the aromatic unit was positioned close to the N-terminal of the peptide backbone near the hydrophobic core, a very strong anionic-π interaction between two components was generated to form nanofibers. The cylindrical morphology of PA was preserved due to the low content of dodecanoic acid that near the hydrophobic core. However, when the aromatic unit was far away from the hydrophobic core, the interaction with dodecanoic acid continued to increase, and the cylindrical structure was destroyed and transformed into a ribbon structure. Moreover, the adjusting of the ratio of dodecanoic acid to PA caused morphological changes of self-assembled PA nanostructures in this binary system.

1.4.2 Solution pH

Peptide self-assembly can be driven by various noncovalent interactions, including hydrogen bond, hydrophobic and electrostatic interactions, as well as others. These weak interactions are easily affected by environmental perturbations, such as the changes in solution pH, ionic strength, and temperature. In addition, some amino acids can provide functional groups for the metal complexation and enzymatic reactions. These intrinsic properties endow the peptide self-assembly with strong responsiveness to external factors. Therefore, the adjusting of solution conditions has become another important strategy to regulate the peptide self-assembly.

The change of solution pH will profoundly affect the side chains of amino acids through the protonation or deprotonation, resulting in the shift in the self-assembly pathway [41]. Ahmed and colleagues reported the self-assembly of Fmoc-protected lysine-based PAs that containing C_{16} chains to form hydrogels [42]. The hydrophobic part of the PA consisted of hexadecyl chains, which provided hydrophobic interactions for the hydrogelation of PA molecules. The free amine of the lysine side chain remained protonated under neutral conditions, thus acting as a hydrophilic group to increase the solubility of PA molecule in water. Under a suitable hydrophilic-lipophilic balance (HLB), driven by the combination of π-π stacking of Fmoc groups and hydrogen-bonding interactions, the PA molecules showed a tendency to aggregate and eventually self-assembled into helical PNFs efficiently. Interestingly, the chirality of the lysine residues affected the helical properties of the formed nanofibers. While the L-analog formed left-handed helical nanofibers, the D-enantiomer formed opposite right-handed helical nanofibers. However, this gelation behavior showed a strong pH dependence. When the pH of the solvent was >8, the free amine of the lysine side chain did not remain protonated, thus affecting the HLB of the PA molecule and the gelation behavior.

Hamley and co-workers investigated the self-assembly behavior of the surfactant-like peptide R_3L_{12} [43]. In aqueous solution, R_3L_{12} was able to self-assemble into nanostructures like nanotubes, tubular networks, and ordered lattice arrays through adjusting the solution pH. These structures consisted of α-helical antiparallel coil-coil peptide dimers that aligned perpendicular to the nanotube axis in a "crossed" nanotube configuration. The aggregation behavior was ascribed to the effects of the dimensionality and the balance of hydrophobic and electrostatic interactions.

Koc et al. explored supramolecular self-assembly of histidine-functionalized PA [44]. It was found that firstly the change of pH affected the noncovalent interactions of chiral PAs, thereby changing final self-assembled nanostructures. In an acidic environment, PAs self-assembled into thin-layer nanosheets mainly driven by hydrophobic interactions. With the change of pH, PA was mainly driven by hydrogen bond interaction and $\pi - \pi$ stacking of imidazole rings in alkaline environment, realizing the transition from thin-layer nanosheets to closed nanotubes. Meanwhile, the chirality of the amino acids also controlled the PA self-assembly behavior to a certain extent.

Cote et al. studied the effect of pH on the self-assembly behavior of the PA molecules carrying palmitoyl groups [1]. The self-assembly behavior of the peptide sequence $CH_3–(CH_2)_{14}CO–NH–I–A_3–E_4–CO–NH_2$ was studied. Affected by strong electrostatic repulsion, that is, under high pH environment, the side chain of glutamic acid was deprotonated, PAs presented a random coil arrangement and self-assembled into spherical micelles. However, affected by weak electrostatic repulsion, the glutamic acid side chain was fully protonated at low pH or in a neutral environment to form nanofibers. Meanwhile, it was found that the deprotonation of glutamic acid residues affected the pH of the solution. When the pH value was in the range of 6–7, the number of hydrogen bonds increased, and the peptide chain realized the conformation transition from random coil to β-sheet, which further affected the self-assembled structures.

Similarly, Liang et al. developed a pH-responsive PA for controllable drug release [45]. The PA has a sequence of DRQIKIWFQNRRMKWKK. At pH 7.4, the PA molecule exhibited a β-sheet conformation, which self-assembled into nanospheres with uniform size. Lysine and arginine were protonated in solution under mildly acidic conditions, for instance when the pH was 5.0. The secondary structure of the PA molecule was transformed, showing an α-helical conformation, and the self-assembled morphology changed from nanospheres to nanofibers. This reorganization of PA nanostructures under the pH stimulation could effectively realize controllable release of drugs into tumor cells and improve the treatment efficiency, showing new ideas for the design and applications of pH-responsive PAs.

1.4.3 Ionic Strength

The ionic strength is an important external factor that can affect the self-assembly of PAs. Iscen et al. explored how to use the Hofmeister effect to control the structure of PA nanofibers through atomic and molecular dynamics simulations [46]. Their results demonstrated that the self-assembly of PAs into supramolecular nanofibers could be tuned by adjusting the salt solution. The charged side-chain amino acid residues on the nanofibers generated strong electrostatic interactions with the anions in the salt solution, and different anions revealed different effects on the structure of PAs. For example, weakly hydrated ions (I^- and Br^-) would interact more easily with non-polar valine and alanine residues, which formed more stable β-sheet hydrophobic residues near the PNF core. However, strong hydrated ions (F^- and Cl^-) could interact with the side-chain amino groups of lysine to form salt bridges between lysine residues in adjacent PA chains. Therefore, the structure of PA nanofibers could be controlled by using different salt solutions, leading to the development of more stable supramolecular nanofibers.

Besides, the folding of PAs can be regulated by the incorporation with metallic ions. Inspired by controllable morphological changes of natural siderophores that secreted by amphiphilic marine bacteria caused by the metal coordination, Knight et al. designed hybrid peptide-polymer conjugates with different morphologies based on the coordination with metallic ions [47], with oligostyrene as hydrophobic tail and hexahistidine as hydrophilic chelating domain, and the thiol-maleimide reaction was used to couple the peptide and oligostyrene. Finally, the self-assembly of the peptide-polymer conjugate in different metal salt solutions was observed. It was found that in the absence of transition metals, PAs self-assembled into vesicles. However, in the presence of Zn^{2+}, due to the coordination reaction, PAs finally aggregated into particles. This situation has also been observed in the salt solution of Co^{2+} and Cu^{2+}. In addition, metal ion-induced micelles were found by adding Ni^{2+} and Cd^{2+}, and vesicles were formed when the PAs were coordinated with Mn^{2+}. Based on the obtained results, it was concluded that the coordination between transition metals and PAs changed the self-assembled structures of PAs, which were affected by the properties of the added transition metals. This study provided great inspiration on the design and synthesis of stimuli-responsive PA-based nanomaterials.

1.4.4 Temperature

Typically, the temperature change can trigger the secondary structure transitions of temperature-responsive peptides, thereby affecting their self-assembled nanostructures. For instance, Han et al. explored the self-assembly of thermo-responsive PAs [48], in which they chose the thermo-sensitive PA, elastin-like peptide (ELP). The ELP molecule composed of the repeating pentapeptide sequence VPGXG, and X was the guest residue, which could be any amino acid except proline. When the temperature of the solution was increased, the ELP molecules were triggered by the temperature, the hydrophobic valine side chain was dehydrated, and the hydrophobic activity increased, which promoted the self-assembly of ELP into spherical micelles in the solution. Silica protein R5 peptide was introduced at the hydrophilic end of the ELP co-block. The silica underwent local aggregation without affecting the spherical micelle structure to form nearly monodispersed hybrid ELP-silica particles. This temperature-controlled hybrid nanoparticle exhibited potential application in biomedical fields.

In a similar study, Cao et al. further introduced the VPGXG peptide sequence into a PA molecule to study the effect of the temperature on the self-assembly behavior of the PA [49]. Studies have shown that the change of temperature not only affected the self-assembly process, but also had a certain control effect on the self-assembly pathway. PA molecules self-assembled into short and smooth nanofibers at 20 °C, while the self-assembled morphology was affected with increasing temperature, transforming into long fibers with a length of several μm at 80 °C. The surface roughness and diameter of the nanofibers generated at both temperatures were quite different, and the increase of temperature led to the formation of more β-sheets, which caused the peptide chain to undergo a conformational transition. It was deduced that the nanofibers may come from different self-assembly pathways. Interestingly, the amount of the introduced VPGXG peptide sequence also significantly affected the self-assembled morphology. Thus, functional peptide materials could be efficiently constructed by introducing thermo-sensitive sequences into the self-assembling peptides.

1.4.5 Other Factors

The introduction of enzymes to peptide systems also induced the self-assembly or dis-assembly of peptides [50]. For instance, Bai et al. reported the design and self-assembly of PA, Ac-VVVVVVKKK-NH_2 (V_6K_3), which underwent a structural transformation induced by plasma amine oxidase (PAO) [51]. The PA consisted of a hydrophobic domain containing six valines and a hydrophilic domain containing three lysines. Its N-terminus was protected by an acetyl group to prevent the degradation by enzymes, such as dipeptidyl peptidase IV. Under the action of hydrogen bonds, van der Waals force, and hydrophobic interactions, the self-assembly of V_6K_3 peptide formed spherical nanoparticles. Under the induction of PAO, the created spherical nanoparticles were destroyed and transformed into nanofibers, as shown in Fig. 1.4a. The secondary structure of V_6K_3 changed and underwent a geometric transformation based on enzymatic reaction. Interestingly, the lysine residues provided positive charges, while the positive charges carried by the PA V6K3 decreased after the addition of PAO. However, the bactericidal performance of V_6K_3 increased instead of decreasing, indicating that the self-assembled structure was also an important factor affecting antibacterial performance of peptide materials.

In another work by Bai et al., an amphiphilic peptide A_6K_2 was designed, which could undergo structural transformation under the catalysis of lysyl oxidase (LO) and PAO [52]. Positively charged A_6K_2 self-assembled into stable nanovesicles in the HEPES buffer. In the presence of enzymes, the lysine residues in the peptide were oxidized by LO or PAO, and the nanovesicles slowly dis-assembled and gradually formed nanofibers, which triggered the conversion of random coil to β-sheet conformation. Lipid-soluble DOX was then encapsulated into the self-assembled

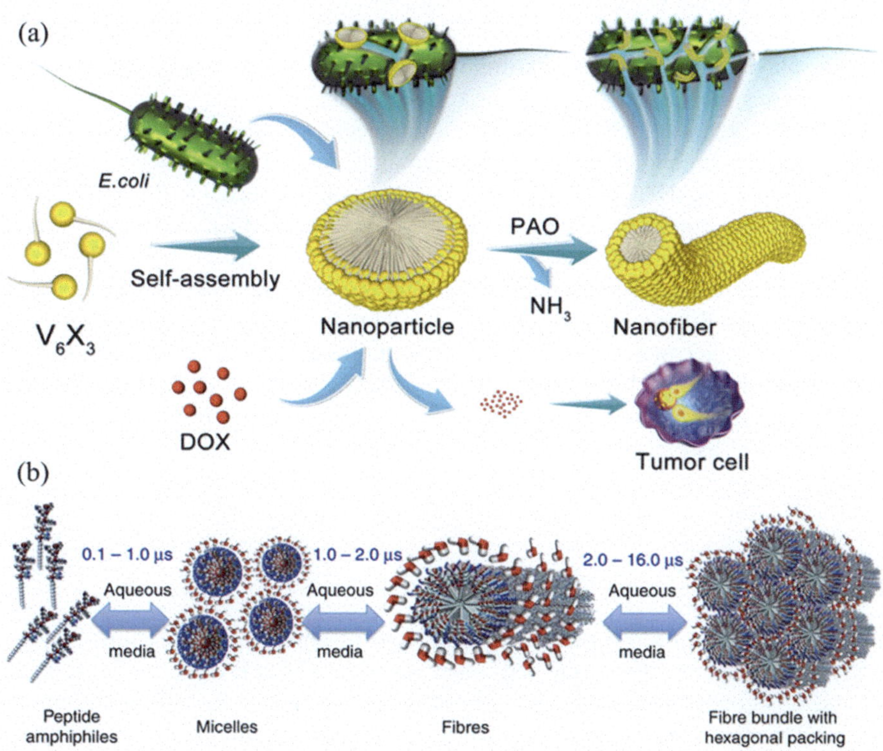

Fig. 1.4 **a** PAO-induced nanoparticle-to-nanofiber geometric transition of V6K3. Reproduced with the permission from Ref. [51]. Copyright 2020, American Chemical Society. **b** Schematic of the various stages involved in PA self-assembly. Reproduced with the permission from Ref. [54]. Copyright 2016, Springer

A_6K_2 nanovesicles as a model drug, and the destruction of the vesicles induced by LO or PAO could promote the drug release.

Nanostructures on different substrates are affected by the local surface concentration and drying method, showing different morphologies. Liao et al. studied the self-assembly behavior of small PA (NapFFKYp) both in solution and on substrates [53]. Studies have shown that PA self-assembled into nanofibers through a nucleation process in solution and further aggregated to form higher-order nanofibers. Subsequently, the self-assembly of PA on mica, highly oriented pyrolytic graphite (HOPG), and polystyrene (PS) films were explored. The local concentrations were different, resulting in the formation of nanosheets with different thicknesses. Meanwhile, the addition of water realized the transformation from nanoflakes to nanofibers. It was clear that the addition of water destroyed the original highly ordered β-sheet layer, and the hydrophilic ends tended to be surrounded by water, eventually forming nanofibers. Finally, molecular dynamics (MD) simulations revealed that the hydrophobicity and ion-ion interactions played the main forces for driving the formation of various nanostructures. The chemical properties and molecular details of water molecules also played a key role in the assembly mechanism and dynamic equilibrium between the phases during the self-assembly of PAs. In addition, the local interfacial ordering of water played a key role in determining the assembly mechanism and dynamic balance among the various stages in the self-assembly process.

Deshmukh et al. explored the role of water molecules in the self-assembly of PAs through multilevel coupled atom-CG MD simulations [54]. The structural formula of the PA was C_{16}–AHL_3K_3–CO_2H, in which C_{16} was connected to the hydrophilic sequence as a hydrophobic tail. Studies have shown that the local ordering of water molecules had a certain influence on early formation of micellar fiber. The order of water molecules near hydrophobic amino acids was broken, and the water cages were broken, which promoted the aggregation of randomly dispersed PAs. The structural order of water molecules near the hydrophilic group was enhanced, driving the transition from micelles to nanofibers, as shown in Fig. 1.4b. During the self-assembly process, water molecules became increasingly ordered, promoting the early hierarchical self-assembly of PAs.

1.5 Conclusion

In summary, we focused on the design and self-assembly of PAs in this chapter. Firstly, some design methods and self-assembly behaviors of PAs were analyzed and discussed. Secondly, we analyzed several strategies to change the amphiphilicity of PAs, such as regulating the number of amino acids in the peptide sequence, changing the relative position and type of amino acids, and changing the chirality of amino acids. Finally, we discussed the influence of environmental factors, such as the interactions, pH, ionic strength, and temperature on the nanostructures of self-assembled PAs.

This chapter provides a reference for designing the sequence of PAs. The design and synthesis of PAs are always performed for specific biomedical and nanotechnological applications. By understanding the preconditions of specific applications and referring to the influences of various factors on the PA self-assembly, PAs with specific structure and function can be designed effectively. We believe that through continuous understanding on the mechanisms of the PA self-assembly, it is possible to develop functional PAs with higher efficacy in the required fields in the future and expand their applications in materials science and nanotechnology.

References

1. Y. Cote, I.W. Fu, E.T. Dobson, J.E. Goldberger, H.D. Nguyen, J.K. Shen, Mechanism of the pH-Controlled self-assembly of nanofibers from peptide amphiphiles. J. Phys. Chem. C **118**, 16272–16278 (2014)
2. A. Manandhar, M. Kang, K. Chakraborty, P.K. Tang, S.M. Loverde, Molecular simulations of peptide amphiphiles. Org. Biomol. Chem. **15**, 7993–8005 (2017)
3. E.R. Draper, D.J. Adams, Low-molecular-weight gels: the state of the art. Chem **3**, 390–410 (2017)
4. A. Dehsorkhi, V. Castelletto, I.W. Hamley, Self-assembling amphiphilic peptides. J. Pept. Sci. **20**, 453–467 (2014)
5. P.A. Korevaar, C.J. Newcomb, E.W. Meijer, S.I. Stupp, Pathway selection in peptide amphiphile assembly. J. Am. Chem. Soc. **136**, 8540–8543 (2014)
6. A.C. Mendes, E.T. Baran, R.L. Reis, H.S. Azevedo, Self-assembly in nature: using the principles of nature to create complex nanobiomaterials. Wires Nanomed. Nanobi. **5**, 582–612 (2013)
7. X. Luan, H. Kong, P. He, G.Z. Yang, D.Z. Zhu, L. Guo, G. Wei, Self-assembled peptide-based nanodrugs: molecular design, synthesis, functionalization, and targeted tumor bioimaging and biotherapy. Small **19**, 2205787 (2022)
8. H.-Q. Song, Y. Fan, Y. Hu, G. Cheng, F.-J. Xu, Polysaccharide-peptide conjugates: a versatile material platform for biomedical applications. Adv. Funct. Mater. **31**, 2005978 (2021)
9. S. Soukasene, D.J. Toft, T.J. Moyer, H.M. Lu, H.K. Lee, S.M. Standley, V.L. Cryns, S.I. Stupp, Antitumor activity of peptide amphiphile nanofiber-encapsulated camptothecin. ACS Nano **5**, 9113–9121 (2011)
10. M. Gontsarik, A. Yaghmur, Q. Ren, K. Maniura-Weber, S. Salentinig, From structure to function: ph-switchable antimicrobial nano-self-assemblies. ACS Appl. Mater. Interfaces **11**, 2821–2829 (2019)
11. M.P. Hendricks, K. Sato, L.C. Palmer, S.I. Stupp, Supramolecular assembly of peptide amphiphiles. Acc. Chem. Res. **50**, 2440–2448 (2017)
12. Y.S. Velichko, S.I. Stupp, M.O. de la Cruz, Molecular simulation study of peptide amphiphile self-assembly. J. Phys. Chem. B **112**, 2326–2334 (2008)
13. A. Dasgupta, D. Das, Designer peptide amphiphiles: self-assembly to applications. Langmuir **35**, 10704–10724 (2019)
14. F. Qiu, Y.Z. Chen, C.K. Tang, X.J. Zhao, Amphiphilic peptides as novel nanomaterials: design, self-assembly and application. Int. J. Nanomed. **13**, 5003–5022 (2018)
15. C. Wang, Z. Wang, X. Zhang, Amphiphilic building blocks for self-assembly: from amphiphiles to supra-amphiphiles. Acc. Chem. Res. **45**, 608–618 (2012)
16. J. Chen, Self-assemble peptide biomaterials and their biomedical applications. Bioact. Mater. **6**, 1790–1790 (2021)
17. H. Cui, M.J. Webber, S.I. Stupp, Self-assembly of peptide amphiphiles: from molecules to nanostructures to biomaterials. Peptide Sci. **94**, 1–18 (2010)

18. W. Zhang, X. Yu, Y. Li, Z. Su, K.D. Jandt, G. Wei, Protein-mimetic peptide nanofibers: motif design, self-assembly synthesis, and sequence-specific biomedical applications. Prog. Polym. Sci. **80**, 94–124 (2018)
19. J. Liang, W.L. Wu, X.D. Xu, R.X. Zhuo, X.Z. Zhang, PH responsive micelle self-assembled from a new amphiphilic peptide as anti-tumor drug carrier. Colloid Surface B **114**, 398–403 (2014)
20. K.M. Cheng, Y.P. Ding, Y. Zhao, S.F. Ye, X. Zhao, Y.L. Zhang, T.J. Ji, H.H. Wu, B. Wang, G.J. Anderson, L. Ren, G.J. Nie, Sequentially responsive therapeutic peptide assembling nanoparticles for dual-targeted cancer immunotherapy. Nano Lett. **18**, 3250–3258 (2018)
21. I. Kim, E.H. Han, J. Ryu, J.-Y. Min, H. Ahn, Y.-H. Chung, E. Lee, One-dimensional supramolecular nanoplatforms for theranostics based on co-assembly of peptide amphiphiles. Biomacromol **17**, 3234–3243 (2016)
22. C. Chen, J. Hu, P. Zeng, Y. Chen, H. Xu, J.R. Lu, High cell selectivity and low-level antibacterial resistance of designed amphiphilic peptide G(IIKK)$_3$I-NH$_2$. ACS Appl. Mater. Interfaces **6**, 16529–16536 (2014)
23. R.H. Zha, S. Sur, S.I. Stupp, Self-assembly of Cytotoxic peptide amphiphiles into supramolecular membranes for cancer therapy. Adv. Healthc. Mater. **2**, 631–631 (2013)
24. D. Xiao, H.Z. Jia, N. Ma, R.X. Zhuo, X.Z. Zhang, A redox-responsive mesoporous silica nanoparticle capped with amphiphilic peptides by self-assembly for cancer targeting drug delivery. Nanoscale **7**, 10071–10077 (2015)
25. H. Ruan, R.S. Xiao, X.H. Jiang, B. Zhao, K. Wu, Z.Z. Shao, Z.J. Zhang, H.Y. Duan, Y.L. Song, Biofunctionalized self-assembly of peptide amphiphile induces the differentiation of bone marrow mesenchymal stem cells into neural cells. Mol. Cell. Biochem. **450**, 199–207 (2019)
26. C. Du, Y.Q. Qi, Y.L. Zhang, Y.Z. Wang, X. Zhao, H. Min, X.X. Han, J.Y. Lang, H. Qin, Q.W. Shi, Z.K. Zhang, X.D. Tian, G.J. Anderson, Y. Zhao, G.J. Nie, Y.M. Yang, Epidermal growth factor receptor-targeting peptide nanoparticles simultaneously deliver gemcitabine and olaparib to treat pancreatic cancer with breast cancer 2 (BRCA2) mutation. ACS Nano **12**, 10785–10796 (2018)
27. I.W. Hamley, Self-assembly of amphiphilic peptides. Soft Matter **7**, 4122–4138 (2011)
28. N. Nandi, K. Gayen, S. Ghosh, D. Bhunia, S. Kirkham, S.K. Sen, S. Ghosh, I.W. Hamley, A. Banerjee, Amphiphilic peptide-based supramolecular, noncytotoxic, stimuli-responsive hydrogels with antibacterial activity. Biomacromol **18**, 3621–3629 (2017)
29. T.J. Moyer, J.A. Finbloom, F. Chen, D.J. Toft, V.L. Cryns, S.I. Stupp, PH and amphiphilic structure direct supramolecular behavior in biofunctional assemblies. J. Am. Chem. Soc. **136**, 14746–14752 (2014)
30. N. Ashwanikumar, J.S. Plaut, B. Mostofian, S. Patel, P. Kwak, C. Sun, K. McPhail, D.M. Zuckerman, S.C. Esener, G. Sahay, Supramolecular self assembly of nanodrill-like structures for intracellular delivery. J. Control. Release **282**, 76–89 (2018)
31. Y.Y. Lin, M.R. Thomas, A. Gelmi, V. Leonardo, E.T. Pashuck, S.A. Maynard, Y. Wang, M.M. Stevens, Self-assembled 2D free-standing janus nanosheets with single-layer thickness. J. Am. Chem. Soc. **139**, 13592–13595 (2017)
32. X.R. Zhou, Y.M. Cao, Q. Zhang, X.B. Tian, H. Dong, L. Chen, S.Z. Luo, Self-assembly nanostructure controlled sustained release, activity and stability of peptide drugs. Int. J. Pharmaceut. **528**, 723–731 (2017)
33. E.T. Pashuck, H. Cui, S.I. Stupp, Tuning supramolecular rigidity of peptide fibers through molecular structure. J. Am. Chem. Soc. **132**, 6041–6046 (2010)
34. H. Xu, Y.M. Wang, X. Ge, S.Y. Han, S.J. Wang, P. Zhou, H.H. Shan, X.B. Zhao, J.A.R. Lu, Twisted nanotubes formed from ultrashort amphiphilic peptide I$_3$K and their templating for the fabrication of silica nanotubes. Chem. Mater. **22**, 5165–5173 (2010)
35. M.T. Jeena, K. Jeong, E.M. Go, Y. Cho, S. Lee, S. Jin, S.W. Hwang, J.H. Jang, C.S. Kang, W.Y. Bang, E. Lee, S.K. Kwak, S. Kim, J.H. Ryu, Heterochiral assembly of amphiphilic peptides inside the mitochondria for supramolecular cancer therapeutics. ACS Nano **13**, 11022–11033 (2019)

36. Y. Sun, Z. Qian, C. Guo, G. Wei, Amphiphilic peptides A6K and V6K display distinct oligomeric structures and self-assembly dynamics: a combined all-atom and coarse-grained simulation study. Biomacromol **16**, 2940–2949 (2015)
37. A. Dasgupta, Exploring architectures at the nanoscale: the interplay between hydrophobic twin lipid chains and head groups of designer peptide amphiphiles in the self-assembly process and application. Soft Matter **12**, 4352–4360 (2016)
38. T. Ehm, H. Shinar, G. Jacoby, S. Meir, G. Koren, M. Segal Asher, J. Korpanty, M.P. Thompson, N.C. Gianneschi, M.M. Kozlov, S. Azoulay-Ginsburg, R.J. Amir, J.O. Rädler, R. Beck, Self-assembly of tunable intrinsically disordered peptide amphiphiles. Biomacromolecules **24**, 98–108 (2022)
39. I.W. Fu, C.B. Markegard, B.K. Chu, H.D. Nguyen, Role of hydrophobicity on self-assembly by peptide amphiphiles via molecular dynamics simulations. Langmuir **30**, 7745–7754 (2014)
40. Z. Yu, A. Erbas, F. Tantakitti, L.C. Palmer, J.A. Jackman, M. Olvera de la Cruz, N.-J. Cho, S.I. Stupp, Co-assembly of peptide amphiphiles and lipids into supramolecular nanostructures driven by anion−π interactions. J. Am. Chem. Soc. **139**, 7823–7830 (2017)
41. Z. Gong, X. Liu, J. Wu, X. Li, Z. Tang, Y. Deng, X. Sun, K. Chen, Z. Gao, J. Bai, PH-triggered morphological change in a self-assembling amphiphilic peptide used as an antitumor drug carrier. Nanotechnology **31**, 165601 (2020)
42. S. Ahmed, J.H. Mondal, N. Behera, D. Das, Self-assembly of peptide-amphiphile forming helical nanofibers and in situ template synthesis of uniform mesoporous single wall silica nanotubes. Langmuir **29**, 14274–14283 (2013)
43. V. Castelletto, J. Seitsonen, J. Ruokolainen, C. Piras, R. Cramer, C.J.C. Edwards-Gayle, I.W. Hamley, Peptide nanotubes self-assembled from leucine-rich alpha helical surfactant-like peptides. Chem. Commun. **56**, 11977–11980 (2020)
44. M. Hatip Koc, G. Cinar Ciftci, S. Baday, V. Castelletto, I.W. Hamley, M.O. Guler, Hierarchical self-assembly of histidine-functionalized peptide amphiphiles into supramolecular chiral nanostructures. Langmuir **33**, 7947–7956 (2017)
45. P.Q. Liang, J.J. Zheng, S.L. Dai, J.Y. Wang, Z.Q. Zhang, T. Kang, C.Y. Quan, PH triggered re-assembly of nanosphere to nanofiber: the role of peptide conformational change for enhanced cancer therapy. J. Control. Release **260**, 22–31 (2017)
46. A. Iscen, G.C. Schatz, Hofmeister Effects on peptide amphiphile nanofiber self-assembly. J. Phys. Chem. B **123**, 7006–7013 (2019)
47. A.S. Knight, J. Larsson, J.M. Ren, R. Bou Zerdan, S. Seguin, R. Vrahas, J. Liu, G. Ren, C.J. Hawker, Control of amphiphile self-assembly via bioinspired metal ion coordination. J. Am. Chem. Soc. **140**, 1409–1414 (2018)
48. W. Han, S.R. MacEwan, A. Chilkoti, G.P. Lopez, Bio-inspired synthesis of hybrid silica nanoparticles templated from elastin-like polypeptide micelles. Nanoscale **7**, 12038–12044 (2015)
49. M.W. Cao, Y. Shen, Y. Wang, X.L. Wang, D.X. Li, Self-assembly of short elastin-like amphiphilic peptides: effects of temperature, molecular hydrophobicity and charge distribution. Molecules **24**, 202 (2019)
50. Z. Feng, X. Han, H. Wang, T. Tang, B. Xu, Enzyme-instructed peptide assemblies selectively inhibit bone tumors. Chem **5**, 2442–2449 (2019)
51. Z.Y. Gong, Y.Y. Shi, H.N. Tan, L. Wang, Z.Q. Gao, B. Lian, G. Wang, H.Y. Sun, P.P. Sun, B.L. Zhou, J.K. Bai, Plasma amine oxidase-induced nanoparticle-to-nanofiber geometric transformation of an amphiphilic peptide for drug encapsulation and enhanced bactericidal activity. ACS Appl. Mater. Interfaces **12**, 4323–4332 (2020)
52. Z.Y. Gong, X.Y. Liu, J.H. Dong, W.F. Zhang, Y.F. Jiang, J.H. Zhang, W.G. Feng, K. Chen, J.K. Bai, Transition from vesicles to nanofibres in the enzymatic self-assemblies of an amphiphilic peptide as an antitumour drug carrier. Nanoscale **11**, 15479–15486 (2019)
53. H.S. Liao, J. Lin, Y. Liu, P. Huang, A. Jin, X.Y. Chen, Self-assembly mechanisms of nanofibers from peptide amphiphiles in solution and on substrate surfaces. Nanoscale **8**, 14814–14820 (2016)

54. S.A. Deshmukh, L.A. Solomon, G. Kamath, H.C. Fry, S.K.R.S. Sankaranarayanan, Water ordering controls the dynamic equilibrium of micelle–fibre formation in self-assembly of peptide amphiphiles. Nat. Commun. **7**, 12367 (2016)

Chapter 2
Self-Assembly of Peptide to One-Dimensional Nanomaterials

Danzhu Zhu and Gang Wei

Abstract Peptide is a type of green and environment-friendly biomacromolecule with both self-assembly and biomimetic functions. It has moderate and controllable self-assembly ability, good biocompatibility, amphoteric performance, and adjustable electrical properties and has attracted widespread attention in the fields of materials science, analytical science, biomedicine, environmental science, and many others. In this chapter, we present the design and synthesis of one-dimensional (1D) peptide nanomaterials, such as nanofibers, nanotubes, nanowires, and nanorods. In addition, the applications of self-assembled 1D peptide nanomaterials in corresponding fields are demonstrated and discussed.

Keywords Peptide · Self-assembly · Nanofibers · Nanotubes · Nanowires · Nanohybrids

2.1 Introduction

In all peptide-based nanostructures, self-assembled peptide nanomaterials can be endowed with special structures and properties through the design and regulation of artificial amino acid sequences, so that they can self-assemble into functional nanomaterials with different dimensions, including zero-dimensional (0D), one-dimensional (1D), two-dimensional (2D), and three-dimensional (3D) structures [1–4]. Peptide nanomaterials with different morphologies have unique functions and can be widely used for biocatalysis, biomedical detection, bioimaging, biosensors, environmental protection, and other fields [5–9].

1D self-assembled peptide nanomaterials include peptide nanofibers (PNFs), nanotubes (PNTs), nanorods (PNRs), and nanowires (PNWs). 1D peptide nanomaterials with ideal size and morphology can be synthesized by designing the peptide sequences and regulating the self-assembly conditions. It can lay a good foundation

D. Zhu · G. Wei (✉)
College of Chemistry and Chemical Engineering, Qingdao University, Qingdao 266071, China
e-mail: weigroup@qdu.edu.cn

G. Wei (ed.), *Peptide Nano-Chemistry and Nanotechnology*,
https://doi.org/10.1007/978-981-96-2387-7_2

for the subsequent synthesis of hybrid materials for drug delivery, antibacterial test, and other biomedical aims [10–13].

2.2 Self-Assembly of Peptide to Nanofibers

Among the numbers of peptide nanomaterials, PNFs have high surface area and volume ratio, which provide more functions and abundant active sites for binding with other molecules and materials. In addition, the nanoscale characteristic of PNFs shows good targeting and retention effects in cells and can be easily captured by cells, which makes PNFs with high potential in controlled drug delivery, tumor targeting, and biosensing [14–16]. Most importantly, PNFs can further form supramolecular hydrogels via nanofiber cross-linking. Therefore, PNFs are considered to be very promising multifunctional biomaterials with outstanding loading capacity, biocompatibility, and biodegradability, showing wide application prospects in materials science, nanotechnology, and biomedicine [17–19].

In the process of the peptide self-assembly (PSA), PNFs can be obtained through the sequence design and simple regulation of the PSA conditions. In the study of Wei and co-workers, a peptide with the sequence of KIIIIKYWYAF was designed, and PNFs with different lengths were synthesized by regulating the PSA [20–22]. The self-assembly of peptide to PNFs was tailored mainly by controlling the incubation solution and the temperature. The synthesized PNFs were then hybridized with other materials to form hybrid materials with specific functions, which showed strong ability in tumor photothermal therapy, colorimetric sensors, electrochemical sensors, and environmental protection.

Meanwhile, the synthesis of PNFs can also be realized by the regulation of the solution pH in the process of the PSA. For instance, Cote et al. described the synthesis of PNFs by amphiphilic peptides under the control of the solution pH [23]. The synthesis of PNFs was achieved by replacing the amino acids in the peptide sequence. In another case, Wychowaniec and co-workers demonstrated the self-assembly of β-sheet PNFs and induced the formation of hydrogels by replacing the flexible phenylalanine residue (F) in the FEFK sequence with rigid phenylglycine (PHG) [24]. The aromatic ring in the resulting peptide PHGEPHGK (PHG4) was conducive to aromatic accumulation, and this motif-designed peptide was conducive to the formation of PNFs. The stable β-sheet PNFs of PHG4 were formed through the π-π stacking of aromatic side chains and hydrogen bonds between main chain amides along the axis of the nanofibers. The formed PNFs exhibited good activity in aqueous medium and incompatible oil phase medium and could further form hydrogels through the intertwining of PNFs. Their results indicated that self-assembled PNFs have great potential for the development of cost-effective, sustainable, and multifunctional biomass hydrogels.

Zou and co-workers demonstrated a facile strategy for the fabrication of a light-triggered platform for the formation of PNFs via PSA, which realized the self-assembly of peptide monomers from nanoparticles to nanofibers in living cells through subtle molecular conformational changes and internal hydrogen bond interaction [25]. The peptide sequence MQIO had photo-responsive effect, and therefore, the light irradiation led to subtle changes in molecular conformation, which significantly and rapidly affected the morphology of self-assembled peptide nanostructure, realizing the transformation from peptide nanoparticles to PNFs. As shown in Fig. 2.1a, the self-assembled PNFs in vitro could not be directly captured into the cell by the cell membrane, while the peptide nanoparticles were well captured by the cell membrane and entered the cell smoothly. After the peptide nanoparticles containing the sequence RGD entered the cell, the conformational change between molecules was triggered by the photo-responsive effect of MQIO. The peptide with the sequence KLVFF self-assembled in the cell through hydrogen bond interaction to form PNFs, which inhibited the cell migration and induced the cell death. The PSA in cells realized the control of PSA in space, did not need chemical reactions, and was harmless to cells, which provided a new way for the application of peptide-based nanomaterials in biomedicine.

In addition to the self-assembly of PNFs through the peptide incubation environment and motif design, the existence of amino, hydroxyl, carboxyl, and aromatic rings make peptides have specific affinities to metallic nanoparticles. Therefore, the self-assembly of peptide monomers could be triggered by the induction of metallic ions [26, 27]. In the work of Wang and colleagues, the intramolecular folding and self-assembly of β-hairpin peptide (CBHH) mediated by platinum ions were realized. As shown in Fig. 2.1b, platinum ions promoted the self-assembly of peptide

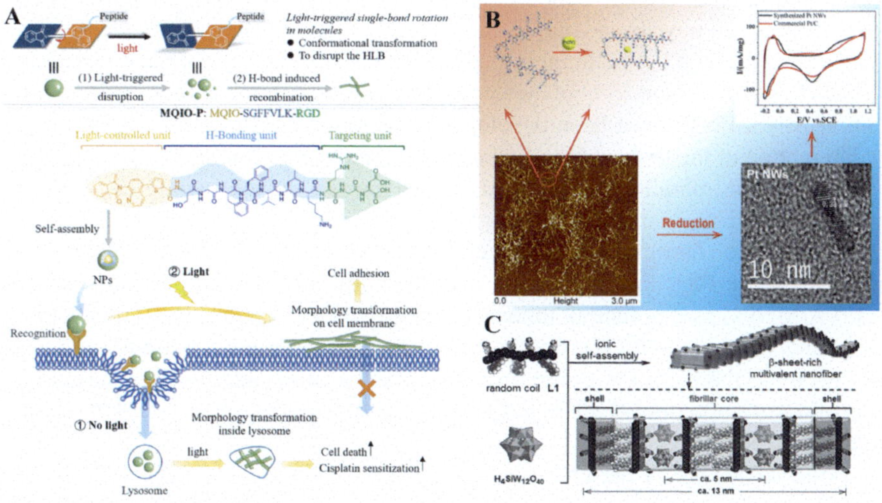

Fig. 2.1 Self-assembly of peptides to PNFs: **a** Design of photo-responsive peptides for the synthesis of PNFs for cancer therapy. Reprinted from Ref. [25] with permission, Copyright 2022 American Chemical Society. **b** Platinum ion-mediated β-hairpin peptide self-assembly into PNFs. Reprinted from Ref. [28] with permission, Copyright 2019 American Chemical Society. **c** pom-driven peptide self-assembly into PNFs. Reprinted from Ref. [29] with permission, Copyright 2016 Wiley VCH

(CBHH) into PNFs by coordinating with the histidine residues. Using the synthesized platinum ion-mediated PNFs as templates, 1D-oriented platinum nanorods were prepared by the in-situ reduction, and the created materials revealed good electrocatalytic activity [28]. In a similar study, Lee et al. used polyoxometalates (POM) as drivers to promote the self-assembly of peptides into PNFs with antibacterial activity [29]. The polyanion properties of POM endowed it with the ability to form complexes with many proteins or peptides containing cationic residues through ionic bonds. POM clusters promoted the self-assembly of short peptides into PNFs with highly concentrated positive charges and good dispersion (Fig. 2.1c). Therefore, POM-driven PNFs showed not only enhanced antibacterial activity, but also improved biological stability.

As a large category of 1D materials, PNFs have broad development prospects in many fields due to their advantages and characteristics. With the in-depth study of the self-assembly processes of PNFs, the development of more peptide-based 1D nanomaterials that beneficial to scientific and technological progress is highly required.

2.3 Self-Assembly of Peptide to Nanotubes

With the in-depth study of the PSA, the synthesis of PNTs has attracted great attention. Nanotubes are formed by the self-assembly of many types of organic molecules. PNTs are a kind of organic nanotubes with a wide range of biological and chemical functions and have broad application prospects in the field of materials science, nanotechnology, and biomedicine [30, 31]. Since Ghadiri *et. al* initiated the design of PNTs based on the stacking of cyclic peptide with even number of D- and L-amino acids in 1993, PNTs have been developed maturely [32]. The main PNTs include the lamination of the β-sheet based on the helix, the lamination of the cyclic peptide dimer of the anti-parallel β-sheet, and the stacking of the alternating D-, L-cyclic peptide [33–35].

In the work of Zarrabi and co-workers, PNTs were created from diphenylalanine peptides via the PSA firstly and then coupled with folic acid/magnetic nanoparticles (FA/MNPs) as a potential anticancer drug delivery system [36]. The coupling compound was obtained by providing covalent bond between the amino group on the PNTs and the carboxyl group on the FA/MNPs. After adding FF solution into 3-hexafluoro-2-propanol (HFIP) and diluting with deionized water, FF PNTs were obtained successfully.

Taking advantage of the ability of FF dipeptides via the self-assembly to form PNTs, Lee and co-workers studied the possibility of coating metal surfaces with self-assembled FF PNTs for releasing flufenamic acid (FnA) [37]. The self-assembly of FF in water and solvent formed nanotubes, which effectively loaded 98% FnA. Meanwhile, mercaptopropionic acid was used to realize stable and orderly arrangement of FF PNTs onto gold surface. The coating of gold surface with FF PNTs made FnA release longer on the surface of gold substrate and had good biocompatibility at the same time. This study provided strong evidence for the potential of self-assembled PNTs as an alternative system for polymer coatings in drug delivery.

In the study of Rho et al., a dual self-assembly method has been developed to stabilize slender peptide nanostructures (Fig. 2.2a) [38]. Cyclic peptides self-assembled through hydrogen bonding to form 1D PNTs. In the process of PSA, a hydrophobic secondary structure was introduced around the peptide as a driving force to stabilize the formed monocyclic PNTs. Due to the formation of this structure, PNTs maintained a single independent morphology. This method could be used to stabilize not only cyclic peptides but also other β-sheet structure, laying a foundation for the applications of supramolecular self-assembly. In another study, artificial transmembrane nanochannels were constructed based on the self-assembly of cyclic peptides into PNTs and the formation of channels in lipid bilayers with intramolecular orientation [39]. As show in Fig. 2.2b, highly hydrophobic cyclic peptide *cyclo*[Gln-(D-Leu-Trp)$_4$-D-Leu] formed PNTs (CPNTs) through intermolecular hydrogen bond, and the formed CPNTs induced the formation of transport channels in the lipid bilayer through intramolecular hollow orientation, which provided the possibility for drug transport across the membrane.

In addition, in the work of Calvelo and colleagues, a kind of transmembrane PNTs with functional lumen was designed [40]. The pH-sensitive functional groups were introduced into the inner cavity of the membrane nanotubes to create a dual adjustable system that was sensitive to the existence of lipid bilayers (controlled by

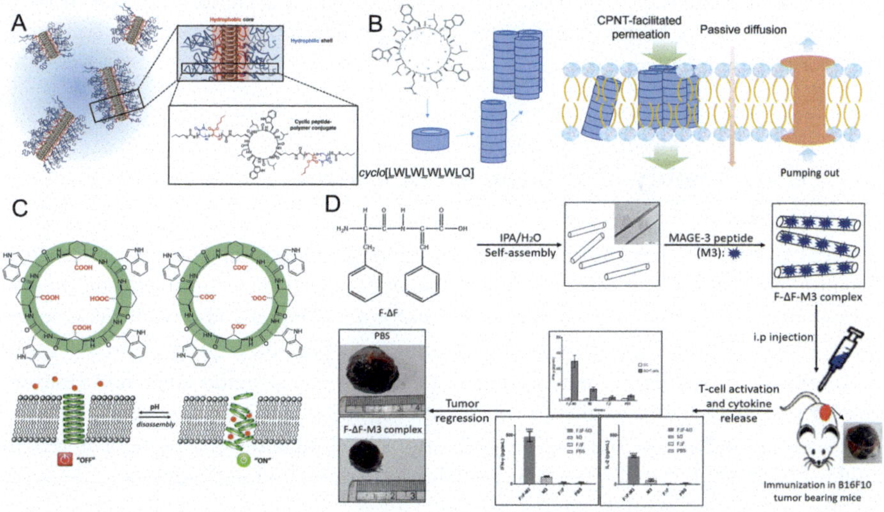

Fig. 2.2 Design and synthesis of PNTs: **a** Self-assembly of cyclic peptide to PNTs. Reprinted from Ref. [38] with permission, Copyright 2019 Springer Nature. **b** Preparation of PNT-based transmembrane nanochannels. Reprinted from Ref. [39] with permission, Copyright 2016 Royal Society of Chemistry. **c** pH-adjustable SCPN assembly/disassembly. Reprinted from Ref. [40] with permission, Copyright 2019 Royal Society of Chemistry. **d** Synthesis of F-ΔF-M3NTs for cancer therapy. Reprinted from Ref. [41] with permission, Copyright 2021 American Chemical Society

the outer surface of the nanotubes and polypeptide side chains) and sensitive to local pH conditions (controlled by internal groups). At low pH value, the protonation of the internal groups (carboxylic acids) of the cyclic peptides made the formed PNTs stable in the lipid environment and permeate to water but not to ions. On the contrary, at higher pH, the deprotonation of these groups seriously changed the properties of the channels, and the structure of the nanotubes was also disintegrated (Fig. 2.2c). This kind of controllable PNTs with pH-responsive properties played important roles in drug delivery and antibacterial materials.

In addition to designing functional peptide sequences through the arrangement and combination of natural amino acids, Verma and colleagues demonstrated that dipeptides containing non-natural amino acid, like α, β-dehydrophenylalanine (ΔF), could self-assemble to form nanotubes, which effectively encapsulated MAGE-3-derived peptides (M3) in the process of self-assembly [41]. As shown in Fig. 2.2d, dipeptide F-ΔF self-assembled to form PNTs and M3 peptides were coated in the prepared PNTs to form F-ΔF-M3 hybrids. It was found that M3 peptides embedded in the F-ΔF exhibited better stability and higher cell uptake. In addition, the formed F-ΔF-M3 injected into mice did effectively inhibit the tumor growth after being ingested by cells. It could be used as a high-performance drug-loaded tumor therapy platform for biomedical application.

As 1D nanomaterials, PNTs have remarkable mechanical properties and a series of potential applications in optoelectronics [42, 43]. The electron transport properties and quantum confinement effects of PNTs provide a better basis for the design of materials with enhanced electronic and optical properties. In addition, the hollow structure of PNTs is highly suitable for coating other substances, which can wrap drugs for targeted drug delivery.

2.4 Self-Assembly of Peptide to Nanowires

Different from PNFs and PNTs, PNWs have longer length and more active sites, showing great advantages in the preparation of electrodes, antibacterial fibers, synthetic templates, micron structures, and biological devices [44–46]. PNWs can be easily obtained through the PSA and the regulation of peptide sequence and assembly environment. PNWs have many active sites, which can provide sufficient binding sites for metals or metal oxides to further synthesize metal nanowires for catalytic, electrochemical, battery, and textile applications [47]. In addition, the micron length enables PNWs to play an important role in the preparation of intelligent nano-devices, crystals, and other functional materials [48].

Men et al. designed self-assembled peptide nanowires (bFPNWs) based on the yeast amyloid protein (Sup35) [49]. As shown in Fig. 2.3a, biotin acceptor peptide (BAP) was modified by spacer peptides on the surface of Sup35 to create the Sup35-BAP, which was then biotinylated by BirA during the in vivo expression, thus realizing the co-transformation of Sup35-BAP and BirA. Through the induction of Sup35-BAP terminal seeds, BirA biotinylated Sup35-BAP self-assembled to

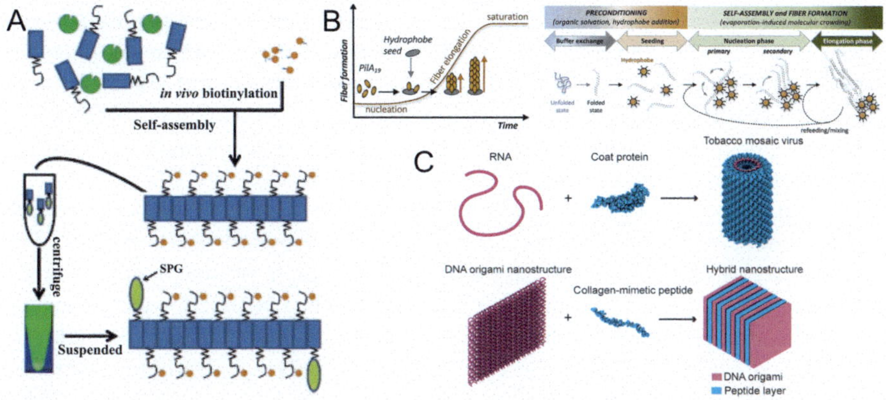

Fig. 2.3 Design and synthesis of PNWs: **a** Design and synthesis of Sup35-SPG PNWs. Reprinted from Ref. [49] with permission, Copyright 2010 Elsevier. **b** BTBT-functionalized PNWs. Reprinted from Ref. [50] with permission, Copyright 2019 American Society for Microbiology. **c** DNA co-assembled with collagen peptides for the formation of hybrid PNWs. Reprinted from Ref. [51] with permission, Copyright 2017 American Chemical Society

form Sup35-SPG PNWs. The self-assembled biotin-integrated bFPNWs realized the detection of commercial diagnostic enzymes and greatly improved the sensitivity of the fabricated immune biosensors.

In the work of Guler and co-workers, highly uniform PNWs with micron length were synthesized via the self-assembly of benzothiophene-benzothiophene (BTBT)-modified peptides in aqueous medium [52]. In the self-assembly process, the peptide modified by BTBT changed the nanostructure of β-sheet and formed micron PNWs that induced by the π-conjugation of BTBT. In the work of Cosert et al., truncated fimbrial protein-based peptides were synthesized by the recombination method, which were then used to fabricate PNWs with the bottom-up strategy [50]. The PilA19 peptide monomer was suspended in organic solvent to stabilize the helical conformation of the peptide, and hydrophobic agent was added as the nucleating seed of fimbrial protein to guide the growth of nanofibers. The controlled evaporation of solvent increased the molecular crowding and promoted the initial nucleation of fimbrial protein on the seed molecule (primary nucleation) and the spontaneous self-assembly of short fibers (secondary nucleation). The addition of hydrophobic molecules to the system provided more nucleating seeds, which improved the length of the fibers and mediated controllable growth of PNWs (Fig. 2.3b).

Large specific surface area can provide rich binding sites for many substances, so the synthesis of other nanowires using PNWs as templates is an environmentally friendly method. For example, in the work of Zhu et al., platinum (Pt) nanowires with ultra-small Pt particles were synthesized using PNWs as templates, which exhibited good stability and high sensitivity in colorimetric and electrochemical dual-mode detection of H_2O_2 [53]. In the work of Wang et al., ferrocene (Fc) was used to modify PNWs to synthesize theFc-PNWs, and then, glucose oxidase (GOx) was immobilized onto the surface of the Fc-PNW [51]. The obtained biological composites showed high sensitivity for the glucose detection [54]. In another study, Jiang and colleagues described the self-assembly of DNA and collagen peptides to form 1D PNWs with a sandwich structure. As shown in Fig. 2.3c, the way of co-assembly of DNA and collagen peptides is the same as that of tobacco mosaic virus. A layer of DNA and a layer of peptides gradually accumulated to form PNWs.

The rich active sites of PNWs provide binding sites for other substances, which is more conducive to the synthesis of multifunctional hybrid nanomaterials. The long structure of PNWs makes it easy to control the length of final materials in the synthesis process. The slender and soft structure can be entangled, so PNWs show wide application prospects for the preparation of functional hydrogels. In addition, the slender structure of PNWs can carry more substances per unit area, which is of great significance in electrochemical detection, biocatalysis, and targeted drug delivery.

2.5 Self-Assembly of Peptide to Nanorods

The shape of PNRs can usually be adjusted by the self-assembly methods via the layer-by-layer stacking, hydrogen bond, and electrostatic interactions [55, 56]. PNRs are different from PNTs and PNFs. PNRs are smaller in size and can enter cells more easily to achieve biomedical applications, such as drug delivery and photothermal therapy. In addition, PNRs with small size and short length are not prone to entanglement, which can greatly reduce the risk of delivery blockage and thrombosis caused by the entanglement compared with PNFs and PNWs.

In the above section, we introduced that FF could easily form PNTs, while in the work of Cong et al., the self-assembly of triphenylalanine FFF into PNRs was reported [57]. Through the simulation and calculation, it was found that FFF was easier to form PNRs, which means that PNRs with different functions could be prepared by designing the peptide sequences with FFF as the core motifs. In the work of Hu et al., it was also found that the self-assembling sequence Fmoc-R $(RCEX)_2$-NH_2 provided the driving force for the self-assembly of the peptide into PNRs [58]. In addition, the self-assembled PNRs could deliver miRNA to cancer cells and then release, which was of great significance in the field of nanomedicine.

Song and colleagues reported a facile strategy to synthesize PNRs as drug delivery carriers [59]. As shown in Fig. 2.4a, the amphiphilic oligopeptide RH3F8 (Arg-His3-Phe8) consists of hydrophilic and hydrophobic ends was designed, in which arginine could penetrate the cell membrane, histidine could induce the escape of endothelial cells and lysosomes, and the hydrophobicity of alanine layed a foundation for the drug entrapment. It was found that RH3F8 self-assembled to form curcumin-embedded PNRs to realize controllable drug delivery, and the formed RH3F8 PNRs with abundant positive charge on their surface were preserved in aqueous phase for up to 35 days with good stability.

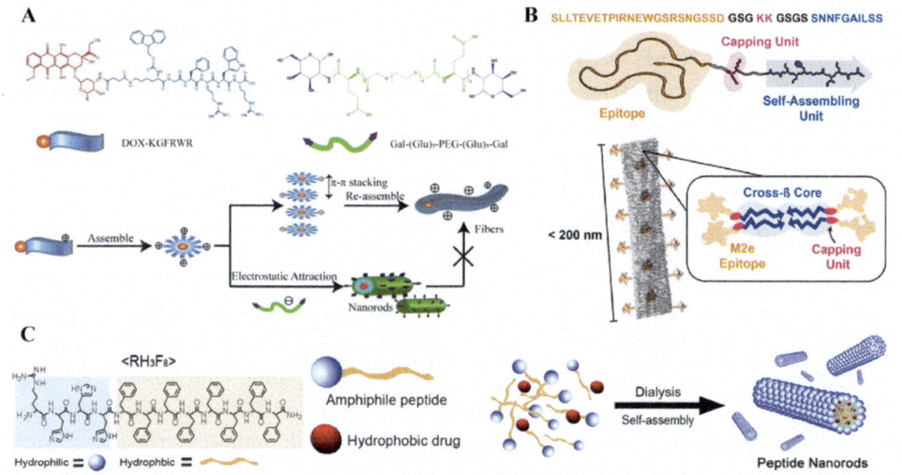

Fig. 2.4 Design and synthesis of PNRs: **a** RH_3F_8 nanorods as drug carriers. Reprinted from Ref. [59] with permission, Copyright 2017 American Chemical Society. **b** Self-assembled M2e-KKI10 PNRs. reprinted from Ref. [60] with permission, Copyright 2021 Elsevier. **c** Self-assembled PDC nanorods. Reprinted from Ref. [61] with permission, Copyright 2016 Wiley

With continuous development biomedicine, the preparation of efficient and safe vaccines using protein and peptide structures has become an important way to resist and prevent influenza, cancer, and other diseases. In the work of Zottig et al., PNRs with uniform length were prepared using the PSA. The formed PNR vaccine delivery platform effectively stabilized the antigen and improved the specific immune response, and revealed a good effect against influenza A H1N1 [60]. As shown in Fig. 2.4b, the designed peptides were connected by the M2e antigen and the self-assembly unit through the packaging unit, in which the package unit KK maintained two units of positive charge to promote the formation of uniform PNRs. In the self-assembly process, PNRs modified by M2e antibody were created without continuous accumulation of two parallel peptides, showing good immune response and affinity.

Peptide-drug conjugate (PDC) has good biocompatibility and is a good platform for biological drug delivery, but usually PDC tends to form fibrous or linear structure under physiological conditions, which greatly increases the risk of the thrombosis caused by the entanglement in the preparation of intravenous injection. Therefore, in the work of Chen and co-workers, the structural transformation from nanofibers to nanorods was realized through controlling the self-assembly of PDC, so as to reach the target position more stably and efficiently [61]. As shown in Fig. 2.4c, the PDC core molecule composed of DOX and peptide KGFRWR. In the process of PDC self-assembly, positively charged micelles were formed by hydrophilic-hydrophobic interaction, and PNRs-embedded DOC was formed by π-π stacking.

The unique rod-like structure and the self-assembly process of PNRs can often achieve drug entrapment and have better encapsulation. Meanwhile, the short rod-like structure of PNRs can better enter the physiological environment, greatly reducing the occurrence of the entanglement of fibers and wire knots. In addition, through the regulation of external environment, the transformation of PNFs to PNRs can be realized. Therefore, better delivery and drug release can be achieved in organisms, which could be a promising research topic in the field of biomedicine.

2.6 Conclusions

In this chapter, we introduced the self-assembly of peptide molecules into 1D nanofibers, nanotubes, nanowires, and nanorods, as well as their applications in biomedicine, drug delivery, sensors, and so on. Compared with other 1D inorganic nanomaterials, 1D peptide nanomaterials exhibited better biocompatibility, designable morphology, and biodegradability. 1D peptide nanomaterials have rich groups on the surface, which can be easily combined with other materials, and the synthesized peptide-based hybrid materials can still ensure good biocompatibility. 1D PNFs and PNWs can provide excellent templates for the synthesis of nanofibers and nanowires of other materials and can be used to synthesize some metal nanofibers/wires or inorganic and organic nanofibers/wires, which greatly reduces the difficulty of synthesis. In addition, 1D PNRs and PNTs are excellent drug-loading/releasing platforms. Through the design of the peptide sequences, peptides can be targeted or coupled with other molecules with specific recognition functions to achieve targeted drug delivery, reducing the damage of full-range drug administration to the healthy physiological tissues. Therefore, it is of great significance to synthesize functional hybrid nanomaterials with specific structure and properties based on self-assembled 1D peptide nanomaterials.

References

1. H. Kong, B. Liu, G.Z. Yang, Y. Chen, G. Wei, Tailoring peptide self-assembly and formation of 2D nanoribbons on mica and HOPG surface. Materials **15**, 310 (2022)
2. N.J. Sinha, M.G. Langenstein, D.J. Pochan, C.J. Kloxin, J.G. Saven, Peptide design and self-assembly into targeted nanostructure and functional materials. Chem. Rev. **121**, 13915–13935 (2021)
3. Y.H. No, N.H. Kim, B. Gnapareddy, B. Choi, Y.-T. Kim, S.R. Dugasani, O.-S. Lee, K.-H. Kim, Y.-S. Ko, S. Lee, S.W. Lee, S.H. Park, K. Eom, Y.H. Kim, Nature-inspired construction of two-dimensionally self-assembled peptide on pristine graphene. J. Phys. Chem. Lett. **8**, 3734–3739 (2017)
4. Q. Song, Z. Cheng, M. Kariuki, S.C.L. Hall, S.K. Hill, J.Y. Rho, S. Perrier, Molecular self-assembly and supramolecular chemistry of cyclic peptides. Chem. Rev. **121**, 13936–13995 (2021)
5. J.J. Han, H.N. Gong, X.K. Ren, X. Yan, Supramolecular nanozymes based on peptide self-assembly for biomimetic catalysis. Nano Today **41**, 101295 (2021)
6. M.J. Webber, C.J. Newcomb, R. Bitton, S.I. Stupp, Switching of self-assembly in a peptide nanostructure with a specific enzyme. Soft Matter **7**, 9665–9672 (2011)

7. Y.M. Liu, E. Naumenko, F. Akhatova, Q. Zou, R. Fakhrullin, X.H. Yan, Self-assembled peptide nanoparticles for enhanced dark-field hyperspectral imaging at the cellular and invertebrate level. Chem. Eng. J. **424**, 130348 (2021)
8. X. Luan, H. Kong, P. He, G.Z. Yang, D.Z. Zhu, L. Guo, G. Wei, Self-assembled peptide-based nanodrugs: molecular design, synthesis, functionalization, and targeted tumor bioimaging and biotherapy. Small **19**, 2205787 (2023)
9. X.D. Xu, B.B. Lin, J. Feng, Y. Wang, S.X. Cheng, X.Z. Zhang, R.X. Zhuo, Biological glucose metabolism regulated peptide self-assembly as a simple visual biosensor for glucose detection. Macromol. Rapid Comm. **33**, 426–431 (2012)
10. P.F. Zou, W.T. Chen, T.Y. Sun, Y.Y. Gao, L. Li, H. Wang, Recent advances: peptides and self-assembled peptide-nanosystems for antimicrobial therapy and diagnosis. Biomater. Sci. **8**, 4975–4996 (2020)
11. Y.X. Chen, W.W. Zhang, C.H. Liang, D.B. Zheng, Y.H. Wang, X.Y. Li, Y. Shi, F.Q. Wang, W.H. Dong, Z.M. Yang, Supramolecular nanofibers with superior anti-angiogenesis and antitumor properties by enzyme-instructed self-assembly (EISA). Chem. Eng. J. **425**, 130531 (2021)
12. Y. Tian, H.V. Zhang, K.L. Kiick, J.G. Saven, D.J. Pochan, Fabrication of one- and two-dimensional gold nanoparticle arrays on computationally designed self-assembled peptide templates. Chem. Mater. **30**, 8510–8520 (2018)
13. W.S. Zhang, X.Q. Yu, Y. Li, Z.Q. Su, K.D. Jandt, G. Wei, Protein-mimetic peptide nanofibers: motif design, self-assembly synthesis, and sequence-specific biomedical applications. Prog. Polym. Sci. **80**, 94–124 (2018)
14. D.B. Cheng, X.H. Zhang, Y.J. Gao, L. Ji, D.Y. Hou, Z.Q. Wang, W.H. Xu, Z.Y. Qiao, H. Wang, Endogenous reactive oxygen species-triggered morphology transformation for enhanced cooperative interaction with mitochondria. J. Am. Chem. Soc. **141**, 7235–7239 (2019)
15. D. Zhang, G.B. Qi, Y.X. Zhao, S.L. Qiao, C. Yang, H. Wang, In situ formation of nanofibers from purpurin18-peptide conjugates and the assembly induced retention effect in tumor sites. Adv. Mater. **27**, 6125–6130 (2015)
16. A. Wagh, J. Singh, S. Qian, B. Law, A short circulating peptide nanofiber as a carrier for tumoral delivery. Nanomedicine **9**, 449–457 (2013)
17. X.Y. Liu, X.B. Sun, G.L. Liang, Peptide-based supramolecular hydrogels for bioimaging applications. Biomater. Sci. **9**, 315–327 (2021)
18. Q. Zhang, Y. Liu, G.Z. Yang, H. Kong, L. Guo, G. Wei, Recent advances in protein hydrogels: from design, structural and functional regulations to healthcare applications. Chem. Eng. J. **451**, 138494 (2023)
19. G.Z. Yang, P. He, D.Z. Zhu, K.M. Wan, H. Kong, X. Luan, L. Fang, Y. Wang, G. Wei, Functional regulation of polymer aerogels by graphene doping and peptide nanofiber-induced biomineralization as sustainable adsorbents of contaminants. Environ. Sci. Nano **9**, 4497–4507 (2022)
20. B. Liu, P. He, H. Kong, D.Z. Zhu, G. Wei, Peptide-induced synthesis of graphene-supported Au/Pt bimetallic nanoparticles for electrochemical biosensor application. Macromol. Mater. Eng. **307**, 2100886 (2022)
21. B. Liu, J.L. Yao, J. Xing, M. Yang, D.Z. Zhu, W.Z. Ren, L.C. Xiang, Y. Wang, A.G. Wu, G. Wei, Design, biomimetic synthesis, and tumor photothermal therapy of peptide-based two-dimensional photothermal conversion nanomaterials. Mol. Syst. Des. Eng. **7**, 1549–1560 (2022)
22. Y. Chen, H. Kong, L. Guo, G. Wei, Biomimetic organic-inorganic hybrid membranes for removal of fluoride ions. Materials **15**, 3457 (2022)
23. Y. Cote, I.W. Fu, E.T. Dobson, J.E. Goldberger, H.D. Nguyen, J.K. Shen, Mechanism of the pH-controlled self-assembly of nanofibers from peptide amphiphiles. J. Phys. Chem. C **118**, 16272–16278 (2014)
24. J.K. Wychowaniec, R. Patel, J. Leach, R. Mathomes, V. Chhabria, Y. Patil-Sen, A. Hidalgo Bastida, R.T. Forbes, J.M. Hayes, M.A. Elsawy, Aromatic stacking facilitated self-assembly of ultrashort ionic complementary peptide sequence: β-sheet nanofibers with remarkable gelation and interfacial properties. Biomacromolecules **21**, 2670–2680 (2020)

25. S. Sun, H.W. Liang, H. Wang, Q.M. Zou, Light-triggered self-assembly of peptide nanoparticles into nanofibers in living cells through molecular conformation changes and H-bond interactions. ACS Nano **16**, 18978–18989 (2022)
26. R.F. Zou, Q. Wang, J.C. Wu, J.X. Wu, C. Schmuck, H. Tian, Peptide self-assembly triggered by metal ions. Chem. Soc. Rev. **44**, 5200–5219 (2015)
27. C. Wallin, J. Jarvet, H. Biverstal, S. Warmlander, J. Danielsson, A. Graslund, A. Abelein, Metal ion coordination delays amyloid-beta peptide self-assembly by forming an aggregation-inert complex. J. Biol. Chem. **295**, 7224–7234 (2020)
28. J. Wang, L. Zhang, J. Yang, H. Yan, X. Li, C. Wang, D. Wang, Y. Sun, H. Xu, Platinum-ion-mediated self-assembly of hairpin peptides and synthesis of platinum nanostructures. Langmuir **35**, 5617–5625 (2019)
29. J.F. Li, Z.J. Chen, M.C. Zhou, J.B. Jing, W. Li, Y. Wang, L.X. Wu, L.Y. Wang, Y.Q. Wang, M. Lee, Polyoxometalate-driven self-assembly of short peptides into multivalent nanofibers with enhanced antibacterial activity. Angew. Chem. Int. Ed. **55**, 2592–2595 (2016)
30. X. Gao, H. Matsui, Peptide-based nanotubes and their applications in bionanotechnology. Adv. Mater. **17**, 2037–2050 (2005)
31. J. Landman, S. Ouhajji, S. Prévost, T. Narayanan, J. Groenewold, A.P. Philipse, W.K. Kegel, A.V. Petukhov, Inward growth by nucleation: Multiscale self-assembly of ordered membranes. Sci. Adv. **4**, eaat1817
32. M.R. Ghadiri, J.R. Granja, R.A. Milligan, D.E. McRee, N. Khazanovich, Self-assembling organic nanotubes based on a cyclic peptide architecture. Nature **366**, 324–327 (1993)
33. C. Valéry, F. Artzner, M. Paternostre, Peptide nanotubes: molecular organisations, self-assembly mechanisms and applications. Soft Matter **7**, 9583–9594 (2011)
34. C. Valéry, M. Paternostre, B. Robert, T. Gulik-Krzywicki, T. Narayanan, J.C. Dedieu, G. Keller, M.L. Torres, R. Cherif Cheikh, P. Calvo, F. Artzner, Biomimetic organization: octapeptide self-assembly into nanotubes of viral capsid-like dimension. Proc. Natl. Acad. Sci. U.S.A. **100**, 10258–10262 (2003)
35. C. Tarabout, S. Roux, F. Gobeaux, N. Fay, E. Pouget, C. Meriadec, M. Ligeti, D. Thomas, M. Ijsselstijn, F. Besselievre, D.-A. Buisson, J.-M. Verbavatz, M. Petitjean, C. Valéry, L. Perrin, B. Rousseau, F. Artzner, M. Paternostre, J.-C. Cintrat, Control of peptide nanotube diameter by chemical modifications of an aromatic residue involved in a single close contact. Proc. Natl. Acad. Sci. U.S.A. **108**, 7679–7684 (2011)
36. G. Emtiazi, T. Zohrabi, L.Y. Lee, N. Habibi, A. Zarrabi, Covalent diphenylalanine peptide nanotube conjugated to folic acid/magnetic nanoparticles for anti-cancer drug delivery. J. Drug. Deliv. Sci. Technol. **41**, 90–98 (2017)
37. T. Zohrabi, N. Habibi, A. Zarrabi, M. Fanaei, L.Y. Lee, Diphenylalanine peptide nanotubes self-assembled on functionalized metal surfaces for potential application in drug-eluting stent. J. Biomed. Mater. Res. Part A **104**, 2280–2290 (2016)
38. J.Y. Rho, H. Cox, E.D.H. Mansfield, S.H. Ellacott, R. Peltier, J.C. Brendel, M. Hartlieb, T.A. Waigh, S. Perrier, Dual self-assembly of supramolecular peptide nanotubes to provide stabilisation in water. Nat. Commun. **10**, 4708 (2019)
39. J. Chen, B. Zhang, F. Xia, Y.C. Xie, S.F. Jiang, R. Su, Y. Lu, W. Wu, Transmembrane delivery of anticancer drugs through self-assembly of cyclic peptide nanotubes. Nanoscale **8**, 7127–7136 (2016)
40. M. Calvelo, J.R. Granja, R. Garcia-Fandino, Competitive double-switched self-assembled cyclic peptide nanotubes: a dual internal and external control. Phys. Chem. Chem. Phys. **21**, 20750–20756 (2019)
41. P. Verma, S. Biswas, N. Yadav, A. Khatri, H. Siddiqui, J.J. Panda, B.S. Rawat, P. Tailor, V.S. Chauhan, Delivery of a cancer-testis antigen-derived peptide using conformationally restricted dipeptide-based self-assembled nanotubes. Mol. Pharmaceut. **18**, 3832–3842 (2021)
42. J.D. Hartgerink, J.R. Granja, R.A. Milligan, M.R. Ghadiri, Self-assembling peptide nanotubes. J. Am. Chem. Soc. **118**, 43–50 (1996)
43. H.S. Kim, J.D. Hartgerink, M.R. Ghadiri, Oriented self-assembly of cyclic peptide nanotubes in lipid membranes. J. Am. Chem. Soc. **120**, 4417–4424 (1998)

44. J. Ryu, C.B. Park, Synthesis of diphenylalanine/polyaniline core/shell conducting nanowires by peptide self-assembly. Angew. Chem. Int. Ed. **48**, 4820–4823 (2009)
45. R.C.G. Creasey, A.B. Mostert, A. Solemanifar, T.A.H. Nguyen, B. Virdis, S. Freguia, B. Laycock, Biomimetic peptide nanowires designed for conductivity. ACS Omega **4**, 1748–1756 (2019)
46. J. Ryu, C.B. Park, High-temperature self-assembly of peptides into vertically well-aligned nanowires by aniline vapor. Adv. Mater. **20**, 3754–3758 (2008)
47. P.P. Li, X. Chen, W.S. Yang, Graphene-induced self-assembly of peptides into macroscopic-scale organized nanowire arrays for electrochemical NADH sensing. Langmuir **29**, 8629–8635 (2013)
48. J. Ryu, S.W. Kim, K. Kang, C.B. Park, Synthesis of diphenylalanine/cobalt oxide hybrid nanowires and their application to energy storage. ACS Nano **4**, 159–164 (2010)
49. D. Men, Z.P. Zhang, Y.C. Guo, D.H. Zhu, L.J. Bi, J.Y. Deng, Z.Q. Cui, H.P. Wei, X.E. Zhang, An auto-biotinylated bifunctional protein nanowire for ultra-sensitive molecular biosensing. Biosens. Bioelectron. **26**, 1137–1141 (2010)
50. K.M. Cosert, A. Castro-Forero, J. Steidl Rebecca, M. Worden Robert, G. Reguera, Bottom-up fabrication of protein nanowires via controlled self-assembly of recombinant geobacter pilins. *mBio* **10**, e02721–02719 (2019)
51. T. Jiang, T.A. Meyer, C. Modlin, X.B. Zuo, V.P. Conticello, Y.G. Ke, Structurally ordered nanowire formation from co-assembly of dna origami and collagen-mimetic peptides. J. Am. Chem. Soc. **139**, 14025–14028 (2017)
52. M.A. Khalily, H. Usta, M. Ozdemir, G. Bakan, F.B. Dikecoglu, C. Edwards-Gayle, J.A. Hutchinson, I.W. Hamley, A. Dana, M.O. Guler, The design and fabrication of supramolecular semiconductor nanowires formed by benzothienobenzothiophene (BTBT)-conjugated peptides. Nanoscale **10**, 9987–9995 (2018)
53. D.Z. Zhu, P. He, H. Kong, G.Z. Yang, X. Luan, G. Wei, Biomimetic graphene-supported ultrafine platinum nanowires for colorimetric and electrochemical detection of hydrogen peroxide. J. Mater. Chem. B **10**, 9216–9225 (2022)
54. J.X. Wang, D. Li, M.H. Yang, Y. Zhang, A novel ferrocene-tagged peptide nanowire for enhanced electrochemical glucose biosensing. Anal. Methods **6**, 7161–7165 (2014)
55. M. von Groning, I. de Feijter, M.C.A. Stuart, I.K. Voets, P. Besenius, Tuning the aqueous self-assembly of multistimuli-responsive polyanionic peptide nanorods. J. Mater. Chem. B **1**, 2008–2012 (2013)
56. L.L. Zhang, P. Zhang, Q.Y. Zhao, Y.C. Zhang, L.Q. Cao, Y.X. Luan, Doxorubicin-loaded polypeptide nanorods based on electrostatic interactions for cancer therapy. J. Colloid. Interf. Sci. **464**, 126–136 (2016)
57. C. Guo, Y. Luo, R.H. Zhou, G.H. Wei, Triphenylalanine peptides self-assemble into nanospheres and nanorods that are different from the nanovesicles and nanotubes formed by diphenylalanine peptides. Nanoscale **6**, 2800–2811 (2014)
58. K. Hu, F. Yin, Z.Y. Zhou, C.S. Lian, Y.H. Liu, C.J. Sun, W.J. Li, J.N. Zhang, Z.G. Li, Directional assembly of a stapled alpha-helical peptide. Chem. Commun. **55**, 10484–10487 (2019)
59. S.J. Song, S. Lee, K.S. Ryu, J.S. Choi, Amphiphilic peptide nanorods based on oligo-phenylalanine as a biocompatible drug carrier. Bioconjugate Chem. **28**, 2266–2276 (2017)
60. X. Zottig, S. Al-Halifa, M. Cote-Cyr, C. Calzas, R. Le Goffic, C. Chevalier, D. Archambault, S. Bourgault, Self-assembled peptide nanorod vaccine confers protection against influenza A virus. Biomaterials **269**, 120672 (2021)
61. R. Chen, J.J. Wang, C. Qian, Y.J. Ji, C.Q. Zhu, L. Wu, W.D. Li, X.L. Bi, Y.T. Wang, G. Cao, Z.P. Chen, From nanofibers to nanorods: nanostructure of peptide-drug conjugates regulated by polypeptide-PEG derivative and enhanced antitumor effect. Adv. Funct. Mater. **29**, 1806058 (2019)

Chapter 3
Synthesis and Properties of Peptide Hydrogels

Guozheng Yang and Gang Wei

Abstract Peptides can form nanofibers of a certain conformation through noncovalent interactions, and these nanofibers are entangled and cross-linked together to form peptide hydrogels with a three-dimensional (3D) porous structure. The controllability and biocompatibility of the peptide self-assembly (PSA) endow peptide hydrogels with a wide range of applications in the fields of biomedicine and tissue engineering. Therefore, the in-depth understanding of the influencing conditions and properties of the PSA to form hydrogels is of great significance for the design and preparation of biomedical nanomaterials. In this chapter, we focus on the factors that induce the PSA to form 3D hydrogels and give a brief overview of the rheological properties and biocompatibility of peptide hydrogels, and look forward to the future development of peptide hydrogels in the field of nanobiotechnology and materials science.

Keywords Peptide · Self-assembly · Nanofibers · Hydrogels · Biomedical applications

3.1 Introduction

Hydrogel is a highly hydrophilic 3D network gel that can exist stably in water and swell by absorbing water [1]. Hydrogel has excellent water absorption and can maintain its original structure after swelling and dehydration. In general, hydrogels can be divided into physically and chemically cross-linked hydrogels [2, 3], in which physically cross-linked hydrogels are formed by physical interactions between molecules, while chemically cross-linked hydrogels are created through chemical reactions. Physically cross-linked hydrogels exhibited weaker mechanical strength than chemically cross-linked ones, but their physical cross-linking characteristics endow them with a more sensitive response to environmental stimuli [4].

There are 20 natural amino acids in nature, which are the building blocks of various peptides and proteins. Each amino acid consists of an alpha carbon atom

G. Yang · G. Wei (✉)
College of Chemistry and Chemical Engineering, Qingdao University, Qingdao 266071, China
e-mail: weigroup@qdu.edu.cn

G. Wei (ed.), *Peptide Nano-Chemistry and Nanotechnology*,
https://doi.org/10.1007/978-981-96-2387-7_3

to which a hydrogen atom, an amino group, a carboxyl group, and a side chain R group are attached. The difference of the R groups leads to the formation of various structures, physicochemical properties, and biological functions of amino acids [5, 6]. An amide bond can be formed by removing a water molecule between two amino acids, thus forming a dipeptide. By analogy, longer peptides can be obtained through the continuous formation of the same amide bond. Peptides can interact each other through noncovalent interactions such as hydrogen bond [7], hydrophobic [8], electrostatic [9], and others to self-assemble, to form zero-dimensional (0D), one-dimensional (1D), two-dimensional (2D) to 3D bionanomaterials with highly ordered structures [10].

In recent years, peptide-based hydrogels have attracted extensive attention in biomedicine and other fields due to their excellent characteristics, including high water-content, microporous structure, tunable mechanical stability, good biocompatibility, and excellent tissue-like elasticity [11]. As a typical biomolecule, peptide hardly needs any cross-linking agent or organic solvent in the process of self-assembly to form hydrogels, so the prepared peptide hydrogels reveal excellent biocompatibility and degradability. In addition, peptide hydrogels with different properties can be obtained by adjusting the conditions of the PSA. For example, it was possible to obtain hydrogel materials with higher stiffness or viscoelasticity by adjusting external stimuli, which could be used as biomedical scaffolds or drug-loading platforms. Some peptide hydrogels exhibited good shear-thinning properties, which enabled them to achieve rapid sol–gel transition at high shear rates and could be used as injectable materials for biomedicine and tissue engineering [12, 13].

In this chapter, we focus on the stimulating factors for inducing the formation of peptide hydrogels and the unique characteristics of peptide hydrogels with various applications.

3.2 Synthesis of Peptide Hydrogels

Peptide can self-assemble into fibrous hydrogels driven by a variety of noncovalent forces. The formation of nanofibers inside the hydrogel and the regulation of the self-assembly of peptide hydrogels can be induced by various means. For example, the PSA can be affected by some stimuli, such as the pH and ionic strength of the peptide solution [14], and the formation of peptide hydrogels can also be promoted by physical stimuli such as light or temperature-induced molecular aggregation [15]. In addition, biomolecules such as enzymes can also affect the self-assembly process of peptides. Currently, various methods based on the gelation for the preparation of stimuli-responsive peptide hydrogels have been developed.

3.2.1 Chemical Stimuli-Induced Formation of Peptide Hydrogels

Due to the difference in the side chain amino groups, amino acids have different isoelectric points. Therefore, under various pH environments, the charge of amino acids will also be affected. Therefore, rationally designed peptide amphiphiles (PAs) can promote the formation of peptide hydrogels by adjusting the solution pH.

Rajagopal et al. proposed a strategy to control the heat-triggered folding, self-assembly, and subsequent hydrogelation of amphipathic β-hairpin peptides in a pH-dependent manner [16]. A β-hairpin structure refers to a β-strand composed of alternating valine and lysine residues. By performing the point substitutions on the lysine residues of the self-assembling peptide MAX1 to change its electrostatic charge, the self-assembly behavior of the peptide at different pH could be tuned. Peptides with lower net positive charges were able to form β-fold conformation and assemble into hydrogels at lower pH. Chen et al. designed a pH-responsive PA system. Represented by alternating arginine and aspartic acid in the surface region, this PA sequence exhibited an alternating pattern of positive and negative charges on the surface of PNFs at the pH values below the pKa of arginine [17]. Therefore, such nanofibers could link together with adjacent interlaced PA molecules through strong interactions, thereby promoting their lateral growth.

Kaur et al. studied the self-assembly behavior of classical non-gelling agents with the ion-complementary peptide sequence FEFK [18]. In PAs, the amino acids with opposite charges can form hydrogels over a wide pH range (2.0–12.0). However, there are obvious differences for the self-assembled nanostructures that formed under different pH environments (Fig. 3.1a). For example, at more extreme pH environments, the gelling agent formed finer nanofibers. In the milder physiological pH range, the nanofibers between the positive and negative charges promoted the lateral association of PNFs, thereby forming thicker fibers. It can be concluded that the change of pH affects the size and positive or negative of the overall charge of the peptide, resulting in different peptide nanostructures, which provides an option for the manufacture of functional nanostructures.

In addition, the adjusting of the ionic strength in the peptide system is also a common means to promote the PSA and the formation of peptide hydrogels. The stiffness and strength of peptide hydrogels can be well improved by adding salt solution to the solvent system to enhance the electrostatic interactions of the whole system. In the report of Zhou et al., the self-assembly of pH and calcium ion dual-responsive PA hydrogels was studied [19]. PAs derived from the core sequence GGXGXDXUX of repeat (RTX) toxins featured dual-responsive self-assembly. In a lower pH environment, after the addition of $CaCl_2$, PA rapidly self-assembled into a β-sheet conformation, resulting in the formation of a dense PNF network. During the self-assembly process, the presence of Ca^{2+} effectively shielded the negative charge that generated by aspartic acid, thereby promoting the self-assembly and hydrogelation of PA.

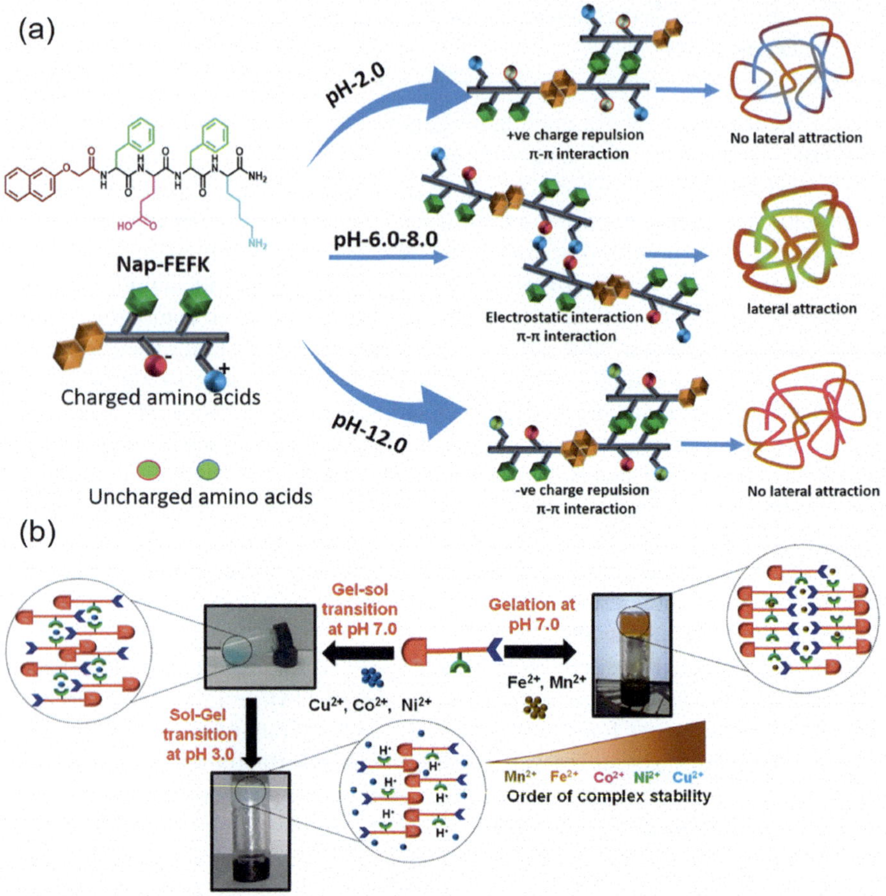

Fig. 3.1 **a** Chemical structure of ion-complementary peptide sequence Nap-FEFK and its self-assembly behavior at different pH. Reproduced with the permission from Ref. [18]. Copyright 2020, American Chemical Society. **b** Changes in the self-assembly behavior of histidine-based PAs in the presence of transition metal salts. Reproduced with the permission from Ref. [21]. Copyright 2019, American Chemical Society

Ozbas et al. reported the effect of salt solution on the self-assembly of MAX1 to form a peptide hydrogel [20]. In a physiological pH environment, due to the primary amino group on the lysine collection, MAX1 revealed a static positive charge at pH 7.4, and the peptide was hard to form the β-sheet conformation due to the electrostatic repulsion that generated between the chains. When a salt solution was added, the positive charge on the lysine residue was shielded by the anion in the solution, so that peptide molecule underwent intramolecular folding to form a β-hairpin structure. Alternatively, an increase in the concentration of the salt solution might also drive the hydrophobic association of residues, thereby promoting the PSA.

In addition, metal ions can also induce the PSA to form different nanostructures. In the work of Sharma et al., the histidine-metal ion interactions were explored (Fig. 3.1b) [21]. Different transition metal ions revealed different effects on the gelation of peptide hydrogels. In the presence of Mg^{2+}, the PSA was mainly driven by the binding of metal ions to the deprotonated carboxyl groups of PA, leading to the formation of robust hydrogels at physiological pH. The binding ability of Fe^{2+} and Mn^{2+} to imidazole N–H was weak, and therefore, the peptide-metal ion complex formed by them was not enough to disrupt the hydrogel formation under physiological conditions. In the presence of Co^{2+}, Ni^{2+}, and Cu^{2+}, the interactions between PA and metal ions led to the transition from gel to sol at physiological pH, which was reversed when the pH was reduced to 3.0. Therefore, the sol-to-gel equilibrium could be achieved by adjusting the pH of the peptide-metal ion complex.

3.2.2 *Physical Stimuli-Induced Formation of Peptide Hydrogels*

In addition to chemical stimuli for the PSA, physical stimuli are also able to induce the formation of peptide hydrogels. The use of physical stimuli to induce the formation of peptide hydrogels can be protected from other reagents in the self-assembly process. Among them, temperature is an important factor for inducing the formation of various peptide hydrogels.

In the work of Pochan et al., a small peptide (MAX3) revealed its ability to self-assemble into network hydrogel thermoreversibly [22]. At pH of 9 and ambient temperature, MAX3 expanded to form a lower viscosity aqueous solution. As the temperature continued to increase, the peptide underwent unimolecular folding to form an amphipathic β-hairpin structure, thereby forming a rich network hydrogel (Fig. 3.2a). Rughani et al. designed a 22-residue peptide MLD that contains eight lysine residues, an N-terminal free amine, and nine hydrophobic valine residues [23]. At pH of 7.4 and lower ionic strength, the side chain of lysine on the MLD was protonated. Therefore, the temperature change was hard to promote the folding of the peptide to form a fibrous structure, but the addition of NaCl to increase the ionic strength of the peptide solution screened the positive charge of lysine, making the thermo-responsive trigger of the PSA possible. Experiments have shown that under the influence of saline solution, the low-temperature environment of the peptide remained unfolded. However, the PSA started at ~ 15 °C, ascribing to the increased hydrophobic effect at elevated temperature. The evolution of this β-sheet structure was completed at 40 °C, and higher concentrated peptide self-assembled and gelled to form peptide hydrogels under the trigger of temperature.

Light is another physical tool that can be used to stimulate the peptide gelation. For instance, Liu et al. proposed a strategy for the self-assembly gelation by adjusting the amphiphilicity of peptides under the irradiation of visible light [15]. The lysine residue of the Fmoc-KK(SP)KF-NH_2 peptide was modified with a nitrosospiropyran

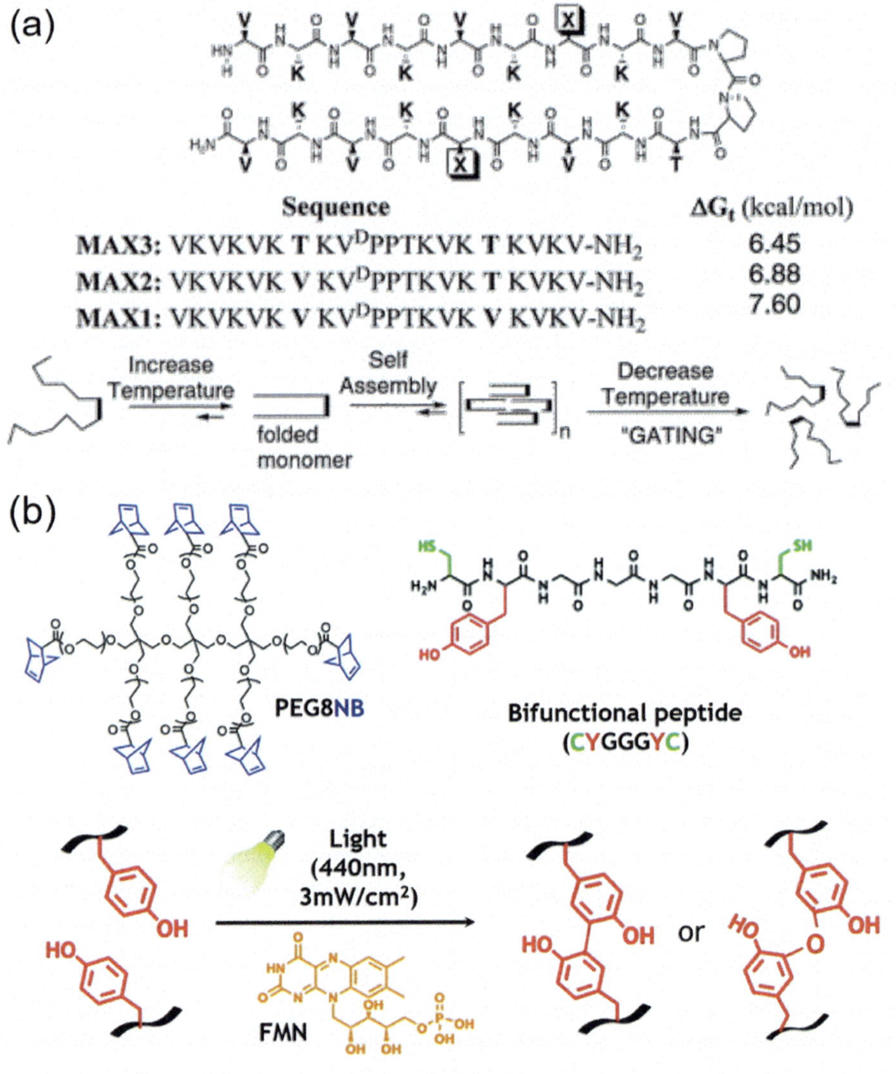

Fig. 3.2 **a** General structure of β-hairpin and heat-triggered folding and self-assembly mechanism of hydrogelation. Reproduced with the permission from Ref. [22]. Copyright 2003, American Chemical Society. **b** Schematic of FMN and visible light-mediated dynamic hydrogel fabrication. Reproduced with the permission from Ref. [24]. Copyright 2018, Wiley

chromophore (SP), which served as a switch to regulate the amphiphilicity of the peptide. Under the irradiation of UV light or increasing temperature, SP will undergo a ring-opening reaction to form of cyanine (MC), which will return to the form of SP after being irradiated with visible light. Therefore, in the trifluoroacetic acid (TFA) system, the presence of TFA induced partial ring-opening of the SP isomer. After being irradiated with visible light, the peptide quickly gelled to form a hydrogel, and the structure of the hydrogel could be regulated by adjusting the irradiation time and the frequency of light.

Lin et al. reported the design and synthesis of a dynamic PEG-peptide hydrogel via the irradiation of visible light and the flavin mononucleotide (FMN)-induced tyrosine dimerization (Fig. 3.2b) [24]. The polymerization of eight-arm PEG-grafted double cysteine-terminated peptides with thiol-norbornene was studied. In the presence of light, FMN induced tyrosyl radicals to bind and lead to bis-tyrosine cross-linking, which resulted in the gelation of peptides to form hydrogels. In the work of Das et al., the azobenzene-functionalized short peptide (Azo-KC) formed hydrogels under the alkaline conditions. After the UV irradiation, the light-triggered trans–cis isomerization of the azobenzene moiety in the peptide hydrogel caused the hydrogel to shrink and expel part of the solvent, which was useful for the adsorption and removal of dye molecules for water purification [25].

3.2.3 Biostimulation-Induced Formation of Peptide Hydrogels

In the self-assembling peptide-based hydrogel systems, enzymes can act as biocatalysts to induce the PSA and the formation of hydrogels, which was called as the enzyme-assisted self-assembly (EASA). In general, hydrogels formed by EASA exhibited excellent chemical stability, robust mechanical properties, and regioselectivity [26].

Bai et al. introduced a fetal bovine serum (FBS) and oxidase (PAO)-induced sol–gel transition of A_9K_2 peptide hydrogels (Fig. 3.3a) [27]. Under natural conditions, the A_9K_2 peptide solution with lysine was a clear free-flowing liquid. After adding a small amount of FBS or PAO, this solution formed a hydrogel after two weeks. This was attributed to the LO in FBS or PAO catalyzed oxidation of lysine residues to form lysine (α-aminoadipate δ-semialdehyde) and the subsequent non-enzymatic condensation reaction. The enzyme-induced peptide hydrogels showed excellent selectivity in a few applications, including promoting the mammalian cell adhesion as well as spreading and the killing of pathogens.

In the work of Gao et al., supramolecular hydrogels were successfully synthesized from hydrophobic compounds by an enzymatic reaction (Fig. 3.3b) [28]. They used N-Fmoc-terminated and C-terminal methyl ester group phosphorylated tyrosine derivatives as precursors and observed the morphologies in aqueous solution and in PBS. It was found that in the presence of phosphatase, this precursor was converted into a hydrophobic compound. Finally, both components were enzymatically converted to form a peptide hydrogel that was stable for at least one month. Interestingly, this enzymatic conversion was the only mechanism by which small hydrophobic peptides formed supramolecular hydrogels.

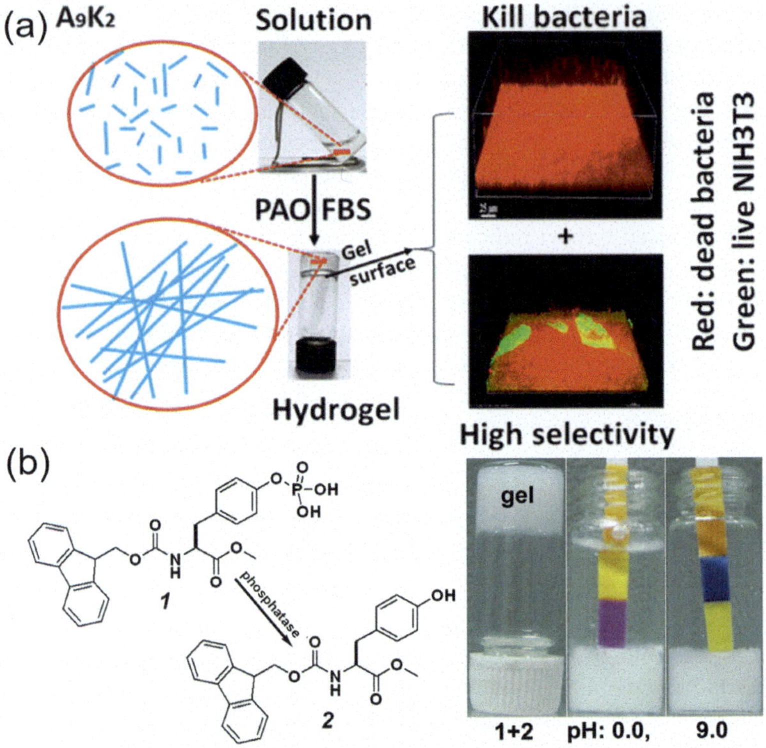

Fig. 3.3 **a** Schematic diagram of A_9K_2 peptide forming a gel and selectively antibacterial under enzymatic action. Reproduced with the permission from Ref. [27]. Copyright 2016, American Chemical Society. **b** Enzyme-induced peptide hydrogels. Reproduced with the permission from Ref. [28]. Copyright 2009, American Chemical Society

In another work, Liang et al. studied the conformation of peptides that self-assembled into peptide hydrogels by the enzymatic method [29]. The peptide derivatives CRB-G^{D}F^{D}Fp^{D}Y and CRB-G^{D}F^{D}F^{D}Y were prepared using chlorambucil (CRB) as a capping group. Among them, the CRB-G^{D}F^{D}Fp^{D}Y was converted into the CRB-G^{D}F^{D}F^{D}Y in the presence of alkaline phosphatase (ALP), and a small amount of ALP was added to the clarified solution of CRB-G^{D}F^{D}Fp^{D}Y at a temperature of 4 °C. The solution was transformed into a transparent hydrogel and was stable for at least three months, while the CRB-G^{D}F^{D}F^{D}Y produced precipitates after a heating–cooling process. To this end, they explored the self-assembled conformation of the peptide under two different conditions. The results demonstrated that in the presence of ALP, the self-assembled conformation of PNFs was α-helix, while the self-assembled nanoparticles obtained by the heating–cooling process was β-sheet. This study indicated that in the presence of enzymes, the peptide conformation could be transformed. Such enzyme-induced α-helical nanofibers revealed better cell membrane fusion ability than the β-sheet peptides and thus exhibited greater application potential in biological fields.

3.3 Properties of Peptide Hydrogels

The unique conformation and network structure endow peptide hydrogels with unique rheological properties, and the nature of peptide molecules determine excellent biocompatibility [30, 31]. Based on the characteristics of the PSA, in this section we focus on the rheological properties and biocompatibility of self-assembled peptide hydrogels.

3.3.1 Rheological Properties

Rheological properties are important factors of self-assembled peptide hydrogels. Generally speaking, the stiffness, viscoelasticity, and shear-thinning behavior of peptide hydrogels are the focus of studying. Peptide hydrogels with different properties have great application advantages in various application fields of biomedicine.

The stiffness is a parameter that describes the ability of a material to resist the deformation. The higher the strength of the network structure of a self-assembled peptide hydrogel, the greater the solidification degree, and the higher the stiffness of the peptide hydrogel [31]. Peptide hydrogels with high stiffness can be used as in vivo scaffolds for the tissue repair and cell growth. For instance, previously it has been reported that the FEFEFKFK peptide self-assembled into antiparallel β-sheet nanofibers and then formed self-supporting hydrogels, which could be utilized for drug delivery and cell culture applications [32]. The potential application of the FEFEFKFK hydrogels as scaffolds for the nucleus pulposus tissue engineering was investigated. Through the characterizations and optimization of the FEFEFKFK hydrogels, it was found that the hydrogels could closely replicate the strength of native nucleus pulposus.

In another work, Hiew et al. designed a short biomimetic peptide GV8 (Ac-GLYGGYGV-NH_2) that derived from squid sucker ring tooth protein, which formed hydrogels in water without adding cross-linkers or chemical modifiers [33]. During the gelation, the helix transformed into antiparallel β-sheets due to interpeptide hydrophobic interactions. The concentration-dependent mechanical response of the GV8 hydrogel allowed its stiffness to vary up to 25 folds, with a maximum value of 35.5 kPa, making it one of the stiffest protein-based hydrogels. This new gelation mechanism extended previous understanding of fibrinopeptide-based hydrogels and offered potential biomedical applications, such as tissue engineering, therapeutic encapsulation, soft tissue adhesives, and stem cell differentiation matrices.

Bairagi et al. developed a histidine-containing peptide that can form a hydrogel by the self-assembly at pH of 5.5–8.5 [34]. They found that the thermal stability and stiffness of this peptide hydrogel could be tuned by incorporating different types of dicarboxylic acids. This two-component hydrogel was applied to 3D cell culture of mouse fibroblasts and showed potential applications in regenerative medicine and tissue engineering. Their study demonstrated the tunability of peptide hydrogel-based soft biomaterials in promoting the cell growth and nutrition, highlighting future prospects in the field of regenerative medicine.

The rheological properties of peptide hydrogels were mainly affected by the number and the fiber length of self-assembled PNFs. Peptide hydrogels with different rheological properties could be obtained by adjusting the properties of PNFs [35]. The variations in the peptide concentration directly affected the number of nanofibers, thereby altering the rheology of the formed peptide hydrogels. In the work of Tang et al., a self-assembled peptide hydrogel based on a short five-residue peptide was synthesized, and the effect of the concentration variation on the mechanical properties of peptide hydrogels was studied [36]. When the concentration of the KYFIL peptide was increased from 0.5 to 3 wt%, the storage modulus of the peptide hydrogel reached from 8000 to 17,000 Pa. Boothroyd et al. incorporated a small amount of peptide FEFEFKFK-GG-FKFKFEFE (FEKII18) into the self-assembling octapeptide FEFEFKFK (FEKII) for the formation of peptide hydrogels. It was found that the elastic modulus of the self-assembled peptide hydrogel continued to increase as the concentration of the incorporated peptide increased, and when the concentration of the incorporated peptide was 0.28 M, the elastic modulus of the formed hydrogel was increased by 30 times [37]. Therefore, in general, an increase of the peptide concentration enhanced the mechanical properties of the created peptide hydrogels.

The change of pH can cause structural change of PNFs, so pH is also another important factor for affecting the rheology of peptide hydrogels. Tang et al. demonstrated that increasing the peptide concentration effectively enhanced the mechanical properties of hydrogels [36]. The change of pH revealed a very obvious effect on the formed hydrogels. According to the obtained results, the mechanical properties of the hydrogels formed by the self-assembly of KYFIL pentapeptide under acidic conditions were very weak ($G' < 80$ Pa). However, the self-assembled hydrogels at pH 7.4 and 10.6 were stiff and exhibited good shear-thinning behavior under the shear flow. In another case, Chen et al. reported a novel self-assembled peptide hydrogel with a pH-tunable structure [38]. The change of pH affected the ionization state of the peptide terminal group. The EIWLK peptide was dissolved, and the change of the self-assembled structure was observed under different pH conditions. When the pH was 2, no obvious PSA occurred, while peptide nanoparticles appeared when the pH was raised to 4–5. As the pH was further increased, the nanofibrous structure appeared. When the pH reached 8–9, the fibrous structure gradually disappeared, and the peptide nanoparticles appeared again (Fig. 3.4a). Based on the obtained results, it can be concluded that the self-assembly of EIWLK could be adjusted via pH to obtain peptide hydrogels with higher strength.

In addition, ionic strength also affected the electrostatic interactions between PNFs, thereby affecting the physical and chemical properties of peptide hydrogels [39]. Morales et al. synthesized nanofibrillar hydrogels based on the co-assembly of charge-complementary peptides, in which four pairs of CATCH peptide variants

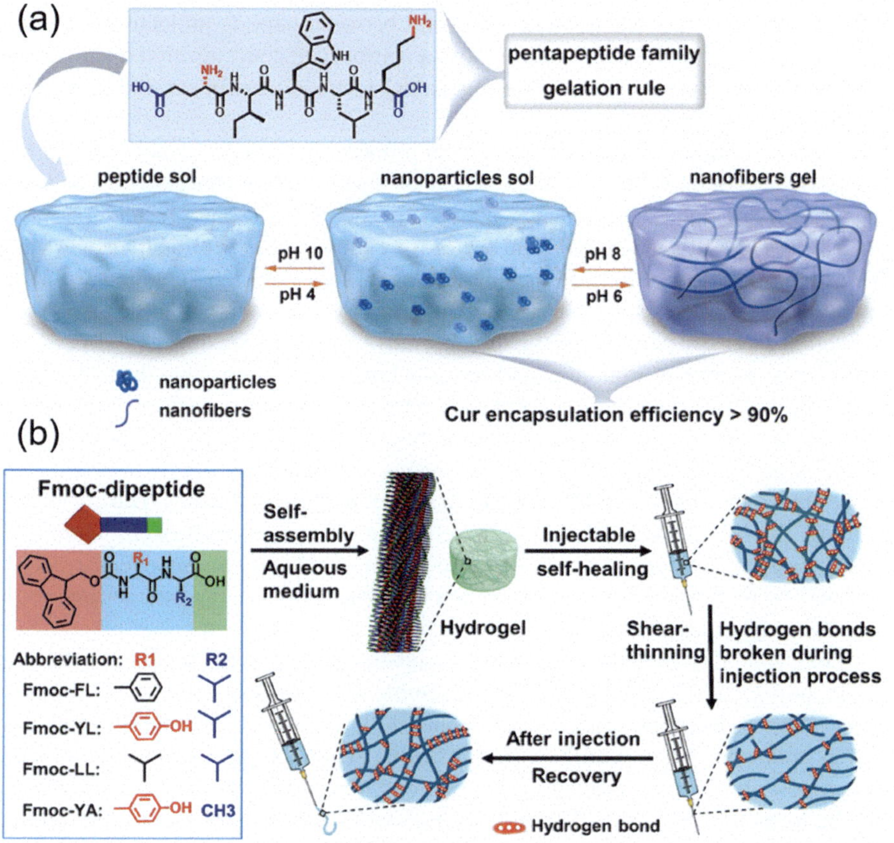

Fig. 3.4 **a** Self-assembled pentapeptide hydrogels under different pH conditions. Reproduced with the permission from Ref. [38]. Copyright 2022, Elsevier. **b** Self-assembled Fmoc-dipeptide hydrogel with shear-thinning and self-healing properties. Reproduced with the permission from Ref. [44]. Copyright 2020, American Chemical Society

underwent different pairings to form a series of peptide hydrogels [40]. It was found that the CATCH(4+/4−) was able to form hydrogels with the largest storage and loss modulus, as well as the smallest pore structure, while the hydrogel formed by the CATCH(6+/6−) was weaker. In addition, they tested the shear-thinning behavior of these hydrogels. Among them, the CATCH(4+/4−) hydrogel exhibited shear-thinning at both low and high shear rates during the first two cycles.

In addition to internal factors, external ionic stimuli can also change the structure and properties of peptide hydrogels. Owczarz et al. studied the effect of the ionic strength on the self-assembly and gelation of PA, RADARADARADARADA (RADA 16-I) [41]. By adjusting the concentration of the saline solution, the critical gel concentration of the self-assembling peptides changed. As the salt concentration increased, the critical gel concentration gradually decreased, which was attributed to the shielding of the electrostatic repulsion between peptide molecules after the addition of salt. Therefore, the addition of an appropriate concentration of salt could stabilize the structure of peptide hydrogels and resulted in stronger viscoelasticity and higher storage modulus.

The shear-thinning is another rheological property of the self-assembling peptide hydrogels that defines the phase transition of viscoelastic materials from solid to low viscosity or liquid at high shear rates [42]. The physical bond formed between self-assembled PNFs is an important factor for endowing the peptide hydrogel with shear-thinning behavior [43]. Ren et al. designed a series of Fmoc-coupled aromatic dipeptides (Fmoc-FL, Fmoc-YL, Fmoc-LL, and Fmoc-YA) to regulate noncovalent interactions between peptides by the substitution of amino acid residues (hydrophobic interactions, hydrogen bonding, and π-π stacking), thus preparing shear-thinning hydrogels with self-healing and tunable mechanical properties (Fig. 3.4b) [44]. The mechanical rigidity of the dipeptide hydrogels was mainly affected by the hydrophobic interaction, and the instantaneous self-healing property was affected by the strong hydrogen bond interaction between peptide molecules. During shear-thinning, the hydrogen bonds between peptide molecules were broken, leading to the destruction of fibers and the fibril entanglement. During the recovery process, the fibers reconnected with each other, and the hydrogen-bonding interactions were instantaneously reestablished.

Clarke et al. designed three aromatic-group-free pentapeptide sequences (IDIDI, DDIII, and DIIID), which could self-assemble into pH-responsive high-strength hydrogels [45]. The peptide sequence contains three aliphatic isoleucine (Ile) residues and an amino acid that readily forms β-sheets. After the protonation, the charge recognition and hydrogen-bonding interactions induced the β-sheet self-assembly and the formation of hydrogels. Studies have shown that the stiffness of the formed peptide hydrogels could be tuned within two orders of magnitude (2–200 kPa) by changing the concentration and charge distribution of the peptide molecules. Hydrogels formed through noncovalent interactions exhibited excellent self-healing properties, and they were able to reassemble through physical cross-linking after the deformation and recover their mechanical properties after high strains. Meanwhile, the hydrogel exhibited typical shear-thinning behavior, and the viscosity decreased linearly with the increase of shear stress.

3.3.2 Biocompatibility

Biocompatible peptide hydrogels have high application prospects in the field of biomedicine and tissue engineering and are often used as bio-scaffolds, drug delivery carriers, and cell growth substrates.

Restu et al. reported a short Phe-rich oligopeptide as an effective supramolecular hydrogel [46]. Especially the peptide P1 (Ac-Phe-Phe-Phe-Gly-Lys-Lys) exhibited a high gelling ability accompanied with high biocompatibility and biodegradability. In their study, the hydrogel formed by oligopeptides was hydrolyzed by proteolytic enzymes, exposing the functional compounds wrapped in them, and different types of functional compounds were released according to the type of peptide hydrogels. The exploration of oligopeptide hydrogels showed broad application prospects in the fields of bioengineering, tissue engineering, and drug delivery. In another report, Liang et al. designed and prepared a novel peptide (Fmoc-FFRGDF) hydrogel containing a bioactive RGD sequence [47]. The formed peptide hydrogel exhibited excellent biocompatibility when injected into the subconjunctival space and anterior chamber of rabbits. It could be used as a carrier for an implantable drug delivery system for the treatment of anterior segment diseases and also as a scaffold for the ocular tissue repair. In addition, the peptide hydrogel was sufficiently viscous and immunologically inert to remain in the anterior segment without causing inflammatory or foreign body reactions. After an appropriate period, the peptide hydrogel was biodegraded to release drugs to treat anterior segment diseases. This study demonstrated the potential application of biocompatible peptide hydrogels in ophthalmology.

Self-assembled peptide hydrogels can also be used as scaffolds for the proliferation and differentiation of stem cells. For instance, Wang et al. constructed supramolecular peptide hydrogels with high biocompatibility, which provided bioactive substrates for the proliferation and differentiation of stem cells [48]. Three highly effective hydrogels from the peptides Fmoc-FF, Fmoc-RGD, and Fmoc-RGE were synthesized successfully. The Fmoc-FF had the ability to self-assemble into nanofibrous hydrogels while the Fmoc-RGD revealed high biocompatibility and good affinity for stem cells. The mixing of Fmoc-FF and Fmoc-RGD formed an anchorage-dependent cell scaffold via co-assembly, which induced the seeding of human mesenchymal stem cells (MSCs) to differentiate into osteoblasts, adipocytes, and chondrocytes. This work provided a unique strategy for developing MSC-based tissue engineering scaffolds for the cancer therapy and regenerative medicine.

3.4 Conclusions

In summary, peptides can self-assemble under the stimulation of certain factors to form functional peptide hydrogels with rich nanofiber networks. This process is basically completed through noncovalent interactions between molecules, and the formed hydrogel has unique rheological properties and excellent biocompatibility. Self-assembled peptide hydrogels have great application potential in tissue engineering, drug delivery, biosensors, and other fields. However, there are still many problems to be solved in the application process of the self-assembled peptide hydrogels. First, the mechanisms by which peptides self-assemble into PNFs with various conformations are still not fully understood, and more experiments are needed to prove the existing mechanisms. Secondly, when used as a tissue engineering scaffold, the mechanical properties of a hydrogel need to be further improved to meet the requirements of special conditions in clinical applications. Thirdly, in the nanocarrier application of drug delivery, how to achieve slow and quantitative drug release under the stimulation of different conditions still needs further exploration.

References

1. Z.Q. Yu, Q. Xu, C.B. Dong, S.S. Lee, L.Q. Gao, Y.W. Li, M. D'Ortenzio, J. Wu, Self-assembling peptide nanofibrous hydrogel as a versatile drug delivery platform. Curr. Pharm. Design **21**, 4342–4354 (2015)
2. E.M. Ahmed, Hydrogel: preparation, characterization, and applications: a review. J. Adv. Res. **6**, 105–121 (2015)
3. P. Patel, P. Thareja, Hydrogels differentiated by length scales: a review of biopolymer-based hydrogel preparation methods, characterization techniques, and targeted applications. Eur. Polym. J. **163**, 110935 (2022)
4. S.C. Lange, J. Unsleber, P. Drucker, H.J. Galla, M.P. Waller, B.J. Ravoo, PH response and molecular recognition in a low molecular weight peptide hydrogel. Org. Biomol. Chem. **13**, 561–569 (2015)
5. X.Z. Hu, M.R. Liao, H.N. Gong, L. Zhang, H. Cox, T.A. Waigh, J.R. Lu, Recent advances in short peptide self-assembly: from rational design to novel applications. Curr. Opin. Colloid. Interf. **45**, 1–13 (2020)
6. G. Wei, Z.Q. Su, N.P. Reynolds, P. Arosio, I.W. Hamley, E. Gazit, R. Mezzenga, Self-assembling peptide and protein amyloids: from structure to tailored function in nanotechnology. Chem. Soc. Rev. **46**, 4661–4708 (2017)
7. J. Wang, K. Liu, L.Y. Yan, A.H. Wang, S. Bai, X.H. Yan, Trace solvent as a predominant factor to tune dipeptide self-assembly. ACS Nano **10**, 2138–2143 (2016)
8. T. Li, X.M. Lu, M.R. Zhang, K. Hu, Z. Li, Peptide-based nanomaterials: Self-assembly, properties and applications. Bioact. Mater. **11**, 268–282 (2022)
9. M. Criado-Gonzalez, D. Wagner, J.R. Fores, C. Blanck, M. Schmutz, A. Chaumont, M. Rabineau, J.B. Schlenoff, G. Fleith, J. Combet, P. Schaaf, L. Jierry, F. Boulmedais, Supramolecular hydrogel induced by electrostatic interactions between polycation and phosphorylated-fmoc-tripeptide. Chem. Mater. **32**, 1946–1956 (2020)
10. Y. Wang, W.S. Zhang, C.C. Gong, B. Liu, Y.D. Li, L.C. Wang, Z.Q. Su, G. Wei, Recent advances in the fabrication, functionalization, and bioapplications of peptide hydrogels. Soft Matter **16**, 10029–10045 (2020)
11. K. Fu, H.G. Wu, Z.Q. Su, Self-assembling peptide-based hydrogels: fabrication, properties, and applications. Biotechnol. Adv. **49**, 107752 (2021)
12. P.K. Nguyen, W. Gao, S.D. Patel, Z. Siddiqui, S. Weiner, E. Shimizu, B. Sarkar, V.A. Kumar, Self-assembly of a dentinogenic peptide hydrogel. ACS Omega **3**, 5980–5987 (2018)
13. Y.K. Zhang, H. Zhang, Q.L. Zou, R.R. Xing, T.F. Jiao, X.H. Yan, An injectable dipeptide-fullerene supramolecular hydrogel for photodynamic antibacterial therapy. J. Mater. Chem. B **6**, 7335–7342 (2018)
14. N. Falcone, H.B. Kraatz, Supramolecular assembly of peptide and metallopeptide gelators and their stimuli-responsive properties in biomedical applications. Chem. Eur. J. **24**, 14316–14328 (2018)
15. M.M. Liu, C.N. Creemer, T.J. Reardon, J.R. Parquette, Light-driven dissipative self-assembly of a peptide hydrogel. Chem. Commun. **57**, 13776–13779 (2021)
16. K. Rajagopal, M.S. Lamm, L.A. Haines-Butterick, D.J. Pochan, J.P. Schneider, Tuning the pH responsiveness of β-hairpin peptide folding, self-assembly, and hydrogel material formation. Biomacromol **10**, 2619–2625 (2009)
17. Y.R. Chen, H.X. Gan, Y.W. Tong, PH-Controlled hierarchical self-assembly of peptide amphiphile. Macromolecules **48**, 2647–2653 (2015)

18. H. Kaur, P. Sharma, N. Patel, V.K. Pal, S. Roy, Accessing highly tunable nanostructured hydrogels in a short ionic complementary peptide sequence via pH trigger. Langmuir **36**, 12107–12120 (2020)
19. X.R. Zhou, R. Ge, S.Z. Luo, Self-assembly of pH and calcium dual-responsive peptide-amphiphilic hydrogel. J. Pept. Sci. **19**, 737–744 (2013)
20. B. Ozbas, J. Kretsinger, K. Rajagopal, J.P. Schneider, D.J. Pochan, Salt-triggered peptide folding and consequent self-assembly into hydrogels with tunable modulus. Macromolecules **37**, 7331–7337 (2004)
21. P. Sharma, H. Kaur, S. Roy, Inducing differential self-assembling behavior in ultrashort peptide hydrogelators using simple metal salts. Biomacromol **20**, 2610–2624 (2019)
22. D.J. Pochan, J.P. Schneider, J. Kretsinger, B. Ozbas, K. Rajagopal, L. Haines, Thermally reversible hydrogels via intramolecular folding and consequent self-assembly of a de Novo designed peptide. J. Am. Chem. Soc. **125**, 11802–11803 (2003)
23. R.V. Rughani, M.C. Branco, D. Pochan, J.P. Schneider, De novo design of a shear-thin recoverable peptide-based hydrogel capable of intrafibrillar photopolymerization. Macromolecules **43**, 7924–7930 (2010)
24. H.-Y. Liu, H.D. Nguyen, C.-C. Lin, Dynamic PEG–peptide hydrogels via visible light and FMN-induced tyrosine dimerization. Adv. Healthc. Mater. **7**, 1800954 (2018)
25. B.K. Das, B. Pramanik, S. Chowdhuri, O.A. Scherman, D. Das, Light-triggered syneresis of a water insoluble peptide-hydrogel effectively removes small molecule waste contaminants. Chem. Commun. **56**, 3393–3396 (2020)
26. J. Gao, J. Zhan, Z.M. Yang, Enzyme-instructed self-assembly (EISA) and hydrogelation of peptides. Adv. Mater. **32**, 1805798 (2020)
27. J. Bai, C. Chen, J. Wang, Y. Zhang, H. Cox, J. Zhang, Y. Wang, J. Penny, T. Waigh, J.R. Lu, H. Xu, Enzymatic regulation of self-assembling peptide A9K2 nanostructures and hydrogelation with highly selective antibacterial activities. ACS Appl. Mater. Interfaces **8**, 15093–15102 (2016)
28. J. Gao, H.M. Wang, L. Wang, J.Y. Wang, D.L. Kong, Z.M. Yang, Enzyme promotes the hydrogelation from a hydrophobic small molecule. J. Am. Chem. Soc. **131**, 11286 (2009)
29. C.H. Liang, D.B. Zheng, F. Shi, T.Y. Xu, C.H. Yang, J.F. Liu, L. Wang, Z.M. Yang, Enzyme-assisted peptide folding, assembly and anti-cancer properties. Nanoscale **9**, 11987–11993 (2017)
30. P.K. Nguyen, B. Sarkar, Z. Siddiqui, M. McGowan, P. Iglesias-Montoro, S. Rachapudi, S. Kim, W. Gao, E.J. Lee, V.A. Kumar, Self-assembly of an antiangiogenic nanofibrous peptide hydrogel. Acs Appl. Bio Mater. **1**, 865–870 (2018)
31. L.J. Yang, C.R. Zhang, C.H. Ren, J.J. Liu, Y.M. Zhang, J. Wang, F. Huang, L. Zhang, J.F. Liu, Supramolecular hydrogel based on chlorambucil and peptide drug for cancer combination therapy. ACS Appl. Mater. Interfaces **11**, 331–339 (2019)
32. S. Wan, S. Borland, S.M. Richardson, C.L.R. Merry, A. Saiani, J.E. Gough, Self-assembling peptide hydrogel for intervertebral disc tissue engineering. Acta Biomater. **46**, 29–40 (2016)
33. S.H. Hiew, H. Mohanram, L.L. Ning, J.J. Guo, A. Sanchez-Ferrer, X.Y. Shi, K. Pervushin, Y.G. Mu, R. Mezzenga, A. Miserez, A short peptide hydrogel with high stiffness induced by 3(10)-helices to beta-sheet transition in water. Adv. Sci. **6**, 1901173 (2019)
34. D. Bairagi, P. Biswas, K. Basu, S. Hazra, D. Hermida-Merino, D.K. Sinha, I.W. Hamley, A. Banerjee, Self-assembling peptide-based hydrogel: regulation of mechanical stiffness and thermal stability and 3D cell culture of fibroblasts. Acs Appl. Bio Mater. **2**, 5235–5244 (2019)
35. F. Koch, M. Muller, F. Konig, N. Meyer, J. Gattlen, U. Pieles, K. Peters, B. Kreikemeyer, S. Mathes, S. Saxer, Mechanical characteristics of beta sheet-forming peptide hydrogels are dependent on peptide sequence, concentration and buffer composition. Roy. Soc. Open Sci. **5**, 171562 (2018)
36. J.D. Tang, C. Mura, K.J. Lampe, Stimuli-responsive, pentapeptide, nanofiber hydrogel for tissue engineering. J. Am. Chem. Soc. **141**, 4886–4899 (2019)
37. S. Boothroyd, A. Saiani, A.F. Miller, Controlling network topology and mechanical properties of Co-assembling peptide hydrogels. Biopolymers **101**, 669–680 (2014)

38. H.M. Chen, T.R. Zhang, Y.Q. Tian, L.J. You, Y. Huang, S.Y. Wang, Novel self-assembling peptide hydrogel with pH-tunable assembly microstructure, gel mechanics and the entrapment of curcumin. Food Hydrocoll. **124**, 107338 (2022)
39. R. Li, C.C. Horgan, B. Long, A.L. Rodriguez, L. Mather, C.J. Barrow, D.R. Nisbet, R.J. Williams, Tuning the mechanical and morphological properties of self-assembled peptide hydrogels via control over the gelation mechanism through regulation of ionic strength and the rate of pH change. Rsc Adv. **5**, 301–307 (2015)
40. B.S. Morales, R.J. Liu, J. Olguin, A.M. Ziegler, S.M. Herrera, K.L. Backer-Kelley, K.L. Kelley, G.A. Hudalla, Injectable nanofibrillar hydrogels based on charge-complementary peptide co-assemblies. Biomater. Sci. **9**, 2494–2507 (2021)
41. M. Owczarz, S. Bolisetty, R. Mezzenga, P. Arosio, Sol-gel transition of charged fibrils composed of a model amphiphilic peptide. J. Colloid Interface Sci. **437**, 244–251 (2015)
42. Y. Yamada, N.L. Patel, J.D. Kalen, J.P. Schneider, Design of a peptide-based electronegative hydrogel for the direct encapsulation, 3D culturing, in vivo syringe-based delivery, and long-term tissue engraftment of cells. ACS Appl. Mater. Interfaces **11**, 34688–34697 (2019)
43. L.X. Mei, K.M. Xu, Z.R. Zhai, S.Y. He, T.T. Zhu, W.Y. Zhong, Doxorubicin-reinforced supramolecular hydrogels of RGD-derived peptide conjugates for pH-responsive drug delivery. Org. Biomol. Chem. **17**, 3853–3860 (2019)
44. P. Ren, J.T. Li, L.Y. Zhao, A.H. Wang, M.Y. Wang, J.L. Li, H.L. Jian, X.O. Li, X.H. Yan, S. Bai, Dipeptide self-assembled hydrogels with shear-thinning and instantaneous self-healing properties determined by peptide sequences. ACS Appl. Mater. Interfaces **12**, 21433–21440 (2020)
45. D.E. Clarke, C.D.J. Parmenter, O.A. Scherman, Tunable pentapeptide self-assembled β-sheet hydrogels. Angew. Chem. Int. Ed. **57**, 7709–7713 (2018)
46. W.K. Restu, Y. Nishida, S. Yamamoto, J. Ishii, T. Maruyama, Short Oligopeptides for biocompatible and biodegradable supramolecular hydrogels. Langmuir **34**, 8065–8074 (2018)
47. L. Liang, X.D. Xu, C.S. Chen, J.H. Fang, F.G. Jiang, X.Z. Zhang, R.X. Zhuo, Evaluation of the biocompatibility of novel peptide hydrogel in rabbit eye. J. Biomed. Mater. Res. B **93b**, 324–332 (2010)
48. Y.L. Wang, S.P. Lin, S.R. Nelli, F.K. Zhan, H. Cheng, T.S. Lai, M.Y. Yeh, H.C. Lin, S.C. Hung, Self-assembled peptide-based hydrogels as scaffolds for proliferation and multi-differentiation of mesenchymal stem cells. Macromol. Biosci. **17** (2017)

Chapter 4
Design and Self-Assembly of Stimuli-Responsive Peptides

Peng He and Gang Wei

Abstract Stimuli-responsive peptides have emerged as promising smart materials for biomedical applications due to their multifunctionality, unique self-assembly ability, high biocompatibility, and ease of customization. Stimuli-responsive peptides have been widely used in tissue engineering, regenerative medicine, diagnostics, and drug delivery systems and could achieve controlled drug release, targeted delivery, and non-invasive operations. The synergistic effects of multiple stimuli-responsive domains within a single peptide system, the design of peptide sequences, and the physicochemical properties for specific functions, as well as the development of rationally designed stimuli-responsive peptide materials are all the hot topics in this promising research field. This chapter focuses on the design, synthesis, and applications of various types of stimuli-responsive peptides, including the thermo-responsive, pH-responsive, light-responsive, metal-ion-responsive, and enzyme-responsive ones, each of them have exhibited unique responses to specific environmental triggers.

Keywords Design · Peptide nanomaterials · Stimuli-responsive · Drug delivery · Biomedicine

4.1 Introduction

The self-assembly property of peptides is one of the important factors that endowing their promising biomedical and nanotechnological applications [1]. Self-assembled peptide structures include the vesicles, nanofibers, nanotubes, nanorods, nanosheets, and hydrogels, which can be induced by the interactions such as ionic bonds, hydrogen bonds, van der Waals forces, and hydrophobic interactions [2, 3]. The composition and properties of these self-assembled peptide structures can be tailored to suit different needs. Peptides can self-assemble into superstructures through noncovalent interactions, and the assembly process of peptide molecules is very sensitive

P. He · G. Wei (✉)
College of Chemistry and Chemical Engineering, Qingdao University, Qingdao 266071, China
e-mail: weigroup@qdu.edu.cn

G. Wei (ed.), *Peptide Nano-Chemistry and Nanotechnology*,
https://doi.org/10.1007/978-981-96-2387-7_4

to environmental conditions, such as the temperature, light, ionic strength, and pH value, and others, which motivate the use of peptide to develop stimuli-responsive nanomaterials [4–6].

Stimuli-responsive peptides can change their shapes and properties according to the changes of both internal and external environments. In addition, the responses and control to different stimuli can be achieved by selecting different amino acid sequences and adjusting their secondary structures, which makes stimuli-responsive peptides ideal candidates for the synthesis of smart materials [7]. When designing stimuli-responsive peptides, many factors need to be considered, such as the amino acid sequence and length, solvent, pH, peptide concentration, and temperature. Different choices of these parameters can be realized through rational design to tailor the function and performance of peptide materials by de novo designing the physicochemical properties, self-assembled structure, specific recognition, and stimuli-response properties of peptide chains [8, 9]. The response mechanisms made the stimuli-responsive peptides widely used in biomedical fields, including tissue engineering, regenerative medicine, cardiovascular therapy, drug delivery, and bio-diagnostics [10–13]. For example, stimuli-responsive peptides could achieve selective and controlled drug release to minimize the drug damage to healthy cells for the cancer therapy [14]. In addition, stimuli-responsive peptides have been used to synthesize injectable hydrogels for minimally invasive surgery, which were applied by the syringe and underwent a transition from solution to condensation triggered by the temperature, pH, ionic strength, oxidative species, or enzymes at the sites of injury [15]. Furthermore, stimuli-responsive peptides have been used for biosensing, that is, the study of the changes of the properties of small chemical or physical sensing materials in response to the stimuli from local environment [16]. These reactive biomaterials contained molecular building blocks that changed at the molecular level, resulting in altered noncovalent interactions that translate into macroscopic reactions.

Stimuli-responsive peptides are very promising smart materials with the advantages of multifunctionality, self-assembly, biocompatibility, and easy customization. With the in-depth understanding of the design and preparation principles of peptides and the continuous improvement of nanotechnology, the applications of peptide nanomaterials in the field of biomedicine and tissue engineering will continue to expand. In this chapter, we focus on the design, synthesis, and applications of stimuli-responsive peptides.

4.2 Thermo-Responsive Peptides

Temperature is a key factor for affecting the self-assembly and conformation transition of the stimuli-responsive peptides. This section will discuss the design and synthesis of thermo-responsive peptides.

Gao et al. reported a novel triple redox/thermo-responsive diselenoamino acid polymer that prepared by the selenocysteine-initiated ring-opening polymerization

(ROP) [17]. Selenides aggregated with peptides and self-assembled into spherical micelles and then formed supramolecular hydrogels in aqueous solution at room temperature (Fig. 4.1a). By changing the molar ratio of selenocysteamine to EG2-Glu-NCA, the molecular weight and thermo-sensitivity of diselenide homopolypeptides could be adjusted. The synthesized peptide micelles showed obvious reversible thermo-sensitivity in five cycles at 10–100 °C and exhibited the characteristics of thermal shrinkage aggregation with the increase of temperature. This study provided new ideas for the design of self-assembled peptide materials with thermo-sensitive response and triple redox properties and provided an inspiration for the further development of multifunctional nanomaterials.

In another work, Shi et al. synthesized the 1-(2-azidoethyl)-4-methylpiperazine-modified peptide by the "click" chemical reaction [18]. By adjusting its degree of polymerization and the block structure, a series of physiologically relevant responsive peptide hydrogels were developed. These hydrogels exhibited fast sol–gel transition

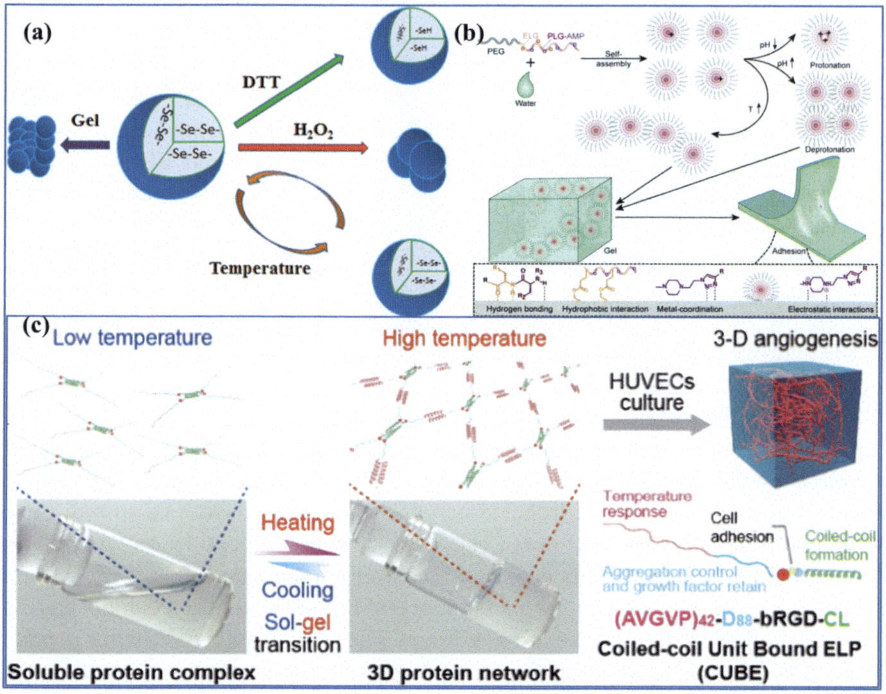

Fig. 4.1 **a** Triple redox/temperature-responsive peptide hydrogels of diselenide-containing homopeptides. Reproduced with the permission from Ref. [17]. Copyright 2018, Wiley–VCH. **b** Gelation and adhesion mechanisms of PPCA hydrogels. Reproduced with the permission from Ref. [18]. Copyright 2021, Royal Society of Chemistry. **c** Thermo-responsive protein-like peptide hydrogel. Reproduced with the permission from Ref. [20]. Copyright 2020, American Chemical Society

behavior and reversible responsiveness, while exhibiting excellent adhesive properties and biocompatibility (Fig. 4.1b). The experimental results showed that the prepared hydrogels had good gel properties in the range of 10–70 °C by optimizing the composition and structure of the peptide. When the concentration of peptide was as low as 1.5% (w/v), the peptide was gelled under different pH and temperature conditions. A reversible sol–gel transition occurred. In addition, in terms of the unique adhesion, the strength of the adhesion properties of the peptide hydrogels was related to the mole fraction of triazole rings and N, N′-substituted piperazine groups. In another study, Wang et al. developed a thermo-responsive peptide fusion strategy that could effectively prolong the drug half-life and improve the pharmacokinetic limitations [19]. In the experiment, a therapeutic protein-interferon α (IFN) was fused with a thermo-responsive elastic peptide (ELP) gene to prepare a thermo-responsive fusion interferon (IFN-ELP). The phase transition temperature (Tt) of the IFN-ELP was lower than the local body temperature. When the hydrogel was injected subcutaneously, the IFN-ELP was phase-separated from the solution to form an IFN-ELP pool.

Mizuguchi et al. developed a genetically engineered thermo-responsive protein-like peptide hydrogel [20]. In the experiment, the ELP peptide $(AVGVP)_n$ was used as a thermo-responsive module, and the polyaspartic acid (poly D) chain and Coil-LL peptide (CL) were fused to the ELP peptide sequence to prepare a hydrogel (Fig. 4.1c). The synthesized coil-coil unit-combined ELPs (CUBEs) exhibited controllable sol–gel transition ability and tunable mechanical properties. The CUBE hydrogel was further modified with the pro-angiogenic activity bone sialoprotein-derived RGD peptide (bRGD) for the synthesis of bRGD-CUBE, which promoted the growth of cells. In addition, angiogenic GFs were immobilized onto the bRGD-CUBEs by electrostatic interaction. It was found that the bRGD-CUBE hydrogel effectively enhanced the angiogenic activity of human umbilical vein endothelial cells (HUVEC).

In addition, the addition of specific functional sequences when designing thermo-responsive peptides is a new strategy for synthesizing stimuli-responsive smart peptide nanomaterials. For instance, Liu et al. demonstrated the use of bioactive peptide sequences to create hydrogel materials that respond to external stimuli, such as the pH, ionic strength, and heat [21]. They designed a peptide containing two antimicrobial amino acid sequences and a central tetrapeptide linker, which could undergo a structural transition from a random coil to a stable unimolecular β-hairpin conformation and then form an elastic hydrogel. The assembly kinetics of this peptide were studied using the time-lapse atomic force microscopy and the Thioflavin T binding assay. The formed peptide hydrogel exhibited intrinsic antibacterial activity against *E. coli*. In addition, it revealed potential applications in drug delivery, tissue engineering, and regenerative medicine.

4.3 pH-Responsive Peptides

The pH-responsive peptides can switch their structures under different pH conditions to adjust their properties and functions, while the presence of basic and acidic amino acids can enable the response mechanisms related to both protonation and deprotonation. Various pH-sensitive peptide systems have been widely used in tissue engineering, regenerative medicine, drug delivery, and bio-diagnostics previously. For instance, Cote et al. discussed the self-assembly of peptide amphiphiles (PAs) that respond to the pH changes for the creation of nanofibers [22]. Computer simulations were utilized to understand the peptide self-assembly (PSA) process, and the results showed that under strong electrostatic repulsion, spherical micelles formed, while under weak or no electrostatic repulsion, the micelles changed into β-sheets between peptide segments to induce the formation of nanofibers. This study also presented a transition between random coils and β-sheets in the pH range 6–7, matching the obtained circular dichroism (CD) data. This study provided insights into the pH-dependent self-assembly of PAs, which is expected to advance the computer-aided design and discovery of pH-responsive peptide nanomaterials.

The hydrophobic/hydrophilic balance plays a key role in the self-assembly and gelation of peptides. Ghosh et al. designed a naturally occurring galectin-1 β chain peptide fragment (Phe-Leu-Leu-Asn-Leu-Gly-Lys-Asp, PEP-1), replacing the Gly motif in the peptide sequence with Ala to increase the tendency of β-sheet folding [23]. The Leu motif in the peptide sequence was replaced by Ala to balance the hydrophobic/hydrophilic interaction, and the PEP-1 self-assembled in aqueous solution. Depending on the pH value, the protonation state of Asp and Lys residues were different, so the PEP-1 had an antiparallel β-sheet conformation secondary structure and its self-assembly was respond to the solution pH. Noncovalent interactions, such as hydrophobic interactions between alkyl groups, intermolecular H-bonds between amide groups, and electrostatic interactions between oppositely charged amino acids, were responsible for the self-assembly of PEP-1 and the formation of nanofibers.

Shimanovich et al. reported that the stimuli-responsive self-assembly of amyloid peptides could be used to construct environment-responsive microgel capsules with release properties [24]. As shown in Fig. 4.2a, amyloid peptides formed fibril hydrogels at pH 8.0, decomposed at lower pH values, and exhibited the best stability near neutral pH. The hydrogel was able to deposit onto the targeted sites and slowly dissolve as the pH changes to 7 under physiological conditions. These microcapsules were biocompatible and degradable and could be used for controlled release of a wide range of biomolecules. In addition, the formation of these microcapsules was reversible, and the release of the encapsulated molecules (such as antibodies) was controllable by adjusting the pH of the solution.

Xue et al. designed a novel biomimetic nanodrug carrier based on a neutrally charged polyglutamic acid lysine-cysteine peptide (p(EK–co–C)) [25]. The tumor-targeting properties of the designed peptide triggered by high pH value of the carrier could effectively prolong the drug circulation time and accelerate the intracellular uptake of the drug (Fig. 4.2b). A part of the thiol group of the cysteine side chain

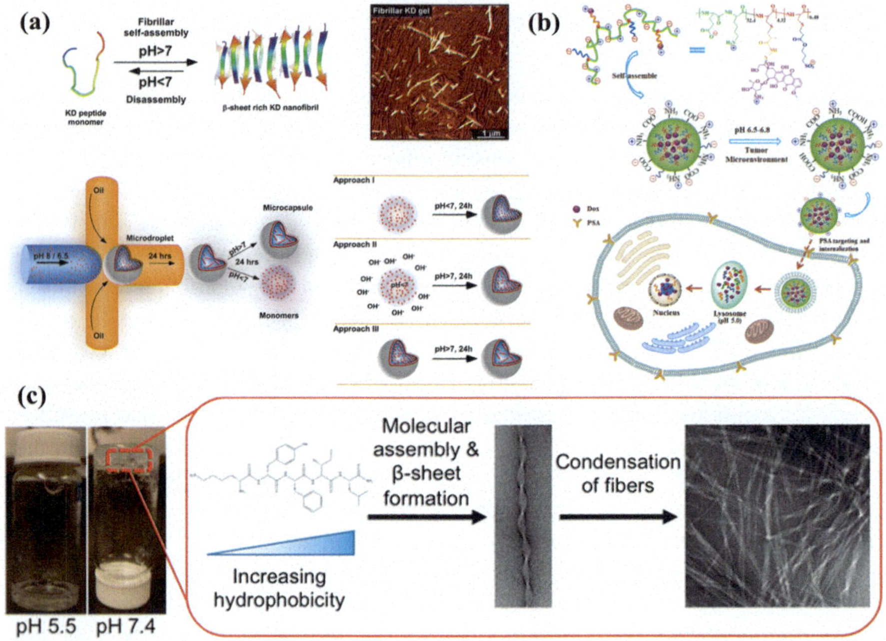

Fig. 4.2 **a** pH-sensitive self-assembly and dis-assembly of PNFs. Reproduced with the permission from Ref. [24]. Copyright 2021, Wiley–VCH. **b** pH-triggered tumor targeting by zwitterionic peptide nanodrug carriers. Reproduced with the permission from Ref. [25]. Copyright 2020, American Chemical Society. **c** Design and structure of RAPID hydrogel. Reproduced with the permission from Ref. [27]. Copyright 2019, American Chemical Society

was changed to a sulfonic acid group, and a hydrazide bond was used to couple the rest with doxorubicin (DOX). It was found that the biomimetic nanomedicine not only prolonged the blood circulation time by reducing the recognition to the reticuloendothelial system, but also effectively targeted solid tumors. The accumulation of the peptide-based nanomedicine in the tumor was significantly higher than that in the liver and kidney, suggesting that the synergistic effect of long circulation time and fast cellular internalization of peptide nanomedicines significantly enhanced the tumor targeting.

In a recent study, Meleties and co-workers explored the gelation properties of helicoid peptide Q-based hydrogels at different pH conditions [26]. At pH 7.4 and 10, the Q peptide self-assembled to form fiber-based hydrogels, and at pH 6, the Q formed polydisperse nanoparticles and could not further self-assemble to form hydrogels. Studies have shown that pH was one of the factors for controlling the Q self-assembly, and hydrogels with different elasticity were obtained according to the pH value, which was helpful for the development of novel scaffolds and functional biomaterials that sensitive to biological stimuli, especially in the fields of drug delivery and disease therapy.

Tang et al. developed a novel short pentapeptide that was capable of self-assembling into hydrogels with nanofibrous structure [27]. In their study, a combined computational and experimental approach was used to design, synthesize, and characterize this pH-triggered self-assembling pentapeptide through evaluating its cytocompatibility under physiological conditions. This pentapeptide formed stable hydrogels at low concentrations with tunable mechanical and structural properties. In the cell culture medium, the peptide self-assembled into long fibrils with a sequence-dependent nanofibrillar structure, only tens of nm in diameter, with a periodicity similar to that of amyloid fibrils (Fig. 4.2c). The properties of this hydrogel system made it suitable as an injectable biomaterial for drug delivery and tissue engineering applications.

4.4 Photo-Responsive Peptides

Photo-responsive peptides can undergo the isomerization, degradation, or dimerization reactions induced by light and have broad application prospects. Photo-responsive peptides are smart materials with controllable light-responsive properties that are easy to manipulate and highly efficient. Among them, the o-nitrophenyl group, as a common photodegradable group, has been widely used for the fabrication of smart hydrogels, nano-templates, and drug nanocarriers. The introduction of light-responsive molecules that can be manipulated non-invasively under highly controlled conditions is of great value for the study of complex biological systems.

Zhang et al. developed a novel phosphinated peptide (FmocFFpS^{C}(oNB)-PEG) that could assemble into a photo-responsive peptide hydrogel under the condition of physiological concentration of calcium ions (Fig. 4.3a) [28]. The composition of the hydrogel, Fmoc-FF motif, o-nitrobenzyl-protected phosphiserine, short PEG chain, and C-terminal carboxylic acid could be dissolved under the irradiation of light to release the encapsulated drug. Phosphorylation provided a way to regulate and induce the charge repulsion between the Fmoc-FF peptides, and light removed (o-nitrobenzyl) oNB from phosphorylated serine, thereby generating the repulsion and disrupting the peptide hydrogel network. The ability of the hydrogel to encapsulate and release DOX using UV light with a wavelength of 365 nm was also tested, and it was found that DOX could be effectively encapsulated and released using this peptide hydrogel. This study demonstrated that the designed FmocFFpS^{C}(oNB)-PEG hydrogel was capable of spatially and temporally controlling the drug release as a smart drug delivery system.

Roth-Konforti et al. designed and synthesized a novel peptide-based hydrogel (Nvoc-FF) for controlled delivery of insulin-fluorescein isothiocyanate (insulin-FITC) [29]. The hydrogel responded to the radiation of UV light with FF peptide connected to the photolyzable 6-nitroveratryloxycarbonyl (Nvoc) of the tail unit as a trigger, and the Nvoc-FF was completely depolymerized under the UV irradiation. Therefore, the insulin-FITC was gradually released, and the amount released was linearly related to the stimulation duration. The created Nvoc-FF exhibited

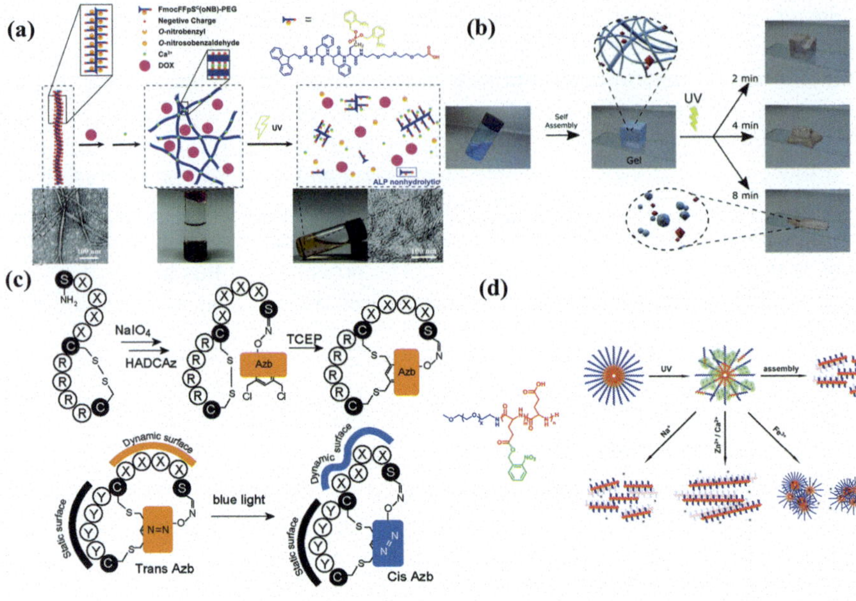

Fig. 4.3 Design and synthesis of photo-responsive peptide nanomaterials: **a** FmocFFpS^{C}(oNB)-PEG hydrogel. Reprinted with the permission from Ref. [28]. Copyright 2020, Royal Society of Chemistry. **b** Nvoc-FF hydrogels. Reprinted with the permission from Ref. [29]. Copyright 2018, Wiley–VCH. **c** bicyclic macrocycles of peptide. Reprinted with the permission from Ref. [30]. Copyright 2018, Royal Society of Chemistry. **d** PEG-b-PNBG copolymer nanomaterials. Reprinted with the permission from Ref. [31]. Copyright 2019, American Chemical Society

robust mechanical properties to self-assemble into 3D hydrogels with high storage modulus and microfibrous structure (Fig. 4.3b). In addition, the Nvoc-FF hydrogels could potentially be used as nano-platforms for the delivery of relatively large active molecules. This novel photo-responsive peptide-based hydrogel offered a promising strategy for the development of new drug delivery platforms.

In another study, Jafari et al. reported a strategy to generate photo-responsive bicyclic peptides from linear peptides using tridentate C2-symmetric hydroxylamines and dichlorobenzene (Hadcaz) linkers containing azobenzene (Fig. 4.3c) [30]. The resulting bicyclic peptide contained two amino acid rings, one of which was constrained by an azobenzene moiety that can change the conformation under visible light. The bicyclic peptides offered enhanced functions, such as improved binding and stability, compared to monocyclic peptides. This study demonstrated the use of the Hadcaz linker in a variety of synthetic bicyclic peptides with loops containing 2–5 amino acids, and the methods used could be used to design double-ring photo-responsive architectures with better binding and stability for various biomedical applications.

Ji et al. demonstrated the use of PEG-block poly(o-nitrophenylglycine) (PEG-b-PNBG) copolymers to construct intelligent peptide-based nanomaterials via the PSA [31]. As shown in Fig. 4.3d, under the irradiation of UV light, this copolymer realized the morphological transition from spherical micelles to cylindrical ones, which was caused by the change of the affinity and hydrophobicity that induced by the cleavage of the o-nitrophenyl group on the side chain. In addition, the study also systematically investigated the effects of the pH, molecular weight, and secondary structure on the self-assembly behavior of block peptide polymers. It was found that the changes in the secondary structure were the determining factors for the spherical–cylindrical transition. This research provided a new approach to realize the response to the biological environment through the polymer self-assembly. The combination of photo-responsiveness and ion-responsiveness enabled the construction of more complex multifunctional nanomaterials with extended applications.

4.5 Metal Ion-Responsive Peptides

The interaction of peptides with metal ions is ubiquitous in nature and can be used to control the PSA to form desired nanomaterials. Selegård et al. reported the dimerization of a peptide (JR2E) induced by Zn^{2+} and the coupling to low molecular weight hyaluronic acid (HA) to form tunable hydrogels (Fig. 4.4a) [32]. Zn^{2+} caused the peptide to form a helical structure, leading to the hydrogel formation and rapid depolymerization by removing Zn^{2+}. Encapsulating hydrolytic enzymes in hydrogels could program their degradation timing. In addition, the hydrogel also exhibited potential antioxidant properties.

Afrasiabi et al. reported the design and synthesis of two new ferrocene-dipeptides, which formed organogels in the organic solvents [33]. The prepared organogels exhibited reversible responses to various external signals such as the heat, sound, redox, and mechanical stress. Both peptides, Boc-L-Phe-L-Lys(Z)-OMe and its analog Boc-L-Lys(Z)-L-Phe-Ome, revealed the ability to form organogels by adding ferrocene carboxylic acid. In a recent study, Zhou and co-workers synthesized a novel dual-responsive PA hydrogel based on the repeat-in-toxin (RTX) motif, which rapidly formed PNFs at the condition of Ca^{2+} and low pH [34]. The PA adopted a β-sheet structure, and the self-assembly of the PA molecule was accompanied by the formation of long interconnected nanofibril superstructure. After the systematic residue substitution, the aspartic acid in the PA was found to be responsible for the selectivity of pH and Ca^{2+}. The environmental responsiveness, secondary structure, morphology, and mechanical properties of this PA made it a good candidate for biomedical and engineering applications.

In another case, Fu et al. developed a strategy to construct nonsteroidal anti-inflammatory drug (NSAID)-conjugated diphenylalanine peptide (NSAID-FF) hydrogels by adding metal ions [35]. The ions of a variety of essential metal elements were introduced into the NSAID-FF aqueous solution, which was used in a wide range of pH values (neutral or alkaline conditions) to form stable peptide hydrogels (Fig. 4.4b). These hydrogels with a 3D nanofibrous network exhibited excellent mechanical properties and thixotropy, as well as excellent responsiveness to a variety of metal ions (Fig. 4.4c). The formation mechanism of the hydrogels was ascribed

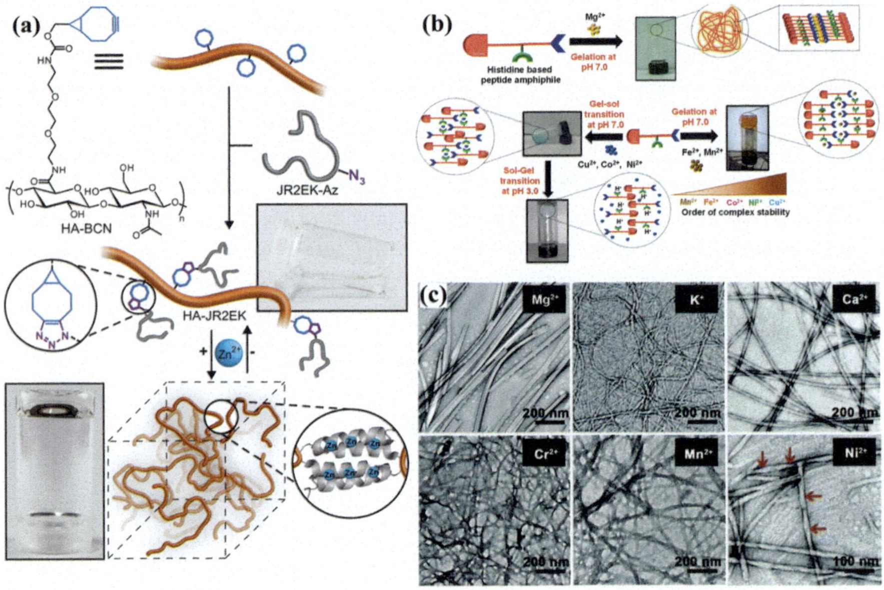

Fig. 4.4 **a** Self-assembly of HA-BCN into hydrogels. Reprinted with the permission from Ref. [32]. Copyright 2017, Nature Publishing Group. **b** Self-assembly behavior of Fmoc-HL-COOH peptide in the presence of metal salt. Reprinted with the permission from Ref. [36]. Copyright 2019, American Chemical Society. **c** Corresponding TEM images of nanofibers in hydrogels. Reprinted with the permission from Ref. [35]. Copyright 2020, Royal Society of Chemistry

to the addition of metal ions, which enhanced the π-π stacking and binding sites of metal ions with the carbonyl/amide groups during the PSA process [36].

4.6 Enzyme-Responsive Peptides

The enzyme-responsive PSA is a new way to treat diseases by making full use of the overexpression of specific enzymes to deliver drugs directly to specific diseased tissues. This approach can minimize non-specific release to healthy tissues and improve therapeutic efficacy, while also being flexible and customizable.

Based on this principle, Carlini et al. reported a cyclic peptide that can achieve low-resistance injection, which self-assembled into hydrogels by the mediation of disease-associated enzymes (Fig. 4.5a) [37]. When the designed cyclic peptide reached the heart, it opened under the linearization action of matrix metalloproteinase (MMP) to form a self-assembling peptide (SAP), which then self-assembled into β-sheet nanofibers and subsequently formed a hydrogel scaffold. Using the enzyme response method to form a hydrogel scaffold solved the problem that the hydrogel could not be delivered through the catheter due to the high viscosity of the material and the short gelation time.

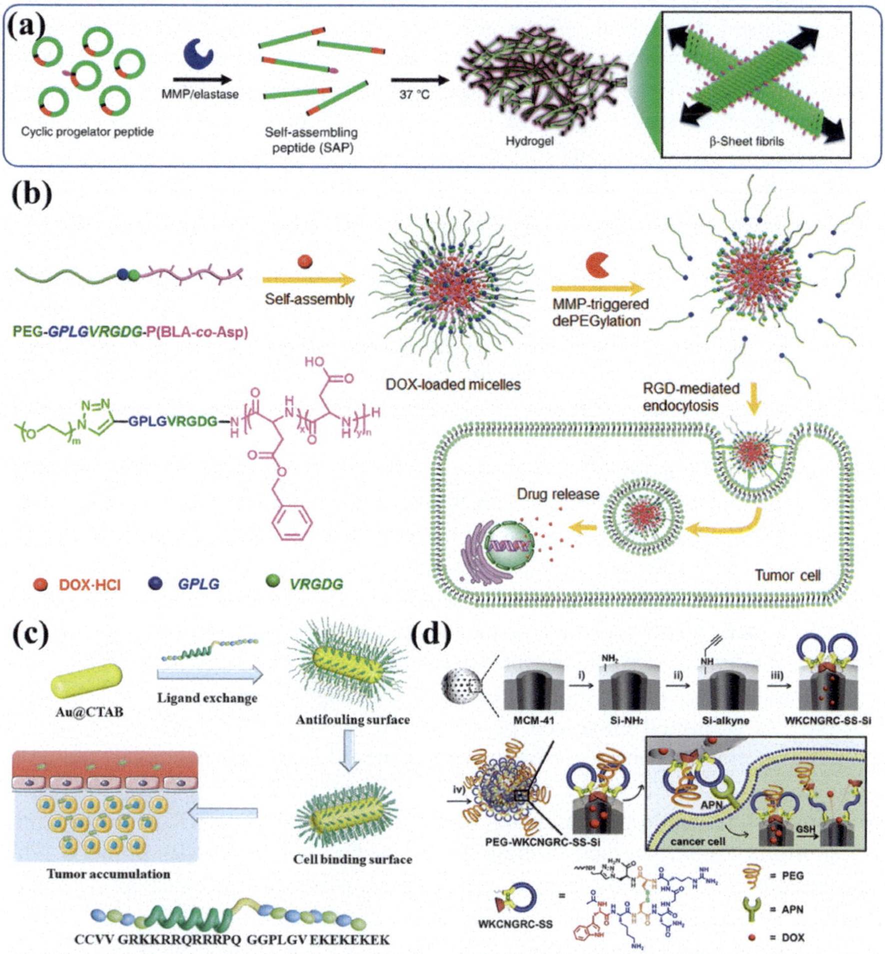

Fig. 4.5 **a** Design of a cyclic enzyme-responsive peptide for activatable gelation. Reprinted with the permission from Ref. [37]. Copyright 2019, Nature Publishing Group. **b** DOX-loaded enzyme (MMP)-responsive block copolymer micelle formation. Reprinted with the permission from Ref. [38]. Copyright 2016, American Chemical Society. **c** Peptide-capped AuNRs for photothermal therapy of cancer. Reprinted with the permission from Ref. [40]. Copyright 2019, Elsevier. **d** GSH-induced conformational transition of peptide self-assembly. Reprinted with the permission from Ref. [41]. Copyright 2017, Wiley–VCH

Ke et al. used the method of click chemistry to prepare the block polymer P3 to encapsulate DOX to achieve the purpose of delivering DOX and improve the effect of tumor treatment [38]. As shown in Fig. 4.5b, a peptide (P3, alkynyl-GPLGVRGDG) was introduced into the PEG termini, followed by the ring-opening polymerization of β-L-aspartic acid N-carboxylic anhydride (BLA-NCA) initiated by the terminal amino group. Under the synergistic effect of the hydrophobicity of the benzyl group and the partial electrophilicity of the carboxyl group, DOX was encapsulated by P3, which improved the stability of the nanodrug particles. When reaching the tumor cells, the drug delivery system was hydrolyzed under the action of MMP to release DOX, so as to achieve the purpose of targeted delivery of drugs. The use of enzyme-responsive peptides to deliver DOX provided useful ideas for the design of other systems for the anticancer drug delivery.

Cabazitaxel is a drug that used to treat metastatic castration-resistant prostate cancer. However, due to poor solubility, lack of targeting ability, and high systemic toxicity, the application of cabazitaxel has been greatly limited. To address this problem, Barve et al. used a biocompatible polymer-peptide micelle to encapsulate cabazitaxel for enhanced drug delivery [39]. This MMP-responsive material greatly improved the solubility of cabazitaxel in water, increased its ability to target cancer cells, and reduced its systemic toxicity. In addition, the encapsulation of Cabazitaxel in peptide micelles increased its activity, which was expected to improve the therapeutic effect toward prostate cancer.

In addition to the peptides with enzyme-responsive effects to deliver drugs, peptides can also be used to deliver AuNRs to achieve targeted therapeutic effects. For instance, Wu et al. developed an enzyme-responsive zwitterionic peptide coating to enhance the efficient accumulation of gold nanorods (AuNRs) with photothermal effects in tumors (Fig. 4.5c) [40]. The peptide contained the cell-penetrating peptide (TAT) sequence, MMP cleavable sequence, and zwitterionic antifouling sequence, which responded to MMP overexpressed in the tumor microenvironment, so that the accumulation of materials in tumor cells was significantly improved. In addition, AuNRs exhibited enhanced photothermal therapy (PTT) effect for tumors. The cell-penetrating peptides effectively penetrated cells and promoted the entry of AuNRs into tumor cells. In addition, two cysteines were designed at the end of the triblock functional peptide, which were firmly bound to AuNRs to form a reliable antifouling interface and prolong the circulation life of AuNRs in blood. This multifunctional hybrid composite of enzyme-responsive zwitterionic peptide-coated AuNRs provided a facile strategy for achieving enhanced PTT of cancer.

Li et al. prepared a mesoporous nanocarrier with a surface-modified stimuli-responsive bifunctional cyclic peptide [41]. As shown in Fig. 4.5d, drug molecules were stored in the pores of monodispersed mesoporous silica, and the surface was then modified with the enzyme-responsive cyclic peptides. Therefore, the modified Asn-Gly-Arg (NGR) bifunctional cyclic peptide targeted aminopeptidase N (APN) that expressed on the cancer cells, which was an overexpression in tumor endothelial cells and various tumor cell lines. In addition, glutathione (GSH) induced a conformational transformation of the peptide and triggered intracellular drug release behavior. GSH in cancer cells reduced the intramolecular disulfide bond of the cyclic peptide to two thiol units, leading to a conformational transition of the peptide gatekeeper from a circular conformation to a linear conformation, thereby releasing the drug molecules within the pores of the mesoporous silica. However, in the absence of GSH in normal cells, the drug was not released. This dual-responsive peptide drug carrier has shown obvious targeting ability and anti-tumor effect.

4.7 Conclusions

In summary, stimuli-responsive peptides are multifunctional, biocompatible, and easily customizable smart materials with broad application prospects. This chapter focused on the design, synthesis, and applications of various types of stimuli-responsive peptides, including the thermo-responsive, pH-responsive, light-responsive, metal ion-responsive, and enzyme-responsive peptides. Each type of peptide exhibited unique responses to specific environmental triggers and could be selected for desired biomedical applications. Thermo-responsive peptides were suitable for the applications requiring controlled drug release because they exhibited reversible thermo-sensitivity. Due to their ability to switch structures under different conditions, the pH-responsive peptides have been increasingly used in tissue engineering, regenerative medicine, bio-diagnostics, and drug delivery systems. The photo-sensitive peptides allowed non-invasive manipulation of materials, making them useful in complex biological systems. The metal ion-responsive peptides utilized the abundance and specificity of metal ions to regulate the PSA, while the enzyme-responsive peptides could target drug to specific diseased tissues with controlled delivery, minimizing the non-specific release effect.

References

1. G.-B. Qi, Y.-J. Gao, L. Wang, H. Wang, Self-assembled peptide-based nanomaterials for biomedical imaging and therapy. Adv. Mater. **30**, 1703444 (2018)
2. P. Katyal, M. Meleties, J.K. Montclare, Self-assembled protein- and peptide-based nanomaterials. Acs Biomater. Sci. Eng. **5**, 4132–4147 (2019)
3. T. Li, X.-M. Lu, M.-R. Zhang, K. Hu, Z. Li, Peptide-based nanomaterials: Self-assembly, properties and applications. Bioact. Mater. **11**, 268–282 (2022)
4. Y. Huo, J. Hu, Y. Yin, P. Liu, K. Cai, W. Ji, Self-assembling peptide-based functional biomaterials. ChemBioChem **24**, e202200582 (2023)
5. L. Wang, C. Gong, X. Yuan, G. Wei, Controlling the self-assembly of biomolecules into functional nanomaterials through internal interactions and external stimulations: a review. Materials **9** (2019)
6. W. Zhang, X. Yu, Y. Li, Z. Su, K.D. Jandt, G. Wei, Protein-mimetic peptide nanofibers: motif design, self-assembly synthesis, and sequence-specific biomedical applications. Prog. Polym. Sci. **80**, 94–124 (2018)
7. R.J. Mart, R.D. Osborne, M.M. Stevens, R.V. Ulijn, Peptide-based stimuli-responsive biomaterials. Soft Matter **2**, 822–835 (2006)
8. A. Shah, M.S. Malik, G.S. Khan, E. Nosheen, F.J. Iftikhar, F.A. Khan, S.S. Shukla, M.S. Akhter, H.-B. Kraatz, T.M. Aminabhavi, Stimuli-responsive peptide-based biomaterials as drug delivery systems. Chem. Eng. J. **353**, 559–583 (2018)
9. N.J. Sinha, M.G. Langenstein, D.J. Pochan, C.J. Kloxin, J.G. Saven, Peptide design and self-assembly into targeted nanostructure and functional materials. Chem. Rev. **121**, 13915–13935 (2021)
10. J. Yang, H.-W. An, H. Wang, Self-assembled peptide drug delivery systems. ACS Appl. Bio Mater. **4**, 24–46 (2021)
11. P. Zou, W.-T. Chen, T. Sun, Y. Gao, L.-L. Li, H. Wang, Recent advances: peptides and self-assembled peptide-nanosystems for antimicrobial therapy and diagnosis. Biomater. Sci. **8**, 4975–4996 (2020)

12. A. Levin, T.A. Hakala, L. Schnaider, G.J.L. Bernardes, E. Gazit, T.P.J. Knowles, Biomimetic peptide self-assembly for functional materials. Nat. Rev. Chem. **4**, 615–634 (2020)
13. W. He, S. Wang, J. Yan, Y. Qu, L. Jin, F. Sui, Y. Li, W. You, G. Yang, Q. Yang, M. Ji, Y. Shao, P.X. Ma, W. Lu, P. Hou, Self-assembly of therapeutic peptide into stimuli-responsive clustered nanohybrids for cancer-targeted therapy. Adv. Funct. Mater. **29**, 1807736 (2019)
14. Q. Guo, Y. Liu, G. Mu, L. Yang, W. Wang, J. Liu, J. Liu, A peptide–drug hydrogel to enhance the anti-cancer activity of chlorambucil. Biomater. Sci. **8**, 5638–5646 (2020)
15. F. Raza, H. Zafar, X. You, A. Khan, J. Wu, L. Ge, Cancer nanomedicine: focus on recent developments and self-assembled peptide nanocarriers. J. Mater. Chem. B **7**, 7639–7655 (2019)
16. A. Karimzadeh, M. Hasanzadeh, N. Shadjou, M. d. l. Guardia, Peptide based biosensors. TrAC, Trends Anal. Chem. **107**, 1–20 (2018)
17. Y. Gao, C.-M. Dong, Triple redox/temperature responsive diselenide-containing homopolypeptide micelles and supramolecular hydrogels thereof. J. Polym. Sci. Part A: Polym. Chem. **56**, 1067–1077 (2018)
18. Y. Shi, D. Li, J. Ding, C. He, X. Chen, Physiologically relevant pH- and temperature-responsive polypeptide hydrogels with adhesive properties. Polym. Chem. **12**, 2832–2839 (2021)
19. Z. Wang, J. Guo, J. Ning, X. Feng, X. Liu, J. Sun, X. Chen, F. Lu, W. Gao, One-month zero-order sustained release and tumor eradication after a single subcutaneous injection of interferon alpha fused with a body-temperature-responsive polypeptide. Biomater. Sci. **7**, 104–112 (2019)
20. Y. Mizuguchi, Y. Mashimo, M. Mie, E. Kobatake, Temperature-responsive multifunctional protein hydrogels with elastin-like polypeptides for 3-D angiogenesis. Biomacromol **21**, 1126–1135 (2020)
21. Y. Liu, Y. Yang, C. Wang, X. Zhao, Stimuli-responsive self-assembling peptides made from antibacterial peptides. Nanoscale **5**, 6413–6421 (2013)
22. Y. Cote, I.W. Fu, E.T. Dobson, J.E. Goldberger, H.D. Nguyen, J.K. Shen, Mechanism of the pH-controlled self-assembly of nanofibers from peptide amphiphiles. J. Phy. Chem. C **118**, 16272–16278 (2014)
23. G. Ghosh, R. Barman, J. Sarkar, S. Ghosh, PH-responsive biocompatible supramolecular peptide hydrogel. J. Phy. Chem. B **123**, 5909–5915 (2019)
24. U. Shimanovich, A. Levin, D. Eliaz, T. Michaels, Z. Toprakcioglu, B. Frohm, E. De Genst, S. Linse, K.S. Åkerfeldt, T.P.J. Knowles, PH-responsive capsules with a fibril scaffold shell assembled from an amyloidogenic peptide. Small **17**, 2007188 (2021)
25. W. Xue, A. Trital, J. Shen, L. Wang, S. Chen, Zwitterionic polypeptide-based nanodrug augments pH-triggered tumor targeting via prolonging circulation time and accelerating cellular internalization. ACS Appl. Mater. Interfaces **12**, 46639–46652 (2020)
26. M. Meleties, P. Katyal, B. Lin, D. Britton, J.K. Montclare, Self-assembly of stimuli-responsive coiled-coil fibrous hydrogels. Soft Matter **17**, 6470–6476 (2021)
27. J.D. Tang, C. Mura, K.J. Lampe, Stimuli-responsive, pentapeptide, nanofiber hydrogel for tissue engineering. J. Am. Chem. Soc. **141**, 4886–4899 (2019)
28. Y.-L. Zhang, R. Chang, H.-Z. Duan, Y.-X. Chen, Metal ion and light sequentially induced sol–gel–sol transition of a responsive peptide-hydrogel. Soft Matter **16**, 7652–7658 (2020)
29. M.E. Roth-Konforti, M. Comune, M. Halperin-Sternfeld, I. Grigoriants, D. Shabat, L. Adler-Abramovich, UV light-responsive peptide-based supramolecular hydrogel for controlled drug delivery. Macromol. Rapid Commun. **39**, 1800588 (2018)
30. M.R. Jafari, H. Yu, J.M. Wickware, Y.-S. Lin, R. Derda, Light-responsive bicyclic peptides. Org. Biomol. Chem. **16**, 7588–7594 (2018)
31. S. Ji, L. Xu, X. Fu, J. Sun, Z. Li, Light- and metal ion-induced self-assembly and reassembly based on block copolymers containing a photoresponsive polypeptide segment. Macromolecules **52**, 4686–4693 (2019)
32. R. Selegård, C. Aronsson, C. Brommesson, S. Dånmark, D. Aili, Folding driven self-assembly of a stimuli-responsive peptide-hyaluronan hybrid hydrogel. Sci. Rep. **7**, 7013 (2017)
33. R. Afrasiabi, H.-B. Kraatz, Stimuli-responsive supramolecular gelation in ferrocene-peptide conjugates. Chem. Eur. J. **19**, 17296–17300 (2013)

34. X.-R. Zhou, R. Ge, S.-Z. Luo, Self-assembly of pH and calcium dual-responsive peptide-amphiphilic hydrogel. J. Pept. Sci. **19**, 737–744 (2013)
35. W. Fu, Z. Farhadi Sabet, J. Liu, M. You, H. Zhou, Y. Wang, Y. Gao, J. Li, X. Ma, C. Chen, Metal ions modulation of the self-assembly of short peptide conjugated nonsteroidal anti-inflammatory drugs (NSAIDs). Nanoscale **12**, 7960–7968 (2020)
36. P. Sharma, H. Kaur, S. Roy, Inducing differential self-assembling behavior in ultrashort peptide hydrogelators using simple metal salts. Biomacromol **20**, 2610–2624 (2019)
37. A.S. Carlini, R. Gaetani, R.L. Braden, C. Luo, K.L. Christman, N.C. Gianneschi, Enzyme-responsive progelator cyclic peptides for minimally invasive delivery to the heart post-myocardial infarction. Nat. Commun. **10**, 1735 (2019)
38. W. Ke, J. Li, K. Zhao, Z. Zha, Y. Han, Y. Wang, W. Yin, P. Zhang, Z. Ge, Modular design and facile synthesis of enzyme-responsive peptide-linked block copolymers for efficient delivery of doxorubicin. Biomacromol **17**, 3268–3276 (2016)
39. A. Barve, A. Jain, H. Liu, Z. Zhao, K. Cheng, Enzyme-responsive polymeric micelles of cabazitaxel for prostate cancer targeted therapy. Acta Biomater. **113**, 501–511 (2020)
40. L. Wu, B. Lin, H. Yang, J. Chen, Z. Mao, W. Wang, C. Gao, Enzyme-responsive multifunctional peptide coating of gold nanorods improves tumor targeting and photothermal therapy efficacy. Acta Biomater. **86**, 363–372 (2019)
41. J. Lee, E.-T. Oh, Y. Han, H.G. Kim, H.J. Park, C. Kim, Mesoporous silica nanocarriers with cyclic peptide gatekeeper: specific targeting of aminopeptidase N and triggered drug release by stimuli-responsive conformational transformation. Chem. Eur. J. **23**, 16966–16971 (2017)

Chapter 5
Hybridization of Peptides and Metallic Nanoparticles

Youyin Xu and Gang Wei

Abstract Peptide nanomaterials via the molecular design and controllable self-assembly provide potential nanoscale building blocks for the synthesis of functional hybrid nanomaterials. Peptide can be hybridized with various nanomaterials, such as nanoparticles (NPs), quantum dots (QDs), nanowires, and two-dimensional (2D) materials (like graphene) through both covalent and noncovalent interactions. In this chapter, we carry out the case studies on the conjugation of self-assembled peptide nanomaterials with various metallic nanoparticles (MNPs), including gold, silver, and others. Meanwhile, the potential applications of the peptide/MNP hybrid nanomaterials for biosensors, bioimaging, drug deliver, and cancer therapy are presented and discussed.

Keywords Peptide nanomaterials · Nanoparticles · Quantum dots · Biomimetic synthesis · Hybrid materials

5.1 Introduction

Biomolecules can bind to the surface of metal nanoparticles (MNPs) through molecular-material interactions [1, 2]. Among these biomolecules, peptides are ideal ones that can be hybridized with MNPs. Although there are only 20 kinds of common amino acids in nature, due to the different types, numbers, arrangement, and spatial structure of the amino acids that make up the peptide sequences, there are countless kinds of peptides that can be used as precursor materials for hybridizing with MNPs [3, 4]. Peptides can form functional materials with wide applications, such as hydrogels, antibacterial membranes, and biological scaffolds. The applications of different properties of these peptide materials have become a hot research topic in recent years [5–7]. In addition, as an active biomolecule, peptide is also one of the most important and basic materials for connecting biology and chemistry as well as

Y. Xu · G. Wei (✉)
College of Chemistry and Chemical Engineering, Qingdao University, Qingdao 266071, China
e-mail: weigroup@qdu.edu.cn

G. Wei (ed.), *Peptide Nano-Chemistry and Nanotechnology*,
https://doi.org/10.1007/978-981-96-2387-7_5

nanotechnology, which reveals good biocompatibility and biodegradability, and can be used as raw materials for biomedical diagnosis and treatment.

With the development of materials science and nanotechnology, hybrid materials have been unanimously pursued by the scientific community. As an important representative of hybrid material, the synthesis of nanoscale hybrids has become a hot topic in the fields of material science, environmental science, biomedicine, and others. Nanoparticles (NPs) can be used as versatile nanoscale building blocks for the synthesis of nanoscale hybrids [8, 9]. Different from bulk metal materials, nanoscale metal materials have great application prospects in biomedicine [10], antitoxic and antibacterial materials [11], disease diagnosis [12], photothermal catalysis [13], and environmental monitoring [14]. Among all the MNPs, gold nanoparticles (AuNPs), silver nanoparticles (AgNPs), and other transition metal nanoparticles (TMNPs) have been synthesized successfully and widely used in last years [15–17]. Nanoscale metal materials can usually reach the scale of dozens of nm or even several nm, showing different morphologies, including NPs, nanorods (NRs), nanostars, and nanoclusters [18–21]. The synthesis of metal nanomaterials with different morphologies and different functions is of great significance for the development of material science, which ensures the development of nanomaterials in multi-level and multi-angle directions, and provides an important guarantee for their subsequent applications in various fields [22].

Different kinds of peptides show various properties and have unique application prospects when hybridized with MNPs. As shown in Fig. 5.1, using the capping effect of mercaptan onto AuNPs, Lévy et al. studied the effects of 58 different peptide sequences on the aggregation of AuNPs [23]. The peptides with different sequences showed specific stability after hybridizing with AuNPs, ascribing to the length, hydrophobicity, and charge of the peptide ligands. The hybridization of peptides and MNPs for the development of novel biomedical nanomaterials has become a hot topic, attracting numerous attentions, and showing broad application prospects in the fields of molecular targeting, cell imaging, drug delivery, and disease therapy [24, 25].

In this chapter, we present the hybridization of peptides with different kinds of MNPs for various applications. Firstly, based on the synthesis methods and main

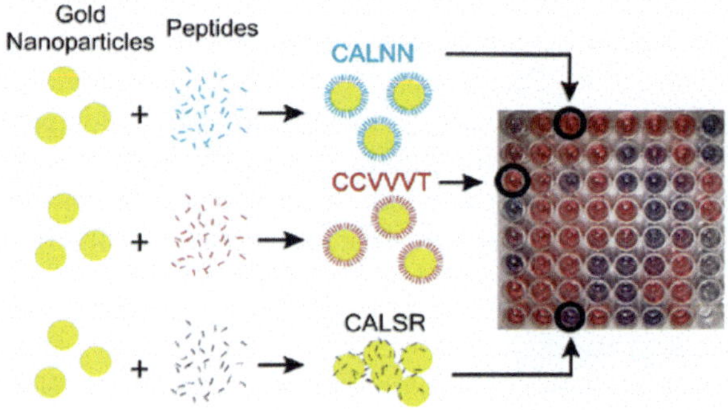

Fig. 5.1 Schematic diagram of AuNPs hybridization with peptides terminated by cysteine in different sequences. Reprinted with the permission from Ref. [23]. Copyright 2004, American Chemical Society

characterization methods, the hybridization of different kinds of peptides with AuNPs is introduced. A variety of applications of these hybrid materials are introduced, including the drug delivery, targeted diagnosis and treatment, and biological receptor detection. Secondly, the applications of the hybrids of antimicrobial peptides (AMP) and AgNPs in antitoxic and antibacterial activities as well as biosensors are briefly introduced. Finally, the hybridization of peptides with other MNPs is introduced, including the platinum nanoparticles (PtNPs) for photothermal therapy (PTT), copper nanoparticles (CuNPs) for the amplification detection, and iron oxide nanoparticles (FeNPs) with certain magnetic properties. Through the presentation of this chapter, readers can understand all kinds of the studies on the hybridization of peptides and MNPs and will obtain the inspiration on the design and synthesis of functional peptide/MNP nanohybrids.

5.2 Hybridization of Peptides with AuNPs

Peptides and AuNPs can be hybridized in many ways. The hybridization of peptides and AuNPs with different types and uses can be obtained by changing the types of peptides, the hybridization conditions, and the particle size of AuNPs [26]. This section mainly focuses on some technical methods for the synthesis of peptide/AuNP hybrids and their applications in drug delivery and disease diagnosis.

5.2.1 Hybridization Methods of Peptide and AuNPs

Alkyl mercaptan has been proved to be useful to combine with Au through the formation of Au–S bonds [27]. Therefore, in order to make peptide interact with AuNPs, the mercaptan group can be introduced into the peptide sequence and then covalently linked with AuNPs through the formation of Au–S bonds. Usually, the design strategy of peptides is to introduce cysteine (C) at the end, and the mercaptan group on the side chain of cysteine can form a covalent bond with the Au surface, so as to achieve the purpose of hybridization of peptides with AuNPs or the stabilization of AuNPs with peptides [28]. The synthesis of stable and uniform sized AuNPs is very important for the hybridization of AuNPs and peptides.

The preparation of AuNPs is usually achieved by the reduction of chloroauric acid with sodium borohydride ($NaBH_4$) and other reducing agents, which has been adopted in the last years, but the prepared AuNPs were easy to aggregate in water. To solve this problem, Cerra et al. used hydrophilic 3-mercapto-1-propane sulfonic acid (3MPS) to stabilize AuNPs, as shown in Fig. 5.2a [29]. In this process, the 3MPS used to coat AuNPs contained a strong negative charge, so the main force of AuNPs-3MPS in the solution was electrostatic repulsion, which made the NPs more dispersed. Subsequently, under the induction of melamine, the dispersed AuNPs

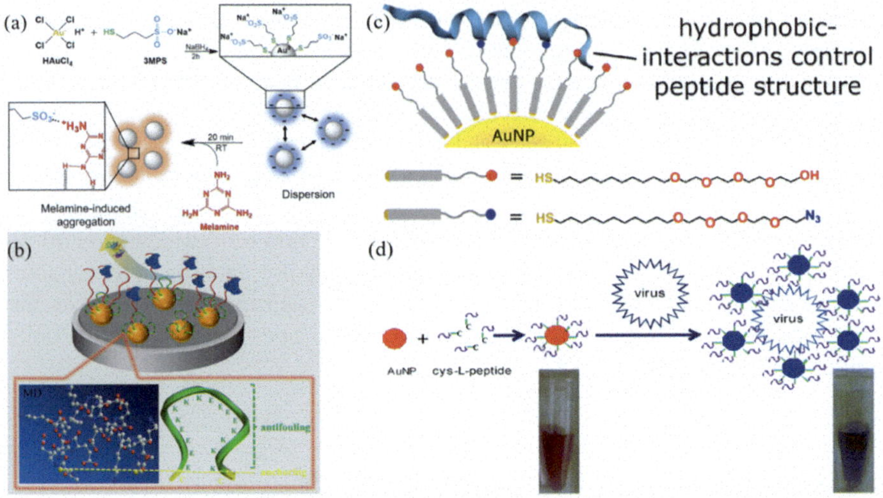

Fig. 5.2 **a** The schematic diagram of using 3MPS to stabilize AuNPs for the detection of melamine. Reprinted with the permission from Ref. [29]. Copyright 2021, Elsevier; **b** Self-assembled cyclic peptide-AuNPs hybrid material for antifouling adaptive sensor. Reprinted with the permission from Ref. [30]. Copyright 2021, American Chemical Society; **c** Oligoethylene glycol-terminated AuNPs induces the conformational transformation of peptides. Reprinted with the permission from Ref. [31]. Copyright 2020, American Chemical Society; **d** AuNP-Cys-peptide hybrids for visual detection of viruses. Reprinted with the permission from Ref. [32]. Copyright 2015, Springer Nature

gathered again, and the presence of melamine was successfully detected in the range of 2.5–500 ppm.

In another work, the peptide sequence (SH-C $(E_4K_4)_2$CSH) was designed by Chen et al., which contained the mercaptan groups at both ends, as shown in Fig. 5.2b. When hybridized with AuNPs, the mercaptan groups at both ends were connected to AuNPs, forming a very stable ring structure [30]. While achieving the goal of stable peptide/AuNPs hybrids, the formed hybrid materials could effectively resist the hydrolysis of self-assembled compounds by enzymes. Their work used the strong interaction between AuNPs and mercaptan to construct hybrid materials of cyclic peptides and AuNPs, which provided guidance and potential application value for the development of various anti-hydrolysis and low-pollution biosensors that suitable for complex biological fluids. In addition, the developed peptide/MNP-based sensors were suitable for organisms and exhibited excellent antifouling performance and high sensitivity, which greatly broadened their practical application for detecting analytes in blood, serum, sweat, or urine.

Wilder and co-workers reported that AuNPs were modified with a multifunctional polymer and then induced to form an irregular peptide with the α-helix configuration, as shown in Fig. 5.2c [31]. The multifunctional polymer consists of two parts, in which one end is an alkyl chain containing mercaptan group that can bind with AuNPs, and the other end is oligomeric ethylene glycol containing –OH or $-N_3$ group that can change the spatial configuration of the peptide. The peptide contained repetitive sequences $(LLKK)_n$ that modified by one or two reactive functional groups. Therefore, the specific sites of the peptides interacted with the –OH or $-N_3$, forming α-helical peptides tangent to AuNPs. This configuration transformation is important for the molecular recognition.

The AuNP-Cys-peptide hybrids based on cysteine have been used for visual detection of viruses, as shown in Fig. 5.2d [32]. The designed peptide containing mercaptan can interact with the glycoproteins on the surface of the Newcastle disease virus (NDV), resulting in the aggregation of AuNPs that linked to the peptides and the change of the color of the solution. In addition, the particle size before and after adding NDV was statistically analyzed. It was found that the particle size increased obviously after adding NDV, indicating that the aggregation of AuNPs was induced, which was also the reason for the change of the solution color.

5.2.2 Peptide/AuNP Hybrids for Drug Delivery and Disease Treatment

AuNPs have good anti-inflammatory effect and play an important role in the immune tolerance. Dul et al. covalently linked an antigen to AuNPs to produce a negatively charged AuNP-peptide complex of less than 5 nm, which were then applied for the antigen-specific immunotherapy [33]. In their study, the clinically approved microneedle delivery system (MicronJet600) was used, which greatly improved the delivery efficiency of NPs in the minimally invasive skin and increased the chance of targeting skin antigen presenting cells. During the test, the AuNP-peptide complex maintained good physical stability when extruded from the microneedles, and the complex injected into the isolated human skin could spread rapidly and widely in the whole dermis.

Peptide-functionalized AuNPs for the drug delivery systems can promote the cell–cell interactions and selectively target the diseased cells. In a typical study, Devi et al. used the peptide (BP100) with the cell-penetration function and the peptide (RGD) with targeting function to hybridize with AuNPs to form two hybrid complexes (BP100@AuNPs and RGD@AuNPs). Meanwhile, Adriamycin DOX was loaded onto both complexes to explore their delivery and therapeutic capabilities [34]. It was found that the peptide BP100 (KKLFKILKYL-amide) containing lysine with the cell-penetration ability exhibited enhanced DOX loading capacity and pH-responsive release ability after hybridizing with AuNPs compared with traditional RGD peptide. In addition, the drug loading, response, release sensitivity, and targeted cells of the BP100@AuNPs-DOX were optimized, which broke through the shackles of traditional targeted peptides in drug delivery, diagnosis, and treatment.

The hybridization of peptides and AuNPs can not only deliver drugs to cells, but also deliver drugs with diagnostic and therapeutic effects to sub-cells such as the lysosomes, mitochondria, and nuclei. For instance, Kang et al. hybridized 30 nm PEG-coated AuNPs with the RGD peptide and nuclear localization signal peptide

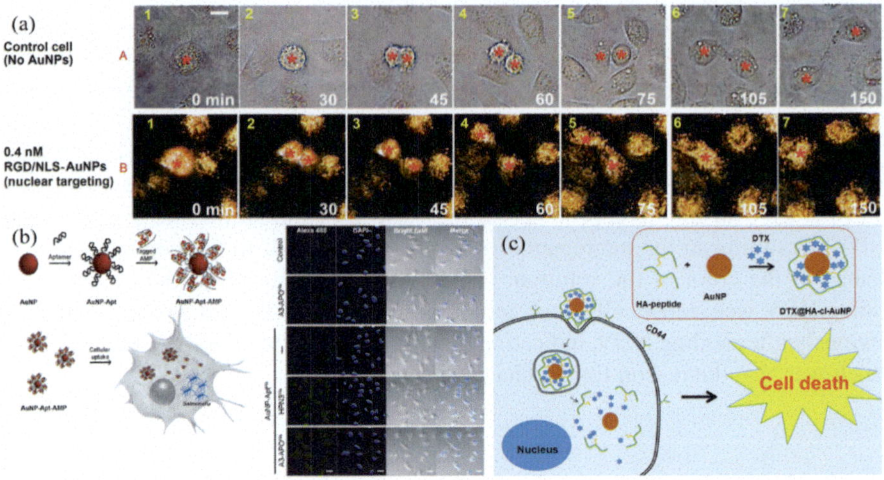

Fig. 5.3 **a** Real-time display image of cancer cell division with or without RGD/NLS-AuNPs. Reprinted with the permission from Ref. [35]. Copyright 2010, American Chemical Society; **b** AuNP-Apt-AMPs assembly and cell delivery. Reprinted with the permission from Ref. [36]. Copyright 2016, Elsevier; **c** HA-peptide and AuNP hybrid loading DTX for inactivation of cancer cells. Reprinted with the permission from Ref. [38]. Copyright 2018, American Chemical Society

(NLS) to form the RGD/NLS-AuNPs hybrid complex [35]. In this hybrid complex, RGD targeted the site on the cell surface, so that the hybrid complex entered the cytoplasm by the endocytosis. The KKKRK sequence in NLS bound specifically to the nuclear adhesin in the cytoplasm and shifted to the nucleus under the action of nuclear adhesin. As shown in Fig. 5.3a, the cell division was significantly stagnated in the presence of 0.4 nM nuclear targeting AuNPs. This stagnation mainly inhibited the division of the cytoplasm, thus interrupting the division and proliferation of cancer cells and reducing the spread of cancer cells.

AMPs are a kind of peptide drugs against bacteria in mammals, which show a good effect on the drug tolerance of diseased cells and can well eliminate *Salmonella typhimurium* in host cells. As shown in Fig. 5.3b, Yeom et al. used a delivery system coupled with AuNPs and DNA aptamer (AuNP-Apt) to efficiently transport AMPs to bacteria in host cells [36]. According to the characterizations of the survival rate of infected host cells, it was found that the delivery system ensured the stability of AMPs in host cells to some extent and reduced the toxicity toward host cells. In addition, the A3-APO (a kind of AMPs) transported by AuNP-Apt could kill *Salmonella typhimurium* invading cells by destroying the bacterial membrane resistant to gram-negative bacteria, thus significantly improving the survival rate of infected host cells.

The hybridization of AuNPs and peptides for entering the cell directly depends on the cell surface recognition ability of the cell-penetrating peptide, and the decisive factor for the destination of AuNPs into the cell is the diameter of AuNPs. In studying the relationship between the diameter of AuNPs and their location into cells, Oh et al. synthesized a series of AuNPs with the sizes ranging from 2.4 to 89 nm, hybridized the AuNPs with methoxy or carboxyl-terminated PEG dithiopentane through the formation of Au–S bonds, and then combined with the cell-penetrating peptides to explore the penetration ability of the complex to cells [37]. In addition, they further used silver staining, fluorescence anti-staining, and dual-mode fluorescence and non-fluorescence scattering to characterize the location of AuNP-peptide conjugates of different sizes in the COS-1 cell model. The results showed that the smallest (2.4 nm) AuNPs were located in the nucleus, while the middle (5.5 and 8.2 nm) AuNPs were partially transported to the cytoplasm, and 16 nm or larger AuNPs did not enter the cell and only located on the periphery of the cell.

It is well known that paclitaxel (DTX) is an effective anticancer drug for the treatment of many kinds of cancer, such as the breast, gastric, prostate, lung, and other cancers. Hyaluronic acid (HA) is a kind of polymer with high biocompatibility, which has been widely used in the fields of medicine and biomedicine. As a receptor of HA on cancer cells, CD44 can be overexpressed in a variety of cancer cells, so many studies have used HA as a carrier to achieve targeted drug delivery. As shown in Fig. 5.3c, Gotov et al. used a short peptide (Gly-Phe-Leu-Gly-Cys, GFLGC) to connect HA to the surface of AuNPs to form a hybrid material, HA-cl-AuNPs [38]. Subsequently, the anticancer drug DTX was loaded with the HA-cl-AuNPs, and HA could accurately locate the cancerous cells, thus achieving the purpose of targeted drug delivery to tumor cells. When the HA-cl-AuNPs loaded with DTX entered the cancerous cell, the short peptide GFLGC used for the binding action was hydrolyzed by cathepsin, thus releasing the anticancer drug. The therapy of mouse cancer cells showed that the HA-cl-AuNPs revealed targeted delivery of DTX in mitochondria to kill cancerous cells by inducing the apoptosis. This method enhanced the therapeutic effect of the drug and reduced the side effects of DTX to normal cells.

5.2.3 *Peptide/AuNPs for Biological Factor Detection*

Vascular endothelial growth factor receptor 1 (Flt-1) is a recognized biomarker for the detection of tumors. Wei et al. proposed a new peptide-based protein detection colorimetric assay for the detection of Flt-1 receptor protein, as shown in Fig. 5.4a [39]. The peptide sequence was WHSDMEWWYLLG-GGGGC, and this designed peptide sequence was highly purposeful. In this case, the cysteine end (C-terminus) containing thiols was used to connect with AuNPs, while the other end was a sequence end (N-terminal) that can bind to Flt-1. The intermediate sequence (GGGG) has a hydrophobic effect that reduces the spatial site resistance between the protein molecule and AuNPs. In addition, cucurbit [8] uril (CB [8]), which can selectively accommodate the two N-terminal aromatic residues of the peptide, was used to induce the aggregation of peptide-modified AuNPs, thus establishing a colorimetric strategy for the detection of proteins. The use of colorimetric method to detect tumors greatly reduced the cost of tumor detection and had special significance for biomedical diagnosis. In addition, this method of using peptides to modify AuNPs increased

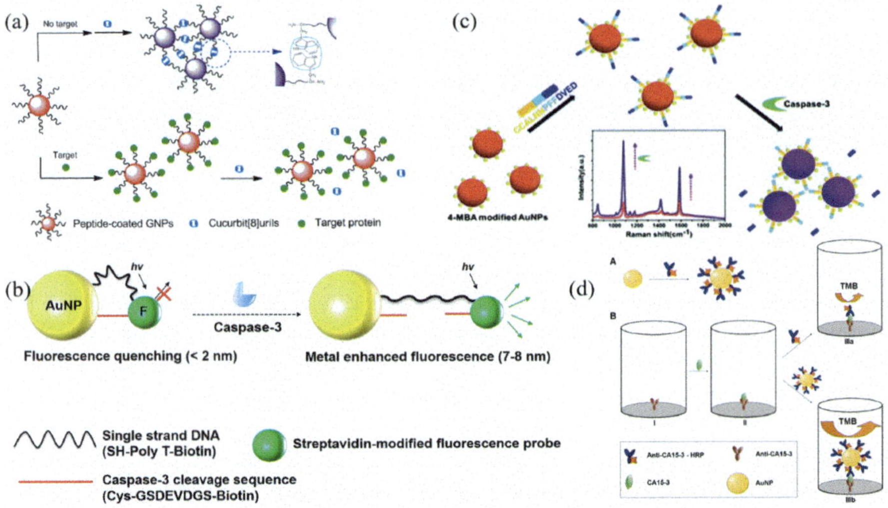

Fig. 5.4 **a** Peptide-coated AuNPs for the detection of Flt-1 receptor protein. Reprinted with the permission from Ref. [39]. Copyright 2015, Elsevier. **b** Peptide/AuNPs hybridization for detection of Caspase-3. Reprinted with the permission from Ref. [40]. Copyright 2020, American Chemical Society. **c** Modification of AuNPs with 4-MBA for the detection of Caspase-3. Reprinted with the permission from Ref. [41]. Copyright 2010, American Chemical Society. **d** Anti-CA15-3-HRP-AuNPs for ELISA procedure. Reprinted with the permission from Ref. [42]. Copyright 2018, Royal Society of Chemistry

the hybridization of peptides with MNPs in a way that allows for a wider range of applications.

There is a correlation between the amount of intracellular protein hydrolases and cancer or neurodegenerative diseases. If the amount of intracellular protein hydrolases is well observed, then the diagnosis of cancer and neurodegenerative diseases can be achieved at an early stage of their development, leading to early treatment and prevention. For this aim, Choi et al. designed a metal-enhanced fluorescence (MEF) bifunctional AuNPs biosensor for the first time, mainly for detecting Caspase-3, an intracellular protein hydrolase that is closely associated with cancer and neurodegenerative diseases [40]. The main construction of the MEF-based Caspase-3 sensor is shown in Fig. 5.4b, in which a single-stranded DNA and a peptide were used to link AuNPs to fluorescein isothiocyanate (FITC) in a form of a double bridge. Due to the short length of the peptide, the distance between AuNPs and FITC was less than 2 nm, and the original FITC fluorescence signal was burst. However, when Caspase-3 catalyzed the hydrolysis of the short peptide that used for the attachment, the distance between AuNPs and FITC increased to 7–8 nm, at which point FTIC exhibited a fluorescence excitation effect. Therefore, the use of MEF to enhance the effect of FITC fluorescence and the hydrolysis of Caspase-3 achieved efficient and accurate detection of Caspase-3 in the intracellular environment or even in the cell-free conformation.

Caspase-3 plays a crucial role in the apoptosis and its activity can be used as a key indicator of the apoptotic process. As shown in Fig. 5.4c, in another study, the hybridization of AuNPs with peptides for the development of probes to detect Caspase-3 was reported by utilizing two peptide sequences (CCALNNPFFDVED and CCALNNKYDDVED) in combination with AuNPs [41]. In the peptide sequence, the CALNN sequence containing thiol groups was used to bind and stabilize AuNPs, and the DVED sequence was used as a substrate for the Caspase-3 protease. In addition, the intermediate PFF or KYD plays an assembly role. When the peptide-modified AuNPs came into contact with the protease Caspase-3, the DVED fragment was recognizably hydrolyzed, then the PFF or KYD fragment was exposed, and the solution formed a colloidal system that could be used as a substrate for surface-enhanced Raman scattering (SERS). Meanwhile, the enhanced Raman signal correlated with the concentration of Caspase-3 with a detection limit of 1.5 ng/mL.

In another study, Ambrosi et al. used AuNPs as signal enhancers and antibody carriers to establish an optical enzyme-linked immunosorbent assay (ELISA) that allows more accurate analysis of CA15-3 antigen, as shown in Fig. 5.4d [42]. CA15-3 is an intracellular glycoprotein, its intracellular level is related to breast cell carcinogenesis or not, and the normal level is usually below 30 U/mL (units/mL). In this study, the effect of the enhanced signal of AuNPs greatly improved the performance of the classical ELISA assay, achieving a higher sensitivity with a detection limit of 30 U/mL, which allows for more timely analysis on the diagnosis of breast cancer.

5.3 Hybridization of Peptides with AgNPs

Peptides have biotherapeutic and target recognition abilities, while AgNPs have certain antibacterial and signaling effects, so the hybridization of peptides and AgNPs can be developed into the systems with biotherapeutic or biosensing effects [43–45]. In this section, we focus on the development of the synthesis of peptide/AgNP hybrids and their applications.

The targeting function of peptides is often used to localize certain specific structures, such as the glycoproteins in the cell membrane and enzymes in the cytoplasm [46, 47]. As shown in Fig. 5.5a, Li and co-workers modified a cyclic RGD peptide onto AgNPs by the click chemistry to develop a therapeutic nanodrug for targeting platelet-rich thrombin, denoted as cRGD-AgNPs [48]. A control experiment using the PEG-AgNPs (without targeting ability) and cRGD-AgNPs revealed that the cRGD-AgNPs inhibited the platelet activation to a certain extent and reduced the number of blood cells adhering to the vessel wall, thus reducing the possibility of thrombosis.

In addition to targeting biomolecules, peptides have antitoxic and antibacterial effects [49–51]. Compared with traditional antibiotics, highly biocompatible AMPs have become a new class of drugs against bacterial infections due to their broad-spectrum antimicrobial properties. As shown in Fig. 5.5b, Jin et al. first loaded

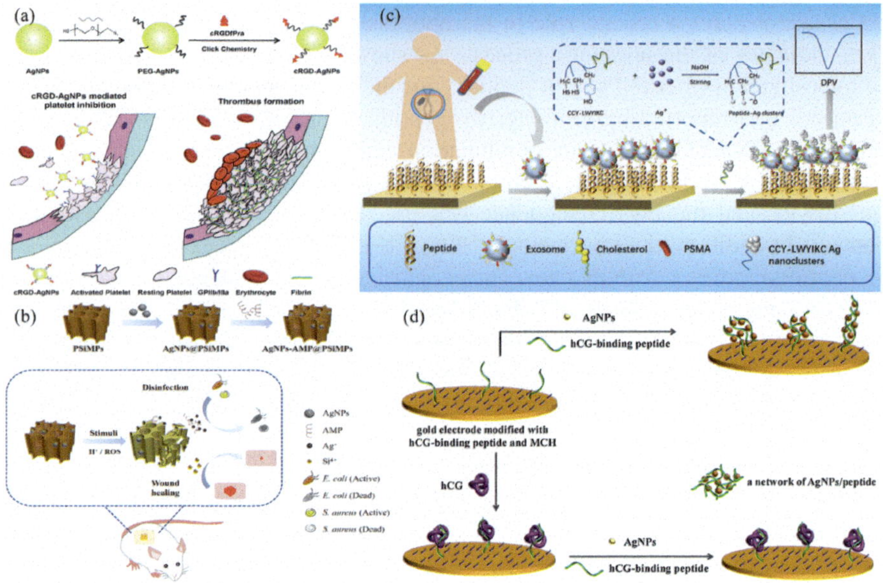

Fig. 5.5 a cRGD-AgNPs complex for thrombosis treatment. Reprinted with the permission from Ref. [48]. Copyright 2022, Elsevier. **b** AgNPs@PSiMPs loaded with AMPs for antimicrobial test. Reprinted with the permission from Ref. [52]. Copyright 2021, American Chemical Society. **c** Peptide/AgNPs for the capture of exosomes. Reprinted with the permission from Ref. [53]. Copyright 2023, Elsevier. **d** hCG-binding peptide induced the binding of AgNPs with electrode for electrochemical sensors. Reprinted with the permission from Ref. [54]. Copyright 2017, Elsevier

AgNPs into the micropores of PSiMPs and then introduced AMPs to form the AgNPs-AMP@PSiMPs complex [52]. When the AgNPs-AMP@PSiMPs reached the wound sites, the carrier PSiMPs slowly fragmented due to the stimulation of acid and reactive oxygen species (ROS), and AMPs and AgNPs released out violently, inactivating the bacteria. The complex exerted a synergistic effect of AgNPs and AMPs, ensuring an improved ability to fight bacterial infections while providing a great improvement in bacterial resistance. In another work, Zhou et al. designed a series of AMPs with osteogenic fragments and complexed the AMPs onto the surface of AgNPs through hydrogen bonding for the infection-triggered antibacterial purposes [11]. It was found that the AMP coating enhanced the ability of bone marrow stem cells (BMSCs) to adhere, spread, and divide, which accelerated the progress of the osseointegration and facilitated the recovery of postoperative patients.

Nanoprobe systems constructed by the hybridization of peptides and AgNPs have the advantages of high biocompatibility, low bacterial infection rate, and low detection limits, which led to the rapid development of biosensors in recent years. As shown in Fig. 5.5c, Cheng and co-workers have developed a sensitive electrochemical biosensor that can detect the prostate cancer-associated exosomes [53]. First, the peptide (CCY-LWYIKC) and silver ions were bound into peptide-Ag nanoclusters under alkaline conditions through the S–Ag bonds and electrostatic interaction force. Second, when exosomes from prostate cancer (pCa) were added to the solution, transmembrane proteins (PSMA) on the prostate cancer cells immobilized on the Au electrode trapped the exosomes and formed a detection array. Finally, when the peptide-Ag nanoclusters were added, the LWYIKC fragment on the peptide bound to the exosomes, generating a strong signaling response that was positively correlated with the number of exosomes within a certain range.

As shown in Fig. 5.5d, Xia et al. designed an electrochemical assay for the detection of human chorionic gonadotropin (HCG) [54]. AgNPs aggregated in response to the specific binding peptide of HCG, but this aggregation disappeared after the addition of HCG. Specific binding peptides of HCG were modified on the electrode as receptors for HCG as well as inducers of the AgNP aggregation, and their bifunctional properties were used to contain the electrochemical signal of HCG solutions and thus determine the level of HCG in human blood/urine. This method was consistent with the results measured by the ELISA technique without using costly antigens, resulting in significant advantage in assay costs.

5.4 Hybridization of Peptides with Other MNPs

In addition to AuNPs and AgNPs, peptides can be hybridized with other MNPs or metal oxide NPs for the synthesis of functional nanomaterials. Taking advantage of the diversity of peptides and the functionality of MNPs, many materials with various applications have been developed by hybridizing peptides with MNPs. [55, 56]

As shown in Fig. 5.6a, Ren et al. demonstrated a strategy for the delivery of intracellular peptides based on the metal–ligand interactions [57]. First, the tripyridine was grafted onto the G5 dendrimer and labeled as GT, which was then coordinated with a series of transition metals to form hybrids. The formed complexes were then used as carriers to transport a series of peptides into living cells via the metal-peptide interactions. It was found that the GT/Mn^{2+} exhibited higher efficiency in delivering peptides into living cells compared to the GT/Zn^{2+}. This method of using the metal ion-coordination polymers to transport peptides into living cells provided a useful idea for designing and developing peptide nanocarriers.

In another work, Li et al. used a mussel-derived peptide (NapGFF-DOPA-RGD) to bind Fe^{3+} to form complexes with intrinsic photothermal and chemical activities [58]. As shown in Fig. 5.6b, firstly the mussel-derived peptide was attached to the cancer-targeting peptide sequence (RGD) at one end, and the naphthyl portion was modified at the other end. Then the naphthalene portion and the FF dipeptide moiety self-assembled into nanofibers by noncovalent interactions. Finally, the formed nanofibers were co-assembled into nanoparticles in the presence of Fe^{3+} (NDFe) and encapsulated with adriamycin (DOX), labeled as NDFeD. The RGD ligand presented on the surface targeted cancer cells, and then, the complex decomposed under the action of light and heat, releasing Fe^{3+} and DOX, both of which killed cancer cells in a synergistic manner.

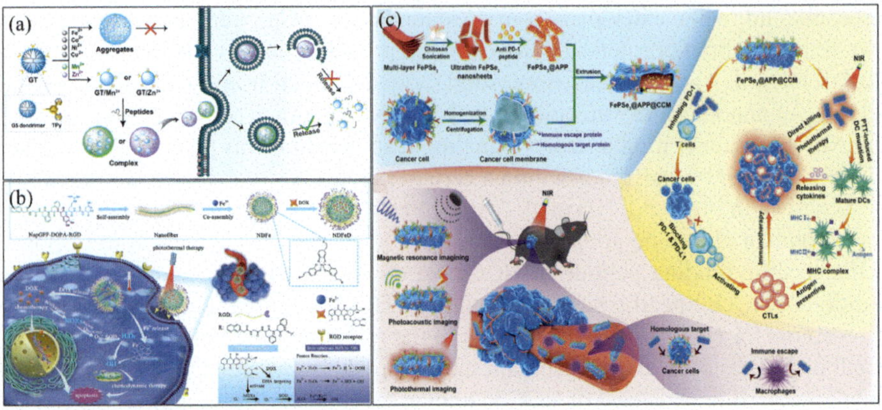

Fig. 5.6 **a** Dendrimer-bound metal ion delivery of peptides into living cells. Reprinted with the permission from Ref. [57]. Copyright 2022, Elsevier. **b** Cancer cell-directed peptides co-assembled with Fe^{3+} into nanofibrous for drug delivery. Reprinted with the permission from Ref. [58]. Copyright 2022, Elsevier. **c** Cancer cell membrane-encapsulated peptide hybridized with FePS3 for cell imaging. Reprinted with the permission from Ref. [62]. Copyright 2021, Wiley Co., Ltd

Photothermal therapy (PTT) has promising applications in the tumor treatment because it causes much less damage to normal tissue cells than other techniques [59, 60]. Peptides were used to stabilize platinum nanoparticles (PtNPs) and thus improve the PTT efficiency of PtNPs. Ma et al. developed PtNPs that encapsulated by the lipophilic triphenylphosphine (TPP) peptide (TPP-Pt) for targeting thermo-sensitive mitochondria, inactivating them under photothermal action to kill cancer cells [17]. Due to the encapsulation of TPP, PtNPs showed good monodispersity, high stability, and enhanced uptake by cancer cells, thus achieving high efficiency of PTT. In another work, Wang et al. developed a novel PTT method to target mitochondria using a small organic photothermal molecule (SOP) instead of the traditional photothermal agent for PTT [61]. Since biotin receptors were expressed much more in tumor cells than in normal tissue cells, they were rightfully the targeting ligands for the active tumors. Meanwhile, the TPP was used to target photothermal agents to thermo-sensitive mitochondria to induce the apoptosis in tumor cells.

In addition to PtNPs that can be used as PTT, metal phosphorus trisulfides (MPX3) as 2D nanomaterials can also be made into efficient photothermal agents for tumor therapy. As shown in Fig. 5.6c, Fang et al. loaded anti-PD-1 peptide (APP) onto chitosan (CS)-modified FePSe3 and then decorated the surface of the complex with cancer cell membranes from CT26 cells to ensure the integrality of the complex [62]. This all-in-one complex not only enabled effective intracellular multimodal imaging of tumors, but also showed high efficiency for synergistic cancer photothermal immunotherapy. Using peptides as capping ends, a shell-like stabilizer was formed around the metal atoms or clusters, thereby controlling the size of the MNPs and the range of the number of atoms in the metal clusters.

Valodkar and co-workers reported a method for the synthesis of copper nanoparticles (CuNPs) with antibacterial activity. Peptides from the latex of Euphorbia grandis plants were used as the capping agents to form peptide-encapsulated CuNPs (LCuNPs) by reducing copper ions [63]. The non-toxic aqueous LCuNPs could be used as a delivery vehicle for cancer therapy and was expected to induce apoptotic destruction of cancer cells.

The tumor formation and spread are closely related to polyadenosine diphosphate ribose polymerase-1 (PARP-1), therefore PARP-1 can be used as a biomarker for the tumor diagnosis, and the exploration of probes that can specifically bind to PARP-1 is essential for the cancer diagnosis. Wang et al. prepared functional CuNPs using a multifunctional peptide as a probe to detect PARP-1. The short peptide sequence CCY on this template peptide could be used for biomineralization, while the arginine (Arg) residue served as a signal marker for the PARP-1 [64]. Thus, not only CuNPs were prepared as reducing agents, but also the PARP-1 was autonomously and specifically recognized and labeled by the guanidine group on top of the peptide. Rather than simply binding to the substrate through non-specific electrostatic interactions, this labeling consisted of peptide-covered CuNPs probes that bind to the PARP-1 with strong covalent interactions, thus acting as a labeling polymerase. During the assay, the PARP-1 oxidized CuNPs to Cu^{2+} and then performed a precise determination of the substrate concentration by dissolution voltammetry. In addition, this peptide provided a template for the synthesis of CuNPs, which improved the efficiency and the selectivity of PARP-1 during the detection process.

5.5 Conclusions

In summary, the hybridization of peptides with MNPs is diverse and versatile. Different peptide sequences, different types of MNPs, and their hybridization methods ensure the richness of multifunctional composites involving peptides. Here, some schemes for hybridization of peptides with common MNPs and biomedical applications of the synthesized hybrid materials were presented. Through an in-depth analysis of the cases, we focused on the hybridization of the peptide-AuNPs and their applications in the disease diagnosis and treatment, as well as the biosensors. In addition, we described the prospects of the peptide-AgNPs for antibacterial applications and the progress of the hybridization of peptides with other MNPs for PTT and bioimaging. The cases introduced in this chapter are some of the more representative works on the hybridization of peptide and MNPs, which hopefully will inspire further design, synthesis, and advanced applications of novel peptide-based hybrid materials.

References

1. Y. Eygeris, S. Patel, A. Jozic, G. Sahay, Deconvoluting lipid nanoparticle structure for messenger RNA delivery. Nano Lett. **20**, 4543–4549 (2020)
2. A.V. Epanchintseva, J.E. Poletaeva, I.S. Dovydenko, B.P. Chelobanov, D.V. Pyshnyi, E.I. Ryabchikova, I.A. Pyshnaya, A lipid-coated nanoconstruct composed of gold nanoparticles noncovalently coated with small interfering RNA: preparation, purification and characterization. Nanomaterials **11**, 2775 (2021)
3. Q. Yang, J.R. Peng, K. Shi, Y. Xiao, Q.Y. Liu, R.X. Han, X.W. Wei, Z.Y. Qian, Rationally designed peptide-conjugated gold/platinum nanosystem with active tumor-targeting for enhancing tumor photothermal-immunotherapy. J. Control. Release **308**, 29–43 (2019)
4. X.J. Liu, J.H. Liu, S.S. Xu, X.F. Li, Z.H. Wang, X.A. Gao, B. Tang, K.H. Xu, Gold nanoparticles functionalized with Au-Se-bonded peptides used as gatekeepers for the off-target release of resveratrol in the treatment of triple-negative breast cancer. ACS Appl. Mater. Interfaces **15**(2), 2529–2537 (2023)
5. G. Fichman, E. Gazit, Self-assembly of short peptides to form hydrogels: design of building blocks, physical properties and technological applications. Acta Biomater. **10**, 1671–1682 (2014)
6. A. Guyomard, E. De, T. Jouenne, J.J. Malandain, G. Muller, K. Glinel, Incorporation of a hydrophobic antibacterial peptide into amphiphilic polyelectrolyte multilayers: a bioinspired approach to prepare biocidal thin coatings. Adv. Funct. Mater. **18**, 758–765 (2008)
7. F. Gelain, Z.L. Luo, M. Rioult, S.G. Zhang, Self-assembling peptide scaffolds in the clinic. Npj Regener. Med. **6**, 9 (2021)
8. S. Shrestha, B. Wang, P. Dutta, Nanoparticle processing: understanding and controlling aggregation. Adv. Colloid Interf. Sci. **279**, 102162 (2020)
9. D. Witzigmann, J.A. Kulkarni, J. Leung, S. Chen, P.R. Cullis, R. van der Meel, Lipid nanoparticle technology for therapeutic gene regulation in the liver. Adv. Drug Deliver. Rev. **159**, 344–363 (2020)
10. D. Zhang, X. L. Ma, Y. Gu, H. Huang, G.W. Zhang, Green synthesis of metallic nanoparticles and their potential applications to treat cancer (Retracted article. See vol. 10, 2023). Front Chem. **8**, 799 (2020)
11. W.H. Zhou, T. Bai, L. Wang, Y. Cheng, D.D. Xia, S. Yu, Y.F. Zheng, Biomimetic AgNPs@antimicrobial peptide/silk fibroin coating for infection-trigger antibacterial capability and enhanced osseointegration. Bioact. Mater. **20**, 64–80 (2023)
12. Z.H. You, Q.M. Qiu, H.Y. Chen, Y.Y. Feng, X. Wang, Y.X. Wang, Y.B. Ying, Laser-induced noble metal nanoparticle-graphene composites enabled flexible biosensor for pathogen detection. Biosens. Bioelectron. **150**, 111896 (2020)
13. C.B. Gao, F.L. Lyu, Y.D. Yin, Encapsulated metal nanoparticles for catalysis. Chem. Rev. **121**, 834–881 (2021)
14. A. Saravanan, P.S. Kumar, S. Karishma, D.V.N. Vo, S. Jeevanantham, P.R. Yaashikaa, C.S. George, A review on biosynthesis of metal nanoparticles and its environmental applications. Chemosphere **264**, 128580 (2021)
15. X.H. Liu, Q.W. Zhang, W.G. Knoll, B. Liedberg, Y. Wang, Rational design of functional peptide-gold hybrid nanomaterials for molecular interactions. Adv. Mater. **32**, 2000866 (2020)
16. Z. Wang, E.S. Xu, C.C. Wang, W. Wei, Y. Liu, S.Q. Liu, High specificity and efficiency electrochemical detection of poly(ADP-ribose) polymerase-1 activity based on versatile peptide-templated copper nanoparticles and detection array. Anal. Chim. Acta **1091**, 95–102 (2019)
17. Z.Y. Ma, Y.F. Zhang, J. Zhang, W.Y. Zhang, M.F. Foda, X.X. Dai, H.Y. Han, Ultrasmall peptide-coated platinum nanoparticles for precise NIR-II photothermal therapy by mitochondrial targeting. ACS Appl. Mater. Interfaces **12**, 39434–39443 (2020)
18. L.C. Liu, A. Corma, Confining isolated atoms and clusters in crystalline porous materials for catalysis. Nat. Rev. Mater. **6**, 244–263 (2021)

19. R. Sardar, A.M. Funston, P. Mulvaney, R.W. Murray, Gold nanoparticles: past, present, and future. Langmuir **25**, 13840–13851 (2009)
20. J. Perez-Juste, I. Pastoriza-Santos, L.M. Liz-Marzan, P. Mulvaney, Gold nanorods: synthesis, characterization and applications. Coord. Chem. Rev. **249**, 1870–1901 (2005)
21. S.M. Mousavi, M. Zarei, S.A. Hashemi, S. Ramakrishna, W.H. Chiang, C.W. Lai, A. Gholami, Gold nanostars-diagnosis, bioimaging and biomedical applications. Drug Metab. Rev. **52**, 299–318 (2020)
22. Y. Shen, X.H. Fu, W.X. Fu, Z.B. Li, Biodegradable stimuli-responsive polypeptide materials prepared by ring opening polymerization. Chem. Soc. Rev. **44**, 612–622 (2015)
23. R. Levy, N.T.K. Thanh, R.C. Doty, I. Hussain, R.J. Nichols, D.J. Schiffrin, M. Brust, D.G. Fernig, Rational and combinatorial design of peptide capping Ligands for gold nanoparticles. J. Am. Chem. Soc. **126**, 10076–10084 (2004)
24. M.J. Kogan, I. Olmedo, L. Hosta, A.R. Guerrero, L.J. Cruz, F. Albericio, Peptides and metallic nanoparticles for biomedical applications. Nanomedicine **2**, 287–306 (2007)
25. E.A. Egorova, M.P. Nikitin, Delivery of theranostic nanoparticles to various cancers by means of integrin-binding peptides. Int. J. Mol. Sci. **23**, 13735 (2022)
26. Z.C. Jin, N. Dridi, G. Palui, V. Palomo, J.V. Jokerst, P.E. Dawson, Q.X.A. Sang, H. Mattoussi, Evaluating the catalytic efficiency of the human membrane-type 1 matrix metalloproteinase (MMP-14) using AuNP-peptide conjugates. J. Am. Chem. Soc. **145**, 4570–4582 (2023)
27. M.-C. Daniel, D. Astruc, Gold Nanoparticles: assembly, supramolecular chemistry, quantum-size-related properties, and applications toward biology, catalysis, and nanotechnology. Chem. Rev. **104**, 293–346 (2004)
28. Z.C. Jin, J. Yeung, J.J. Zhou, Y. Cheng, Y. Li, Y. Mantri, T.Y. He, W. Yim, M. Xu, Z.H. Wu, P. Fajtova, M.N. Creyer, C. Moore, L. Fu, W.F. Penny, A.J. O'Donoghue, J.V. Jokerst, Peptidic sulfhydryl for interfacing nanocrystals and subsequent sensing of SARS-CoV-2 protease. Chem. Mater. **34**, 1259–1268 (2022)
29. S. Cerra, T.A. Salamone, F. Sciubba, M. Marsotto, C. Battocchio, S. Nappini, F.A. Scaramuzzo, R.L. Voti, C. Sibilia, R. Matassa, A.M. Beltran, G. Familiari, I. Fratoddi, Study of the interaction mechanism between hydrophilic thiol capped gold nanoparticles and melamine in aqueous medium. Colloid Surf. B **203**, 111727 (2021)
30. M. Chen, R. Han, W.Q. Wang, Y. Li, X.L. Luo, Antifouling aptasensor based on self-assembled loop-closed peptides with enhanced stability for CA125 assay in complex biofluids. Anal. Chem. **93**, 13555–13563 (2021)
31. L.M. Wilder, P.R. Handali, L.J. Webb, R.M. Crooks, Interactions between oligoethylene glycol-capped AuNPs and attached peptides control peptide structure. Bioconj. Chem. **31**, 2383–2391 (2020)
32. B. Sajjanar, B. Kakodia, D. Bisht, S. Saxena, A.K. Singh, V. Joshi, A.K. Tiwari, S. Kumar, Peptide-activated gold nanoparticles for selective visual sensing of virus. J. Nanopart. Res. **17**, 1–9 (2015)
33. M. Dul, T. Nikolic, M. Stefanidou, M.A. McAteer, P. Williams, J. Mous, B.O. Roep, E. Kochba, Y. Levin, M. Peakman, F.S. Wong, C.M. Dayan, D. Tatovic, S.A. Coulman, J.C. Birchall, E.-A. Consortium, Conjugation of a peptide autoantigen to gold nanoparticles for intradermally administered antigen specific immunotherapy. Int. J. Pharm. **562**, 303–312 (2019)
34. N. Devi, P. Singh, R. Sharma, M. Kumar, S.K. Pandey, R.K. Sharma, N. Wangoo, A lysine-rich cell penetrating peptide engineered multifunctional gold nanoparticle-based drug delivery system with enhanced cellular penetration and stability. J. Mater. Sci. **57**, 16842–16857 (2022)
35. B. Kang, M.A. Mackey, M.A. El-Sayed, Nuclear targeting of gold nanoparticles in cancer cells induces DNA damage, causing cytokinesis arrest and apoptosis. J. Am. Chem. Soc. **132**, 1517–1519 (2010)
36. J.H. Yeom, B. Lee, D. Kim, J.K. Lee, S. Kim, J. Bae, Y. Park, K. Lee, Gold nanoparticle-DNA aptamer conjugate-assisted delivery of antimicrobial peptide effectively eliminates intracellular Salmonella enterica serovar Typhimurium. Biomaterials **104**, 43–51 (2016)

37. E. Oh, J.B. Delehanty, K.E. Sapsford, K. Susumu, R. Goswami, J.B. Blanco-Canosa, P.E. Dawson, J. Granek, M. Shoff, Q. Zhang, P.L. Goering, A. Huston, I.L. Medintz, Cellular uptake and fate of pegylated gold nanoparticles is dependent on both cell-penetration peptides and particle size. ACS Nano **5**, 6434–6448 (2011)
38. O. Gotov, G. Battogtokh, Y.T. Ko, Docetaxel-loaded hyaluronic acid-cathepsin B-cleavable-peptide–gold nanoparticles for the treatment of cancer. Mol. Pharm. **15**, 4668–4676 (2018)
39. L.M. Wei, X.Y. Wang, C. Li, X.X. Li, Y.M. Yin, G.X. Li, Colorimetric assay for protein detection based on "nano-pumpkin" induced aggregation of peptide-decorated gold nanoparticles. Biosens. Bioelectron. **71**, 348–352 (2015)
40. J.-H. Choi, J.-W. Choi, Metal-enhanced fluorescence by bifunctional au nanoparticles for highly sensitive and simple detection of proteolytic enzyme. Nano Lett. **20**, 7100–7107 (2020)
41. A. Ambrosi, F. Airo, A. Merkoci, Enhanced gold nanoparticle based ELISA for a breast cancer biomarker. Anal. Chem. **82**, 1151–1156 (2010)
42. C.L. Zhao, L.H. Qiu, P. Lv, A.L. Han, G.Z. Fang, J.F. Liu, S. Wang, AuNP-peptide probe for caspase-3 detection in living cells by SERS. Analyst **144**, 1275–1281 (2019)
43. H.D. Beyene, A.A. Werkneh, H.K. Bezabh, T.G. Ambaye, Synthesis paradigm and applications of silver nanoparticles (AgNPs), a review. Sustain. Mater. Technol. **13**, 18–23 (2017)
44. W.Z. Wang, Z.Y. Hu, Targeting peptide-based probes for molecular imaging and diagnosis. Adv. Mater. **31**, 1804827 (2019)
45. W.S. Kuan, I. Ibrahim, S.P. Chan, Z. Li, O.W. Liew, C. Frampton, R. Troughton, C.J. Pemberton, J.P.C. Chong, L.L. Tan, W. Lin, S.B.S. Ooi, A.M. Richards, Mid-regional pro-adrenomedullin outperforms N-terminal pro-B-type natriuretic peptide for the diagnosis of acute heart failure in the presence of atrial fibrillation. Eur. J. Heart Failure **22**, 692–700 (2020)
46. C.D. Buckley, D. Pilling, N.V. Henriquez, G. Parsonage, K. Threlfall, D. Scheel-Toellner, D.L. Simmons, A.N. Albar, J.M. Lord, M. Salmon, RGD peptides induce apoptosis by direct caspase-3 activation. Nature **397**, 534–539 (1999)
47. J. Auernheimer, C. Dahmen, U. Hersel, A. Bausch, H. Kessler, Photoswitched cell adhesion on surfaces with RGD peptides. J. Am. Chem. Soc. **127**, 16107–16110 (2005)
48. Y.Y. Li, Y.Y. Jin, X.F. He, Y.H. Tang, M. Zhou, W.J. Guo, W.J. Miao, Cyclo(RGD) peptide-decorated silver nanoparticles with anti-platelet potential for active platelet-rich thrombus targeting. Nanomed. Nanotechnol. **41**, 102520 (2022)
49. M. Magana, M. Pushpanathan, A.L. Santos, L. Leanse, M. Fernandez, A. Ioannidis, M.A. Giulianotti, Y. Apidianakis, S. Bradfute, A.L. Ferguson, A. Cherkasov, M.N. Seleem, C. Pinilla, C. de la Fuente-Nunez, T. Lazaridis, T.H. Dai, R.A. Houghten, R.E.W. Hancock, G.P. Tegos, The value of antimicrobial peptides in the age of resistance. Lancet Infect. Dis. **20**, E216–E230 (2020)
50. C. Wang, T.T. Hong, P.F. Cui, J.H. Wang, J. Xia, Antimicrobial peptides towards clinical application: delivery and formulation. Adv. Drug Deliver. Rev. **175**, 113818 (2021)
51. P. Tan, H.Y. Fu, X. Ma, Design, optimization, and nanotechnology of antimicrobial peptides: from exploration to applications. Nano Today **39**, 101229 (2021)
52. Y. Jin, Y.D. Yang, W. Duan, X.T. Qu, J.M. Wu, Synergistic and on-demand release of Ag-AMPs loaded on porous silicon nanocarriers for antibacteria and wound healing. ACS Appl. Mater. Interfaces **13**, 16127–16141 (2021)
53. W. Cheng, Y. Sun, G. Zhao, A. Khan, J. Zhang, Z. Zhang, Y. Yi, D. Kong, J. Li, A novel peptide-templated AgNPs nanoprobe for theranostics of prostate cancer. Biosens. Bioelectron. **223**, 114978 (2023)
54. N. Xia, Z.H. Chen, Y.D. Liu, H.Z. Ren, L. Liu, Peptide aptamer-based biosensor for the detection of human chorionic gonadotropin by converting silver nanoparticles-based colorimetric assay into sensitive electrochemical analysis. Sens. Actuators B Chem. **243**, 784–791 (2017)
55. Z. Song, Y. Li, H. Teng, C.F. Ding, G.Y. Xu, X.L. Luo, Designed zwitterionic peptide combined with sacrificial Fe-MOF for low fouling and highly sensitive electrochemical detection of T4 polynucleotide kinase. Sens. Actuators B Chem. **305**, 127329 (2020)
56. D. Lukacs, M. Nemeth, L. Szyrwiel, L. Illes, B. Pecz, S.H. Shen, J.S. Pap, Behavior of a Cu-Peptide complex under water oxidation conditions—molecular electrocatalyst or precursor to nanostructured CuO films? Sol. Energy Mater. Sol. Cells **201**, 110079 (2019)

57. L.F. Ren, Y. Gao, Y.Y. Cheng, A manganese (II)-based coordinative dendrimer with robust efficiency in intracellular peptide delivery. Bioact. Mater. **9**, 44–53 (2022)
58. J. Li, S.J. Wu, X. Tian, X.M. Li, Fabrication of a multifunctional nanomaterial from a mussel-derived peptide for multimodal synergistic cancer therapy. Chem. Eng. J. **446**, 136837 (2022)
59. X.Y. Huang, Y. Lu, M.X. Guo, S.Y. Du, N. Han, Recent strategies for nano-based PTT combined with immunotherapy: from a biomaterial point of view. Theranostics **11**, 7546–7569 (2021)
60. D. Zhang, M. Wu, Y.Y. Zeng, L.J. Wu, Q.T. Wang, X. Han, X.L. Liu, J.F. Liu, Chlorin e6 conjugated poly(dopamine) nanospheres as PDT/PTT dual-modal therapeutic agents for enhanced cancer therapy. ACS Appl. Mater. Interfaces **7**, 8176–8187 (2015)
61. H.Y. Wang, J.J. Chang, M.W. Shi, W. Pan, N. Li, B. Tang, A Dual-targeted organic photothermal agent for enhanced photothermal therapy. Angew. Chem. Int. Ed. **58**, 1057–1061 (2019)
62. X.Y. Fang, X.L. Wu, Z.D. Li, L.J. Jiang, W.S. Lo, G.M. Chen, Y.J. Gu, W.T. Wong, Biomimetic anti-PD-1 peptide-loaded 2D $FePSe_3$ nanosheets for efficient photothermal and enhanced immune therapy with multimodal MR/PA/thermal imaging. Adv. Sci. **8**, 2003041 (2021)
63. M. Valodkar, R.N. Jadeja, M.C. Thounaojam, R.V. Devkar, S. Thakore, Biocompatible synthesis of peptide capped copper nanoparticles and their biological effect on tumor cells. Mater. Chem. Phys. **128**, 83–89 (2011)
64. J. Zhao, L.L. Yang, Y.H. Dai, Y.Y. Tang, X.Q. Gong, D.S. Du, Y. Cao, Peptide-templated multifunctional nanoprobe for feasible electrochemical assay of intracellular kinase. Biosens. Bioelectron.Bioelectron. **119**, 42–47 (2018)

Chapter 6
Hybridization and Applications of Peptides and Two-Dimensional Materials

Lei Guo

Abstract Peptides exist widely in the biological world and reveal high biological activity. Because of their excellent properties, high selectivity, and low toxicity, peptides have been widely used as biomedical nanodrugs, drug carriers, tissue engineering materials, and nutritious foods. At present, it is well known that peptides can interact with various nanomaterials to regulate the structure or function of peptide-based hybrids, which can improve the characteristics and functions of raw peptide nanomaterials. In this paper, we present recent advance in the combination of peptides and two-dimensional materials (2DMs). In addition, potential applications of the peptide/2DM hybrids in materials science, biomedicine, and nanotechnology are demonstrated and discussed.

Keywords Hybridization · Peptide nanomaterials · Two-dimensional materials · Biomedicine · Nanotechnology

6.1 Introduction

With the development of molecular biology and biochemical technology, the research on the synthesis and functionalization of peptides has made a breakthrough since the twentieth century [1, 2]. Peptide is a compound naturally produced by all organisms and an intermediate product of the protein hydrolysis, which is a promising material for the synthesis of functional nanomaterials [3, 4]. Peptide has high biological activity, and many active substances in human body exist in the form of peptide. At present, tens of thousands of endogenous peptides, including hormones and growth factors, have been found.

L. Guo (✉)
Institute of Biomedical Engineering, College of Life Science, Qingdao University, Qingdao 266071, China
e-mail: guoleibio@qdu.edu.cn

G. Wei (ed.), *Peptide Nano-Chemistry and Nanotechnology*,
https://doi.org/10.1007/978-981-96-2387-7_6

Peptide compounds have a wide range of physiological, immunomodulatory, and wound healing functions, unique physical and chemical properties, and some protective functions for microorganisms and tumor cells [5]. It can bind toxic compounds, heavy metals, and carcinogens that can enter the human body, and promote their excretion. In addition, it has the function of neutralizing and detoxifying, and is an excellent raw material for the design of drugs [6]. Since insulin was introduced nearly a century ago, peptides have gradually become a research hotspot. Up to now, there are hundreds of peptides in the world that can be used as drugs for marketing or clinical trials to treat many diseases, including diabetes, liver disease, and osteoporosis [7]. Because of their high selectivity and low toxicity, more and more functional peptide molecules have been developed and widely used in biomedical fields, such as the cancer treatment, bioimaging, drug screening, and biomolecular detection. However, due to the limitations of peptides, such as low membrane permeability, protein degradation sensitivity, and short half-life, how to realize efficient utilization of peptides in practical applications is still a challenge [8].

It is an effective strategy to combine peptides with other nanomaterials to improve their efficient utilization. The regulation and modification of nanomaterials by peptides can effectively avoid the shortcomings of natural peptides in practical applications and also make the created materials to obtain better performance. For example, there are covalent interactions between the amino side chain groups of antibacterial peptides (ABPs) and the carboxyl groups of reduced graphene oxide (rGO), and therefore, the ABP-functionalized rGO not only revealed significantly improved antibacterial ability, but also showed selective antibacterial ability, which were attributed to the synergistic effects of both ABPs and rGO [9]. Using functional organic molecules to decode biological functions in organisms plays a very important role in the past few decades. Through the coupling of metal chelate, porphyrin-NTA (nitrilotriacetic acid) hybrid has been effectively coordinated with peptides, and the created porphyrin-peptide complex exhibited the ability to penetrate and accumulate in cancer cells and presented potential use as an anticancer agent [10]. In addition, the polyurethane material with peptide coating indicated that the functionalized peptide reduced the accumulation of inflammatory cells in the wound, and could effectively reduce the bacterial infection when it was used as a wound dressing [11].

In this chapter, we introduce recent progress on the combination of peptides and 2D materials (2DMs) to regulate their structure and functions for specific applications. The surface modification of 2DMs by peptides can change the physical and chemical properties of the original materials. The strategies for the conjugation of 2DMs, such as tungsten disulfide (WS_2) [12], graphene [13], molybdenum dioxide [14], and others with peptides, are introduced and discussed. The formed peptide/2DM hybrids for the applications in food detection, medicine, and environmental science are also presented.

6.2 Hybridization of Peptides with 2DMs

6.2.1 Hybridization with WS_2

WS_2 is a fine crystal or powder with metallic luster, belonging to the hexagonal crystal system, semiconductor, and diamagnetism [15]. WS_2 has a wide range of applications due to its large specific surface areas and unique optical, electronic, and catalytic properties. Layered WS_2 nanosheets are one of the newly emerging 2D transition-metal dichalcogenides (TMDCs), which have attracted intensive attentions in the field of catalysts, field-effect transistors (FET), lithium-ion batteries (LIBs), and others [16].

Recently, Pacheco et al. [17] mixed sodium dodecyl sulfate with WS_2 and MoS_2 for the synthesis of functional 2DMs. Then, peptide was used to bind with the formed 2D hybrids via the hydrophobic and electrostatic interactions between peptides and 2DMs. Finally, a Janus micromotor-based fluorescent detection method was developed for the determination of endotoxins in the "OFF" state. The fluorescently labeled peptide was bound to the 2DMs, WS_2 or MoS_2, which were encapsulated in the micromotor, resulting in quenching of its native fluorescence. In the presence of the specific endotoxin, the high affinity of peptide to the analyte promoted the recovery of fluorescence inside the micromotor, as shown in Fig. 6.1a. The increase of the fluorescent intensity was directly proportional to the concentration of the analyte, which could be used for quantitative detection of endotoxin. This micromotors took advantage of the nature of peptides and 2DMs and exhibited high selectivity and sensitivity.

In another study, Zuo et al. [12] designed a WS_2 nanosheet-based sensing platform by employing a dye-labeled peptide substrate for the BACE1 detection. The biosensing procedure is shown in Fig. 6.1b, in which the 6-carboxyfluorescein (FAM)-labeled peptide substrate was adsorbed onto the surface of WS_2 nanosheets to quench the fluorescence in the absence of enzymes. However, in the presence of BACE1, an irreversible emission of the designed long peptide substrate occurred through the BACE1-induced enzymatic hydrolysis, releasing short peptide fragments. The short FAM-linked peptide fragments were not adsorbed on the surface of WS_2 nanosheet, resulting in weak fluorescence quenching. Thus, the fluorescence signal was restored. When the BACE1 activity was inhibited, the hydrolysis of BACE1 was inhibited, and most FAM-labeled peptide substrates were still adsorbed on the surface of the nanosheet, resulting in weak fluorescence intensity. The stronger the inhibitor, the weaker the fluorescence intensity. The proposed sensing platform was sensitive, fast responsive, and convenient and could be easily extended by changing the peptide substrate sequences, providing potential applications for the drug screening.

Sun et al. [18] found that WS_2 nanosheets displayed a distinct adsorption for Arg amino acid, particularly the Arg-rich peptides. Therefore, Arg-rich peptides were used as probes to construct novel WS_2- and MoS_2-based biosensing platforms for the collagen detection, as shown in Fig. 6.1c. The results showed that the surface of

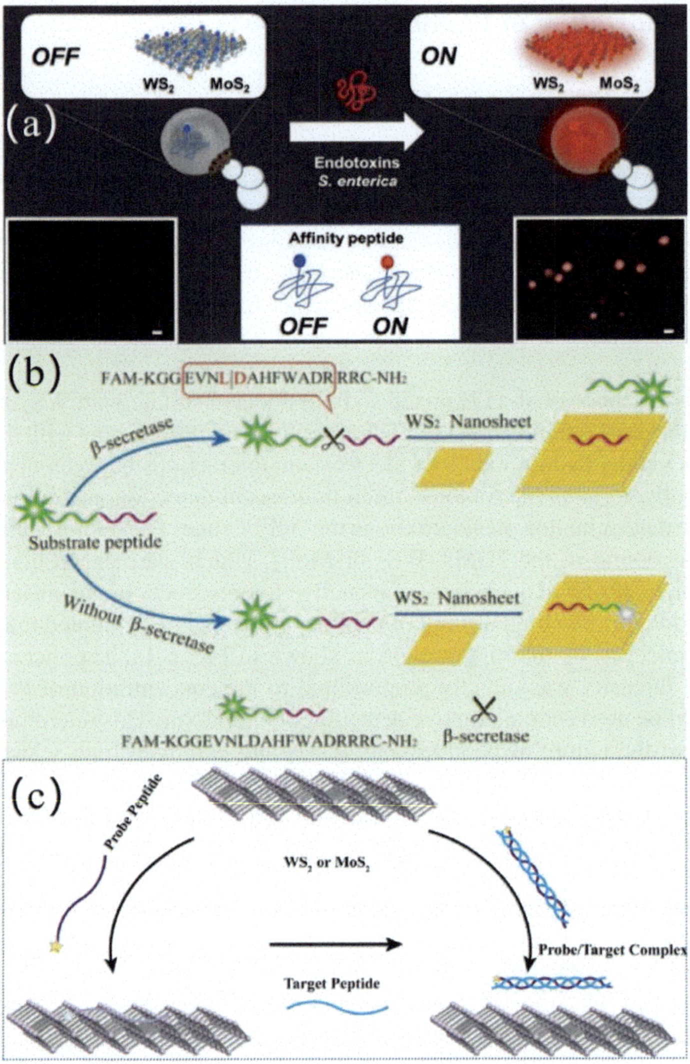

Fig. 6.1 Hybridization of peptides with WS_2: **a** Micromotors prepared from peptides and WS_2 or MoS_2 for the endotoxin detection [17]. **b** WS_2 nanosheet-based sensing platform for BACE1 [12]. **c** WS_2- and MoS_2-based biosensing platforms using peptides as probes [18]

WS_2 nanosheets was easily modified by the thiolated compounds through a simple self-assembly process. In addition, Arg showed obvious adsorption onto the WS_2 nanosheets, and the addition of Arg promoted the interactions of peptide with WS_2 and MoS_2 nanosheets, leading to more significant fluorescence quenching, while the hybridization between probe peptide and target collagen sequence restored the fluorescence. This peptide- and WS_2-based sensor platform was easy to prepare and manipulate, highly specific to the targets, virtually undisturbed by other proteins, and had great potential in biomedical applications.

Wang et al. [19] developed a WS_2-based DNA detection method based on the hybridization between peptide nucleic acid (PNA) and WS_2 nanosheets, as well as the high binding affinity and specificity of PNA to DNA. The method was fast, simple, stable, and sensitive. More recently, Liu et al. [20] developed an electrochemical sensor system based on the molecular imprinted polymer (MIP) for measuring C-reactive protein (CRP) to express inflammatory response. MIP composition was optimized using various concentrated imprinting templates (CRP peptide). Furthermore, as a dopant in the recognition polymer, WS_2 was doped into MIPs, which not only enhanced the electrochemical response associated to the recognition of the template molecules, but also reduced the detection limit toward CRP. The fabricated sensor was proved to be sensitive and accurate on the detection of real serum samples when compared to the ELISA results, which was cost-effective for future home-care diagnostic systems. In another case, Song et al. developed a novel hybrid by conjugating the WS_2-Ga^{3+}-PEG with a short peptide with favorable specificity and affinity to the GPC3 protein and fabricated a sensitive sensor platform for the detection of the tumor biomarker, GPC3 [21].

6.2.2 Hybridization with MXene

MXene is a metal carbide or nitride with a 2D-layered structure, which has good hydrophilicity and biocompatibility and has been used as an ideal material to improve the sensitivity of biosensors [22]. Recently, Reddy et al. reported the fabrication of an ultra-sensitive peptide-based biosensor using the nickel oxide (NiO)-rGO/MXene nanocomposites to detect active influenza viruses (H1N1 and H5N2) and viral proteins [23]. The formed NiO-rGO/MXene nanocomposites were deposited onto the surface of glass carbon electrode (GCE) to obtain the electrochemical biosensors. The NiO-rGO/MXene nanocomposites had a large surface area and were more conducive to the peptide binding. The developed biosensors were capable of highly sensitive and specific detection of influenza viruses and viral proteins with a very low detection limit (3.63 nM for H1N1 and 2.39 nM for H5N1) and good recovery, which provided high potential for clinical applications.

In another study, Lee et al. added different concentrated CRP peptides as templates to the equimolar mixture of aniline (AN) and m-aminobenzenesulfonic acid (MSAN) monomers to form the MIP [24]. A very small percentage of MXene was then added into the above mixture to prepare titanium carbide-doped peptide-imprinted polymers (TCPIPs) using peptide templates in the monomer mix or non-imprinted poly(AN-co-MSAN) (TCNIPs)-coated electrodes, as shown in Fig. 6.2a. The effects of the addition of MXene on the electrochemical response were investigated, and it was found that the electrochemical reaction of the peptide template with the addition of MXene could amplify the signal by about 1.1–1.3 times. In addition, the upper detection range was increased from 100 fg/mL to 10 pg/mL, which significantly improved the performance of the MIP electrode. This study demonstrated that the sensors produced using the MIP technology and the doping with MXene reveal great promise for the sensitive detection of biomarkers.

Wu et al. prepared 2D nanoreactors with cascade catalytic activity for glucose oxidation and hydrogen peroxide (H_2O_2) decomposition via immobilizing glucose oxidase (GO*x*) onto Ti_3C_2 MXene nanosheets [25]. As shown in Fig. 6.2b, Ti_3C_2

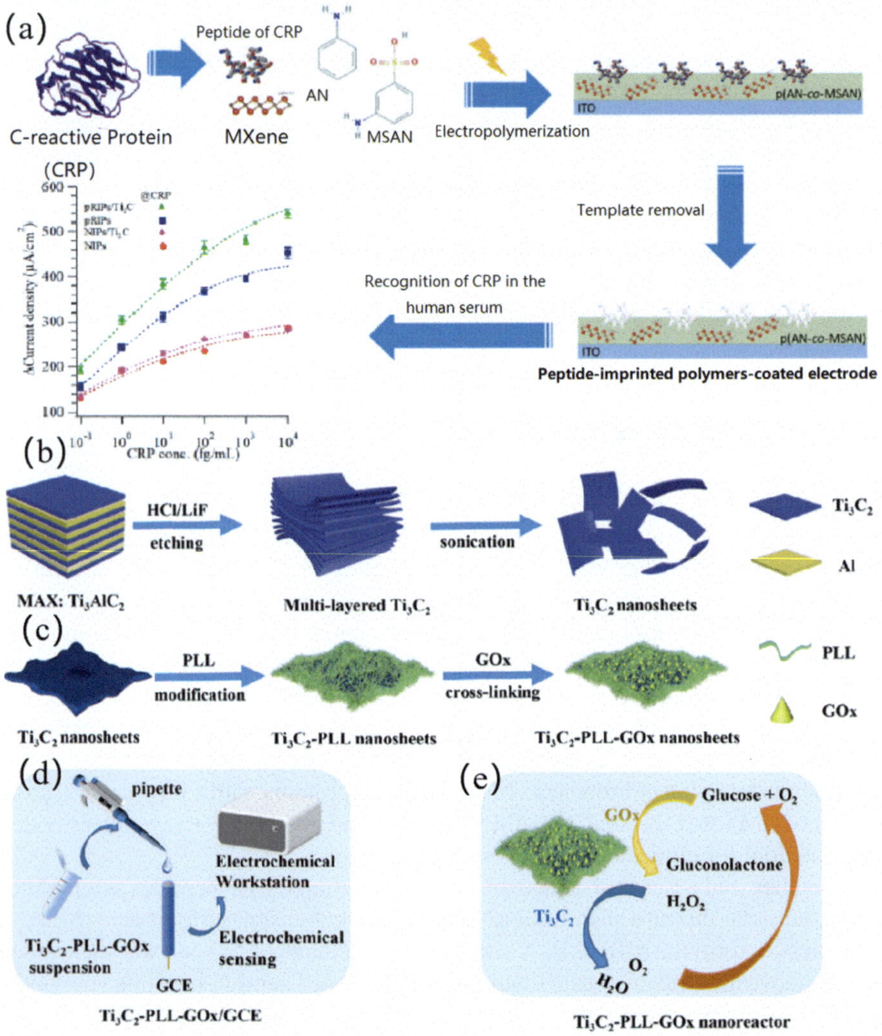

Fig. 6.2 Hybridization of peptides with 2D MXene: **a** MXene-doped CRP peptide-imprinted electrodes [24]. **b** Synthesis of MXene nanosheets, **c** Synthesis of Ti_3C_2-PLL-GO*x*, **d** Ti_3C_2-PLL-GO*x* for cascade glucose oxidation, and **e** Ti_3C_2-PLL-GO*x*-based electrochemical glucose sensor [25]

powder was used as the precursor to synthesize Ti_3C_2 nanosheets. Next, poly-L-lysine (PLL) was added into the Ti_3C_2 suspension for the modification. As a positively charged peptide, PLL provided the electrostatic adsorption of the negatively charged GO*x*. The Ti_3C_2-PLL-GO*x* nanosheets were obtained, as shown in Fig. 6.2c. The obtained result proved that the Ti_3C_2-PLL-GO*x* nanosheets showed superior enzymatic activities than the activities of GO*x* immobilized on an inert porous silica substrate. The Ti_3C_2 nanosheets catalyzed the decomposition of H_2O_2 that generated from the glucose oxidation and improved the reaction rate, as shown in Fig. 6.2d. Moreover, high-performance glucose sensing electrode was constructed (Fig. 6.2e). This work not only enriched the technology of the GO*x* immobilization, but also provided a general and efficient strategy for the preparation and stabilization of various enzyme-loaded MXene with high enzyme content and homogeneous enzyme distribution.

In another study, Xu et al. developed a universal strategy for the construction of ratiometric antifouling electrochemical biosensors based on multifunctional peptides and MXene loaded with AuNPs and methylene blue (MB) [17]. The MXene-Au-MB nanocomposite exhibited excellent conductivity, where AuNPs were able to capture biomolecules containing sulfhydryl terminus, and the MB molecules were used to generate electrochemical signal. Electrochemical biosensors based on the MXene-Au-MB and multifunctional peptides possessed high selectivity, accuracy, and sensitivity. In another study, Wang et al. synthesized chimeric peptide-functionalized titanium carbide (Pep-Ti_3C_2) for the activity assay of enzymes by the integration with carboxypeptidase Y (CPY)-mediated peptide cleavage [26]. The Pep-Ti_3C_2 was self-assembled by chimeric peptide on the surface of Ti_3C_2, and MXene was used as a fluorescent nanoprobe to detect the activity of the PTM enzyme. This work provided important information for the application of MXene in biosensors and clinical diagnosis.

6.2.3 Hybridization with Black Phosphorus (BP)

As a new functional nanomaterial, BP has been widely used in biological imaging, fluorescence sensing, cancer therapy, and others in recent years. For example, recently Liu et al. selected cationic polymer polyethylenimine (PEI) to modify BP quantum dots (QDs) to improve the biocompatibility, and then, the targeting peptide motif, Arg-Gly-Asp (RGD), was linked to PEI to achieve better cancer targeting capability [27]. Under acidic condition, the PEG layer underwent charge flip, releasing positively charged PEI and RGD-modified nanosystem BPQDs@PEI + RGD, as shown in Fig. 6.3a. Moreover, the BPQDs exhibited a spherical shell structure (Fig. 6.3b). Experiments had proved that due to their enhanced permeability and retention, BPQDs could effectively penetrate to tumor blood vessels and accumulate in tumor sites to enhance the drug efficacy and reduce toxic and side effects. Moreover, BPQDs revealed bioimaging capability and biocompatibility, highlighting their great potential in cancer treatment [28]. This study provided a new method for efficiently and accurately transporting BPQDs as novel photosensitive nanomaterials into tumor cells for photodynamic therapy (PDT) of cancer.

In another study, Zhang et al. introduced the binding mechanism of bovine serum albumin (BSA) and bovine hemoglobin (BHB) with 2D BP nanosheets (BP NSs) [29]. Their results showed that the combination of BP NSs with proteins resulted in the

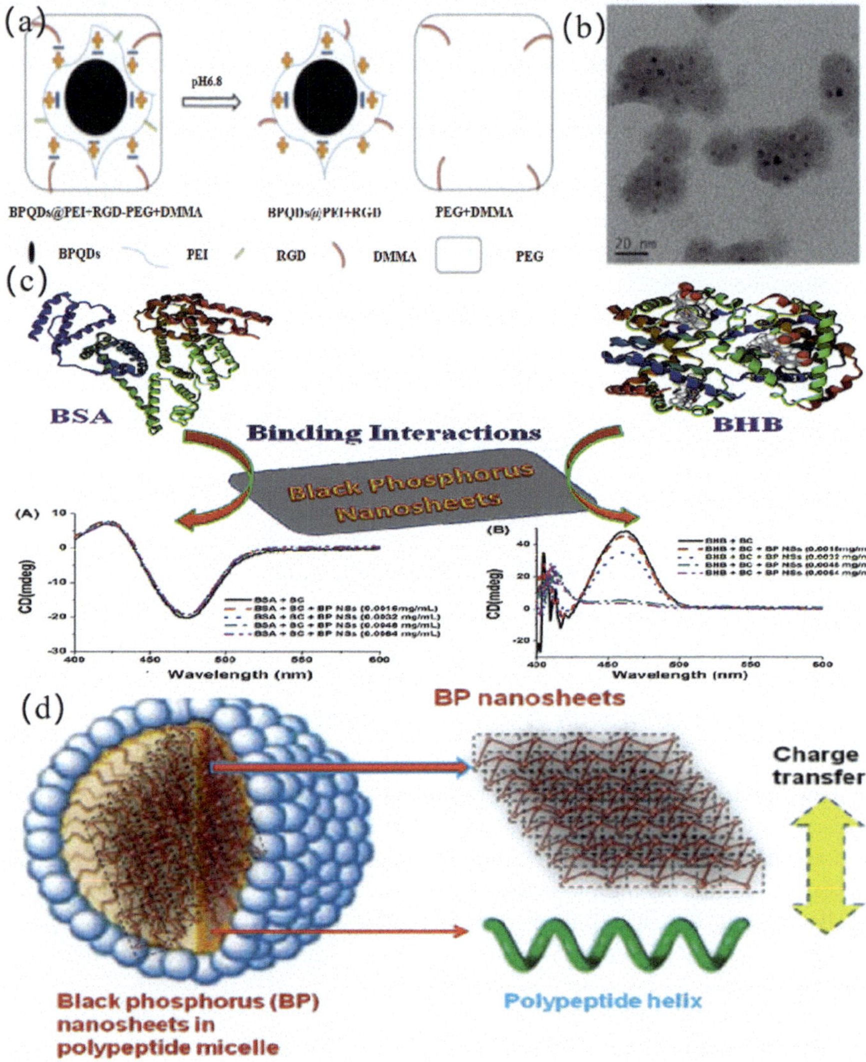

Fig. 6.3 Hybridization of peptides with BP: **a** Schematic illustration of structure of BPQDs@PEI+RGD-PEG+DMMA, **b** TEM of BPQDs [27]. **c** Diagram of BP nanosheet binding to BSA and BHB [29]. **d** Schematic representation of BP encapsulated in a peptide polymer [18]

fluorescence quenching of BSA (25.7%) and BHB (48.5%), indicating that BSA and BHB had different binding ability with BP NPs. Moreover, it was observed that the decrease of the negative ellipticity at 209 and 221 nm indicated that the interactions of BP NSs with BSA/BHB partly induced the extension of the peptide chains, as shown in Fig. 6.3c. This work pointed out that the physicochemical properties of BSA and BHB changed upon association with BP NSs, which would be helpful to understand the fundamental interactions of these biocompatible 2DMs with various biomacromolecules.

In an earlier study, Kumar et al. presented that the as-obtained BP nanosheets were passivated and encapsulated simultaneously by using an artificial peptide with the ability to form micelles [30]. In the first step, BP nanosheets were obtained by the sonication. Next, helical copolymer based on PEG and poly(phenyl isocyanide-peptide) blocks was then allowed to blend with the suspension of BP nanosheets for their inclusion in micelles, as shown in Fig. 6.3d. The size of nanosheets was reduced after their encapsulation inside the polymeric micelleS. The peptide-based copolymer improved the biocompatibility and protected BP nanosheets from the oxidation, showing promising application for bioimaging and drug delivery.

6.2.4 Hybridization with $MoSe_2$

$MoSe_2$ nanosheets and QDs have shown great applications for the fabrication of various chemical and biological sensors due to their unique structure and properties [31]. In a recent study, Ye et al. have developed a novel "signal-on" electrochemical biosensor based on peptide-conjugated hemin/G-quadruplex (DNAzyme-peptide) hybrid and rosebud-like $MoSe_2$@rGO nanocomposites, as shown in Fig. 4a [32]. The multifunctional peptide was composed of cationic and prostate-specific antigen (PSA)-recognition sequences. Specifically, the DNAzyme-peptide hybrid was labeled with Fe_3O_4-COOH to form a magnetic DNAzyme-peptide-Fe_3O_4 probe, which was cleaved to separate DNAzyme and peptide from Fe_3O_4 by simple magnetic separation. The functionalized electrode was prepared (Fig. 6.4b) by immobilizing the DNAzyme-peptide hybrid onto an electrode. The cleaved DNAzyme-peptide was then captured to the electrode surface to generate an electrochemical signal. The electrocatalytic enhancement of H_2O_2 by DNAzyme produced a significant signal, and the $MoSe_2$@rGO contributed to the signal enhancement. This sensor could be used to detect PSA with high sensitivity, which provided an effective tool for early diagnosis of cancer.

Recently, Lin et al. constructed Bacitracin (an antimicrobial peptide, AMP)-modified dextran-$MoSe_2$ nanosheets (AMP/dex-$MoSe_2$ NSs) and further studied the activity of nanosheets in vitro [33]. The results showed that dex-$MoSe_2$ was combined with the AMP through electrostatic adsorption. Low-dose AMP/dex-$MoSe_2$ had antibacterial effect on the Gram-negative bacteria *E. coli*, and the growth inhibition rate was up to 99%, indicating that the antibacterial effect was good, and the modification of AMP enhanced the binding ability of the composites to bacteria. Meanwhile,

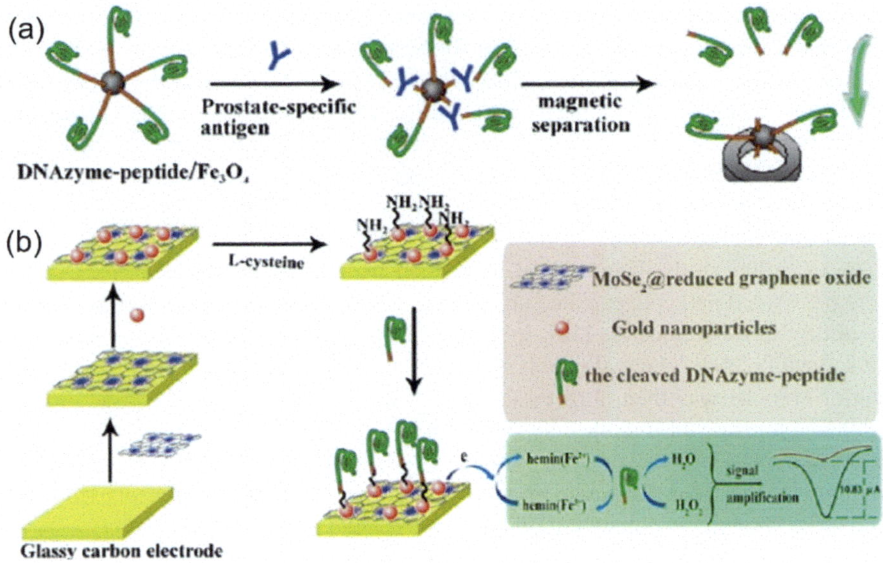

Fig. 6.4 Hybridization of peptides with $MoSe_2$: **a** DNAzyme-peptide/Fe_3O_4 probe for electrochemical biosensor of PSA and **b** functionalization of GCE [32]

the cytotoxicity and hemolysis of the created AMP/dex-$MoSe_2$ were weak in a relatively safe range, which were more suitable for safe and efficient antibacterial application. In another case, Madhu et al. conducted a comparative study regarding the very-few-layer (VFL) MoS_2 and VFL-$MoSe_2$ NS-induced conversion of glutathione (GSH) to GSH disulfide (GSSG) in live cells [34]. It was found that the VFL-$MoSe_2$ converted more efficiently GSH to GSSG than the VFL-$MoSe_2$ NSs. These findings could be attributable to the relatively large-sized selenium vacant sites in the VFL-$MoSe_2$ NSs. The conjugation of selenocysteine on the VFL-$MoSe_2$ NSs suppressed their catalytic activity, which were good triggers for the GSH dimerization in solution and the GSH oxidation in live cells.

6.2.5 Hybridization with GO and rGO

Biocompatibility, mechanical properties, responsiveness, and specificity of biomaterials are key factors for the design of materials for specific biomedical applications. Surface modification of biomaterials by grafting peptides and proteins is a useful strategy to tailor the physical and chemical properties of the original materials. Adibi-Motlagh et al. proposed a method based on the Ugi four-component assembly process (Ugi 4-CAP) to covalently immobilize cell adhesion peptides onto the surface of graphene (G) [35]. The modified adhesive motif peptide KKGDRGDG was used as the amine component, and water was used as the green solvent to perform the four-component reaction in the presence of formaldehyde, cyclohexyl isocyanate, and carboxyl-graphene (G-COOH), and then, the peptide-G biomaterial was prepared. The cell survival proliferation assays proved that the created peptide-G biomaterial had high biocompatibility and enhanced adhesion of human mesenchymal stem cells (hMSCs).

Due to the interaction between peptides and graphene/GO, the formed nanohybrids can also be applied for the treatment of Alzheimer's disease. The aggregation of amyloid beta (Aβ) peptides into fibrillary oligomers is the cause of Alzheimer's disease, one of the most common forms of neurodegenerative disease. Low molecular weight Aβ (1–42) trimers and their higher-order oligomers have been reported to be responsible for the cytotoxicity in neuronal cells, leading to the death of cells. Yang et al. found that graphene nanosheets could penetrate and extract a large amount of peptides from pre-formed amyloid fibrils through the molecular dynamic simulation [36]. These effects appeared to be related to unusually strong dispersive interactions between peptide and graphene, which were further enhanced by strong π-π stacking between aromatic residues of the extracted Aβ peptide and the graphene surface. The cytotoxicity of Aβ peptide amyloid was indeed attenuated by GO, as confirmed by the AFM images and thioflavin fluorescence, as well as the cell viability and ROS measurements.

The peptide/graphene nanohybrids can also be utilized for the detection of virus. In a typical example, Fu et al. demonstrated the modification of GO with 5-carboxyl tetramethylrhodamine (5-TAMRA)-labeled peptide probes for the detection of virus [37]. Three peptide fragments, RS5, QY7, and ED17, were used to construct peptide probes by the co-assembly of GO and three fluorescent peptide ligands (T-RS5, T-QY7, and T-ED17) for the detection of three corresponding viruses, including Ebola virus (EBOV), Marburg virus (MARV), and receptor generalized vesicular stomatitis virus (VSV). The fluorescence spectrum analysis showed that the removal of peptide probes from the GO surface led to the fluorescence recovery of TAMARA by forming the peptide/virus complexes, achieving the detection of various virus. This approach could become a simple and general strategy for rapidly discriminating viruses of different origin. In another similar work, Bolat et al. designed a biosensor based on self-assembled peptide nanotubes (PNT) combined with GO on the pencil graphite electrodes (PGEs), to achieve sensitive impedance analysis of mirNA-192 [38]. As a miniaturized diagnostic tool, the high specificity of the label-free mirNA-192 gene sensor for mirNA-192 and the recovery of peripheral blood serum samples demonstrated the capability of this sensor in the diagnostics of diabetes and cancer.

Puah et al. first prepared multilayer graphene oxide (m-GO) films by a simple drop-casting method [39] and then dropped the peptide solution onto the surface of m-GO films to prepare the peptide/m-GO films, as shown in Fig. 6.5a. The prepared peptide/m-GO films were further characterized with AFM, and the images are shown in Fig. 6.5b–f. Due to the presence of peptide, the m-GO films exhibited better cell binding sites for extracellular matrix. As shown in Fig. 6.5g and h, the osteogenic differentiation of cells on the m-GO and peptide/m-GO films was evaluated in basal medium (BM) and osteogenic medium (OM), and the results proved that the staining

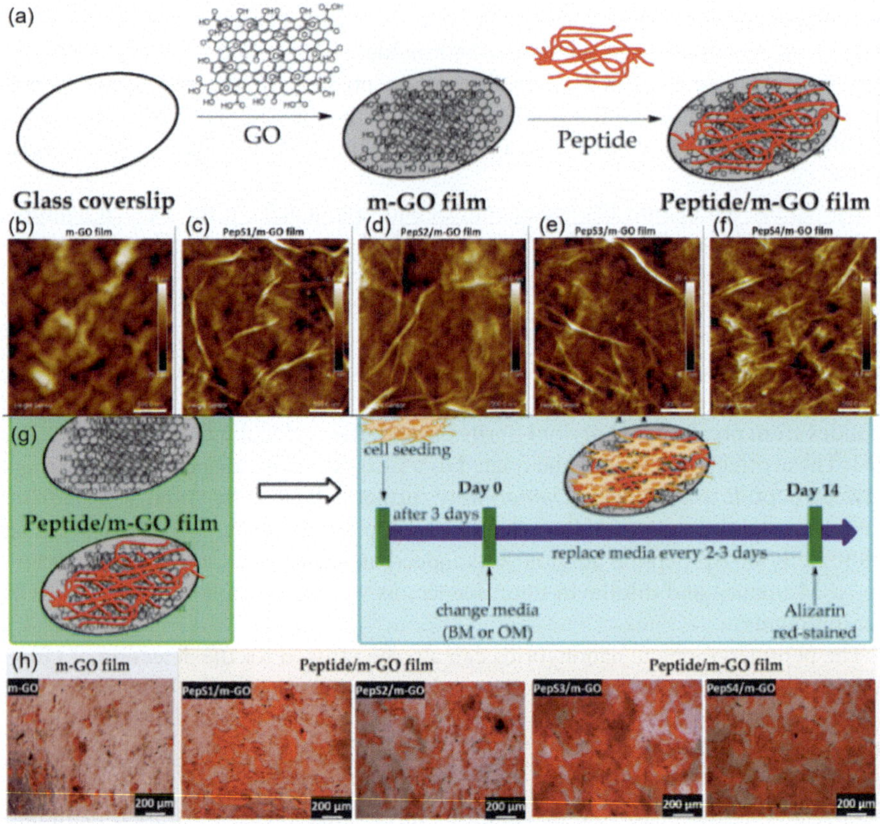

Fig. 6.5 **a** Preparation of conjugated peptide on m-GO membrane, **b–f** AFM images of various membranes, **g** differentiation of osteoblast cell lines on m-GO and peptide/m-GO membranes, **h** images of peptide/m-GO membranes after incubation with osteogenic medium (OM) for 14 days [39]

area of all peptide/m-GO films was more intense than that of the maternal m-GO films. The peptide/m-GO films together with OM provided a biocompatible platform to promote the cell–substrate interactions, thereby enhancing the osteogenic differentiation of cells. This study provided groundbreaking evidence for the use of the peptide/m-GO films as potential biomaterials for mediating the growth and induction of osteogenic differentiation of cells.

6.2.6 Hybridization with Boron Nitride (BN)

The hybridization of peptides, such as BP1, BP7, and B3D, with 2D BN nanosheet produced highly unique biological coupling interface systems [40]. By binding a peptide to the nanosheet, the properties of the final material can be precisely controlled. To identify the affinity of BP7 to h-BN, Brljak et al. used the BP7 peptides to investigate the effects of the fatty acid incorporation sites and carbon chain lengths on the binding with h-BN [41]. It was found that although the fatty acid coupling did not affect the affinity of the resulting biomolecules, it changed the viscoelasticity of the biomolecular coating on the surface of h-BN. The simulation results showed that the C-terminal conjugation promoted the enhancement of the upright adsorption state compared with the N-terminal case. In a further study, they revealed the importance of tyrosine in the peptide/h-BN interactions through a combination of experimental approaches and modeling [42]. The tyrosine residue was identified as a key anchoring material that was exploited through the bio-coupling of BP7 to fatty acids to generate highly viscoelastic interfaces.

Mudedla et al. used classical molecular dynamics simulations and density functional theory (DFT) calculations to study the interactions between α-helical polyalanine peptide and the BN-based nanomaterials, such as boron nitride nanotubes (BNNT) and boron nitride sheets (BNS) [43]. The interactions between polyalanine and BNNT in convex (outer body mode) and concave (inner body mode) surface and between PA and BNS plane were studied, respectively. The results showed that the helical structure of the peptide underwent conformational changes depending on the properties of the surface curvature.

However, few peptides are known to be specific binders of h-BN, and a recent study by Jin et al. indicated that dodecapeptide revealed stronger binding to h-BN than BP7 [44]. This hypothesis was tested using a solute tempered molecular dynamics simulation of replication exchange, which showed a significant increase in surface contact between the two proposed BP7-derived sequences. This work provided a reasonable, economical, and general method to propose, design, and examine new material-binding peptides based on their composite properties.

The interaction between BN and peptide can be applied for the treatment of Alzheimer's disease. For instance, Sorout et al. first investigated the effect of the BN nanotubes on the secondary structure of Aβ (1–42) during amyloid oligomerization [45, 46]. They monitored the helical stability of the BN nanotubes by studying their effect on hydrogen bonds and counting the total number of hydrogen bonds in amyloid-β trimers. The secondary structural stability of the BN nanotubes provided a possibility for the design of different nanomaterials as therapeutic agents for the treatment of amyloid genesis by adjusting the hydrophobicity.

6.3 Conclusions

Based on the above introduction and analysis, it can be concluded that the structure and function of 2DM matrix can be adjusted by adding the peptides, which greatly improve the practical practicability of the prepared hybrid peptide/2DM materials. Firstly, the biocompatibility of 2DMs can be improved by adding peptides. Secondly, the addition of peptides makes 2DMs with high selectivity and specificity, which can broaden the applications of materials in many fields. Thirdly, the addition of antimicrobial peptides enhanced the antibacterial properties of 2DMs, providing a new direction for the treatment of drug-resistant bacteria. Finally, peptides are more available for the combination with some 2DMs, such as WS_2, rGO, BN, and MXene also in order to enhance their structural specificity and multiple functions, thus promoting their practical applications in clinical diagnosis, food detection, biomedicine, and environmental monitoring.

References

1. K. Tsuchiya, K. Numata, Chemoenzymatic synthesis of polypeptides for use as functional and structural materials. Macromol. Biosci. **17**, 1700177 (2017)
2. Y. Chen, G. Yang, B. Liu, H. Kong, Z. Xiong, L. Guo, G. Wei, Biomineralization of ZrO_2 nanoparticles on graphene oxide-supported peptide/cellulose binary nanofibrous membranes for high-performance removal of fluoride ions. Chem. Eng. J. **430**, 132721 (2022)
3. L. Guo, Y. Hu, Y. Lei, H. Wu, G. Yang, Y. Wang, G. Wei, Vitrification of petrochemical sludge for rapid, facile, and sustainable fixation of heavy metals. J. Environ. Chem. Eng. **10**, 108812 (2022)
4. Q. Zhang, Y. Liu, G. Yang, H. Kong, L. Guo, G. Wei, Recent advances in protein hydrogels: from design, structural and functional regulations to healthcare applications. Chem. Eng. J. **451**, 138494 (2023)
5. Y. Liscano, J. Onate-Garzon, J.P. Delgado, Peptides with dual antimicrobial-anticancer activity: strategies to overcome peptide limitations and rational design of anticancer peptides. Molecules **25**, 4245 (2020)
6. X. Luan, H. Kong, P. He, G. Yang, D. Zhu, L. Guo, G. Wei, Self-assembled peptide-based nanodrugs: molecular design, synthesis, functionalization, and targeted tumor bioimaging and biotherapy. Small **19**, 2205787 (2023)
7. Y. Ji, D. Liu, H. Zhu, L. Bao, R. Chang, X. Gao, J. Yin, Unstructured polypeptides as a versatile drug delivery technology. Acta Biomater. **164**, 74–93 (2023)
8. J.L. Lau, M.K. Dunn, Therapeutic peptides: Historical perspectives, current development trends, and future directions. Bioorg. Med. Chem. **26**, 2700–2707 (2018)
9. A.L.T. Zheng, H.N. Farrag, S. Sabidi, T. Kato, T. Maeda, Y. Andou, Accessing the antimicrobial activity of cyclic peptide immobilized on reduced graphene oxide. Mater. Lett. **304**, 130621 (2021)
10. E. Glymenaki, M. Kandyli, C.P. Apostolidou, C. Kokotidou, G. Charalambidis, E. Nikoloudakis, S. Panagiotakis, E. Koutserinaki, V. Klontza, P. Michail, A. Charisiadis, K. Yannakopoulou, A. Mitraki, A.G. Coutsolelos, Design and synthesis of porphyrin-nitrilotriacetic acid dyads with potential applications in peptide labeling through metallochelate coupling. ACS Omega **7**, 1803–1818 (2022)
11. A.C. Stromdahl, L. Ignatowicz, G. Petruk, M. Butrym, S. Wasserstrom, A. Schmidtchen, M. Puthia, Peptide-coated polyurethane material reduces wound infection and inflammation. Acta Biomater. **128**, 314–331 (2021)
12. X. Zuo, H. Dai, H. Zhang, J. Liu, S. Ma, X. Chen, A peptide-WS_2 nanosheet based biosensing platform for determination of beta-secretase and screening of its inhibitors. Analyst **143**, 4585–4591 (2018)
13. S.S. Kim, Z. Kuang, Y.H. Ngo, B.L. Farmer, R.R. Naik, Biotic-abiotic interactions: factors that influence peptide-graphene interactions. ACS Appl. Mater. Interfaces **7**, 20447–20453 (2015)
14. R. Wu, X. Ou, L. Zhang, F. Wang, L. Liu, Interfacial interactions within amyloid protein corona based on 2D MoS_2 nanosheets. ChemBioChem **23**, e202100581 (2022)

15. Z.L. Lei, B. Guo, 2D material-based optical biosensor: status and prospect. Adv. Sci. **9**, e2102924 (2022)
16. X. Xiang, J. Shi, F. Huang, M. Zheng, Q. Deng, J. Xu, MoS_2 nanosheet-based fluorescent biosensor for protein detection via terminal protection of small-molecule-linked DNA and exonuclease III-aided DNA recycling amplification. Biosens. Bioelectron. **74**, 227–232 (2015)
17. M. Pacheco, B. Jurado-Sánchez, A. Escarpa, Transition metal dichalcogenide-based Janus micromotors for on-the-fly Salmonella detection. Microchim. Acta **189**(5), 1–9 (2022)
18. X. Sun, J. Fan, C. Fu, L. Yao, S. Zhao, J. Wang, J. Xiao, WS_2 and MoS_2 biosensing platforms using peptides as probe biomolecules. Sci. Rep. **7**, 10290 (2017)
19. S. Wang, Y. Zhang, Y. Ning, G.J. Zhang, A WS_2 nanosheet-based platform for fluorescent DNA detection via PNA-DNA hybridization. Analyst **140**, 434–439 (2015)
20. K.H. Liu, H.Y. Lin, J.L. Thomas, C.Y. Chen, Y.T. Chen, C.Y. Chen, C.H. Yang, M.H. Lee, Sensing of C-reactive protein using an extended-gate field-effect transistor with a tungsten disulfide-doped peptide-imprinted conductive polymer coating. Biosensors **12**, 31 (2022)
21. X.D. Song, W.T. Shang, L. Peng, H.M. Jiang, K. Wang, C.H. Fang, J. Tian, Novel GPC3-binding WS_2-Ga^{3+}-PEG-peptide nanosheets for in vivo bimodal imaging-guided photothermal therapy. Nanomedicine **13**, 1681–1693 (2018)
22. M. Naguib, M. Kurtoglu, V. Presser, J. Lu, J. Niu, M. Heon, L. Hultman, Y. Gogotsi, M.W. Barsoum, Two-dimensional nanocrystals produced by exfoliation of Ti_3AlC_2. Adv. Mater. **23**, 4248–4253 (2011)
23. Y.V. Manohara Reddy, J.H. Shin, J. Hwang, D.H. Kweon, C.H. Choi, K. Park, S.K. Kim, G. Madhavi, H. Yi, J.P. Park, Fine-tuning of MXene-nickel oxide-reduced graphene oxide nanocomposite bioelectrode: sensor for the detection of influenza virus and viral protein, Biosens. Bioelectron. **214**, 114511 (2022)
24. M.H. Lee, K.H. Liu, J.L. Thomas, C.Y. Chen, C.Y. Chen, C.H. Yang, H.Y. Lin, Doping of MXenes enhances the electrochemical response of peptide-imprinted conductive polymers for the recognition of C-Reactive protein. Biosens. Bioelectron. **200**, 113930 (2022)
25. M. Wu, Q. Zhang, Y. Fang, C. Deng, F. Zhou, Y. Zhang, X. Wang, Y. Tang, Y. Wang, Polylysine-modified MXene nanosheets with highly loaded glucose oxidase as cascade nanoreactor for glucose decomposition and electrochemical sensing. J. Colloid. Interf. Sci. **586**, 20–29 (2021)
26. L. Pan, M.H. Ai, C.Y. Huang, L. Yin, X. Liu, R.R. Zhang, S.B. Wang, Z. Jiang, X.W. Zhang, J.J. Zou, W.B. Mi, Manipulating spin polarization of titanium dioxide for efficient photocatalysis. Nat. Commun. **11**, 418 (2020)
27. Z. Liu, Z. Xie, X. Wu, Z. Chen, W. Li, X. Jiang, L. Cao, D. Zhang, Q. Wang, P. Xue, H. Zhang, PH-responsive black phosphorus quantum dots for tumor-targeted photodynamic therapy. Photodiagn. Photodyn. Ther. **35**, 102429 (2021)
28. M. Wang, Y. Liang, Y. Liu, G. Ren, Z. Zhang, S. Wu, J. Shen, Ultrasmall black phosphorus quantum dots: synthesis, characterization, and application in cancer treatment. Analyst **143**, 5822–5833 (2018)
29. H. Zhang, Q. Han, X. Yin, Y. Wang, Insights into the binding mechanism of two-dimensional black phosphorus nanosheets-protein associations. Spectrochim. Acta A **227**, 117662 (2020)
30. A. Kumar, Simultaneous passivation and encapsulation of black phosphorus nanosheets (phosphorene) by optically active polypeptide micelles for biosensors. ACS Appl. Nano Mater. **2**, 2397–2404 (2019)
31. L. Lin, D. Chen, C. Lu, X. Wang, Fluorescence and colorimetric dual-signal determination of Fe^{3+} and glutathione with $MoSe_2$@Fe nanozyme. Microchem. J. **177**, 107283 (2022)
32. Z. Ye, G. Li, L. Xu, Q. Yu, X. Yue, Y. Wu, B. Ye, Peptide-conjugated hemin/G-quadruplex as a versatile probe for "signal-on" electrochemical peptide biosensor. Talanta **209**, 120611 (2020)
33. T. Lin, G. Jiang, D. Lin, Y. Lai, L. Hou, S. Zhao, Bacitracin-functionalized dextran-$MoSe_2$ with peroxidase-like and near-infrared photothermal activities for low-temperature and synergetic antibacterial applications. ACS Appl. Bio Mater. **5**, 2347–2354 (2022)
34. M. Madhu, Q.-Y. Xue, W.-B. Tseng, T.-H. Chen, A.S. Krishna Kumar, W.-L. Tseng, MoS_2 and $MoSe_2$ nanosheets as triggers for glutathione dimerization in solution and glutathione oxidation in live cells, ACS Appl. Nano Mater. **5**, 10583–10595 (2022)

35. B. Adibi-Motlagh, A.S. Lotfi, A. Rezaei, E. Hashemi, Cell attachment evaluation of the immobilized bioactive peptide on a nanographene oxide composite. Mater. Sci. Eng. C **82**, 323–329 (2018)
36. Z. Yang, C. Ge, J. Liu, Y. Chong, Z. Gu, C.A. Jimenez-Cruz, Z. Chai, R. Zhou, Destruction of amyloid fibrils by graphene through penetration and extraction of peptides. Nanoscale **7**, 18725–18737 (2015)
37. M.Q. Fu, X.C. Wang, W.T. Dou, G.R. Chen, T.D. James, D.M. Zhou, X.P. He, Supramolecular fluorogenic peptide sensor array based on graphene oxide for the differential sensing of ebola virus. Chem. Commun. **56**, 5735–5738 (2020)
38. G. Bolat, O. Akbal Vural, Y. Tugce Yaman, S. Abaci, Label-free impedimetric miRNA-192 genosensor platform using graphene oxide decorated peptide nanotubes composite. Microchem. J. **166**, 106218 (2021)
39. P.Y. Puah, P.Y. Moh, C.S. Sipaut, P.C. Lee, S.E. How, Peptide conjugate on multilayer graphene oxide film for the osteogenic differentiation of human Wharton's jelly-derived mesenchymal stem cells. Polymers **13**, 3290 (2021)
40. T.R. Walsh, M.R. Knecht, Biomolecular material recognition in two dimensions: peptide binding to graphene, h-BN, and MoS_2 nanosheets as unique bioconjugates. Bioconjug. Chem. **30**, 2727–2750 (2019)
41. N. Brljak, M.R. Knecht, T.R. Walsh, Controlling the orientation and viscoelasticity of materials-binding peptides on hexagonal boron nitride using fatty acids. J. Phys. Chem. B **125**, 10621–10628 (2021)
42. N. Brljak, R. Jin, T.R. Walsh, M.R. Knecht, Selective manipulation of peptide orientation on hexagonal boron nitride nanosheets. Nanoscale **13**, 5670–5678 (2021)
43. S.K. Mudedla, K. Balamurugan, V. Subramanian, Unravelling the structural changes in α-helical peptides on interaction with convex, concave, and planar surfaces of boron-nitride-based nanomaterials. J. Phys. Chem. C **120**, 28246–28260 (2016)
44. R. Jin, T.R. Walsh, Modeling-led materials-binding peptide design for hexagonal boron nitride interfaces. Adv. Mater. Interfaces **9**, 2102397 (2022)
45. N. Sorout, A. Chandra, Effects of boron nitride nanotube on the secondary structure of abeta(1–42) trimer: possible inhibitory effect on amyloid formation. J. Phys. Chem. B **124**, 1928–1940 (2020)
46. N. Sorout, A. Chandra, Interactions of the A beta(1–42) peptide with boron nitride nanoparticles of varying curvature in an aqueous medium: different pathways to inhibit beta-sheet formation. J. Phys. Chem. B **125**, 11159–11178 (2021)

Chapter 7
Peptide Nanomaterials for Cancer Therapy

Yi Cao, Pin Wang, Xiaojiao Ge, Xiaolin Duan, Yuanyuan Wei, Li Sun, Aiguo Wu, and Juan Li

Abstract Various nanomaterials have been developed for biomedical applications, which can be classified into organic, inorganic, and organic–inorganic hybrid nanomaterials. Peptide-based nanomaterials have demonstrated unique functions and advantages in tumor diagnosis and therapy, such as specific tumor targeting for accurate tumor diagnosis and excellent tumor selectivity for efficient tumor therapy. Moreover, peptides can be easily integrated into a variety of different nanomaterials to develop superior integrated multifunctional nanomaterials for tumor diagnosis and therapy. Therefore, the peptide-based multifunctional nanomaterials will greatly promote the development of antitumor research and further enhance the transformation from scientific research to clinical applications.

Keywords Peptide nanomaterials · Hybrid materials · Cancer therapy · Functional regulation

7.1 Peptide-Modified Organic Nanomaterials

Peptide-modified organic nanomaterials refer to the modification of peptides on the surface of organic nanomaterials to confer them specific biological properties and functions. Peptide-modified organic nanomaterials used in drug delivery have achieved targeted drug delivery and improved the bioavailability and therapeutic effects. Polymers are suitable candidates for targeted drug delivery vehicles with substantial doses of chemotherapeutic drugs and therapeutic genes, among which

Y. Cao · P. Wang · X. Ge · X. Duan · Y. Wei · L. Sun · A. Wu (✉) · J. Li
Ningbo Institute of Materials Technology and Engineering, Chinese Academy of Sciences, Ningbo, China
e-mail: aiguo@nimte.ac.cn

Y. Cao · P. Wang · X. Ge · X. Duan · Y. Wei · L. Sun · A. Wu · J. Li (✉)
Ningbo Cixi Institute of Biomedical Engineering, Ningbo, China
e-mail: lijuan@nimte.ac.cn

G. Wei (ed.), *Peptide Nano-Chemistry and Nanotechnology*,
https://doi.org/10.1007/978-981-96-2387-7_7

poly (lactic-co-glycolic acid) (PLGA) and poly (ethylene glycol) (PEG) are FDA-approved medical polymers. In addition, bioactive polysaccharides, proteins, and cell membranes have also shown considerable application potential.

7.1.1 *Peptide-Modified Polymer Nanomaterials*

7.1.1.1 PLGA

PLGA polymers are commonly used in biomedicine as they are hydrolyzed in vivo and can produce lactic acid and glycolic acid, which are biocompatible monomers and can take part in the citric acid cycle process. Their autocatalysis of the degradation also accelerated the drug release.

Cerundolo et al. developed a PLGA-based nanoparticles (NPs) for carrying the immunogenic cancer germline antigen NY-ESO-1 and an α-GalCer analog IMM60, as a novel iNKT cell agonist and dendritic cell transactivator [1]. Three peptide sequences derived from immunodominant regions of NY-ESO-1 were selected for a wide HLA coverage and efficiently processed by dendritic cells via various HLA subtypes. Sarmento and co-workers reported the functionalization of PLGA with a cell-penetrating peptide (p28) from the bacterial protein azurin [2]. The p28 could potentiate the selective interaction of these nanosystems and target at lung cancer cells actively. In addition, the formed p28-NPs could effectively penetrate the membranes of cancer cells to deliver drags, providing a new strategy for lung cancer therapy. Moku et al. [3] reported a NP surface-bound targeting ligand that could target at monocarboxylate transporter 1 (MCT1) receptor-positive cancer cells and angiogenic endothelial cells. As shown in Fig. 7.1, PLGA NPs encapsulated with the drug paclitaxel (PTX) were synthesized. The CPK peptide was then conjugated onto the surface of NPs via the maleimide-thiol chemistry. The polymeric NPs with MCT1 receptor-selective, PTX-encapsulated, and CPK peptide-decorated could effectively deliver drugs to cancer and enhance therapeutic efficacy.

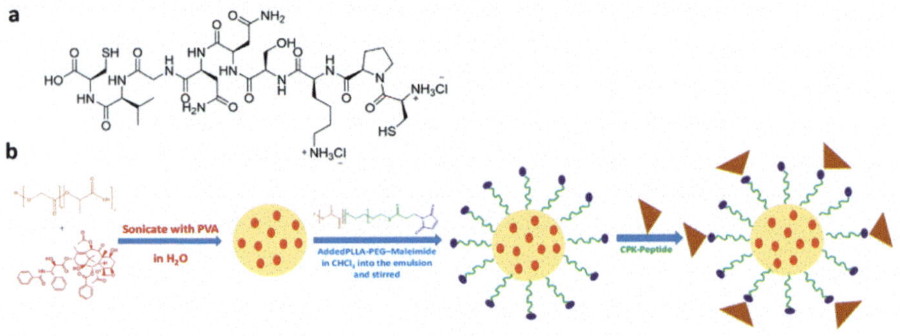

Fig. 7.1 Schematic representation of NP preparation and characterization

7.1.1.2 PEG

PEG can significantly improve the drug accumulation and penetration, as well as high cell selectivity and uptake for their prolonged circulation time. Yu et al. [4] chose an E-selectin-binding peptide as a targeting ligand and metastasis inhibitor, and constructed a system of the PEG-Pep-SN38, which achieved glutathione (GSH)-responsive chemotherapy by inhibiting and killing the tumor and endothelial cells. Zhong et al. [5] constructed tandem nanomicelles to avoid non-specific penetration during the blood circulation, providing a specific way for highly efficient and specific antiglioma chemotherapy. As shown in Fig. 7.2, Zhang et al. synthetized PD-L1 blocked PEG-PLGA NPs, with the surface decoration of PD-L1 binding peptides [6]. The synthetic NPs could be loaded with either chemotherapy or photothermal therapy (PTT) agents to harness the synergistic effects from the combo-therapy. The experimental results showed that the NPs achieved great antitumor efficiency in the metastasis model of colorectal cancer.

7.1.1.3 Polysaccharides

Natural polysaccharides are a class of macromolecular active substances composed of monosaccharides linked by the glycosidic bonds, which not only have unique immunomodulatory effects but also possess good biocompatibility. Polysaccharide-based nanocarriers are widely used for the drug delivery and in the disease research. Chitosan nanoparticles (CHNPs) can be used as drugs, gene delivery, and control carriers precisely because of the good bioadhesive, absorption-promoting, and enzyme-inhibiting properties of chitosan. As a novel drug carrier, chitosan has become a hot spot for research.

Kundu et al. [7] investigated the effect of the pH and RGD peptide-based targeting on the efficacy of CHNPs (RGD-CHNPs) in breast cancer. The RGD-CHNPs had higher stability in the tumor pH environment and were able to enhance cellular uptake. In addition, the RGD peptide enhanced the in vitro uptake of CHNPs by $\alpha_v\beta_3$ integrin-expressed breast cancer cells and induced higher apoptosis. In conclusion, the RGD-CHNPs enhanced the therapeutic effects of chemotherapeutic agents in breast cancer through the synergistic effects of both pH-responsiveness and tumor-specific targeting. Through the co-precipitation of chitosan (CN) and collagen peptide (CP) as shown in Fig. 7.3, Raichur et al. [8] prepared monodispersed pH-responsive CPCN NPs. The potentiality of CPCN NPs to encapsulate and release of doxorubicin (DOX) for drug delivery was also investigated. The proposed research not only expanded the knowledge on the formation of CPCN NPs but also offered a contributing drug delivery system for the encapsulation and release of water-soluble drugs.

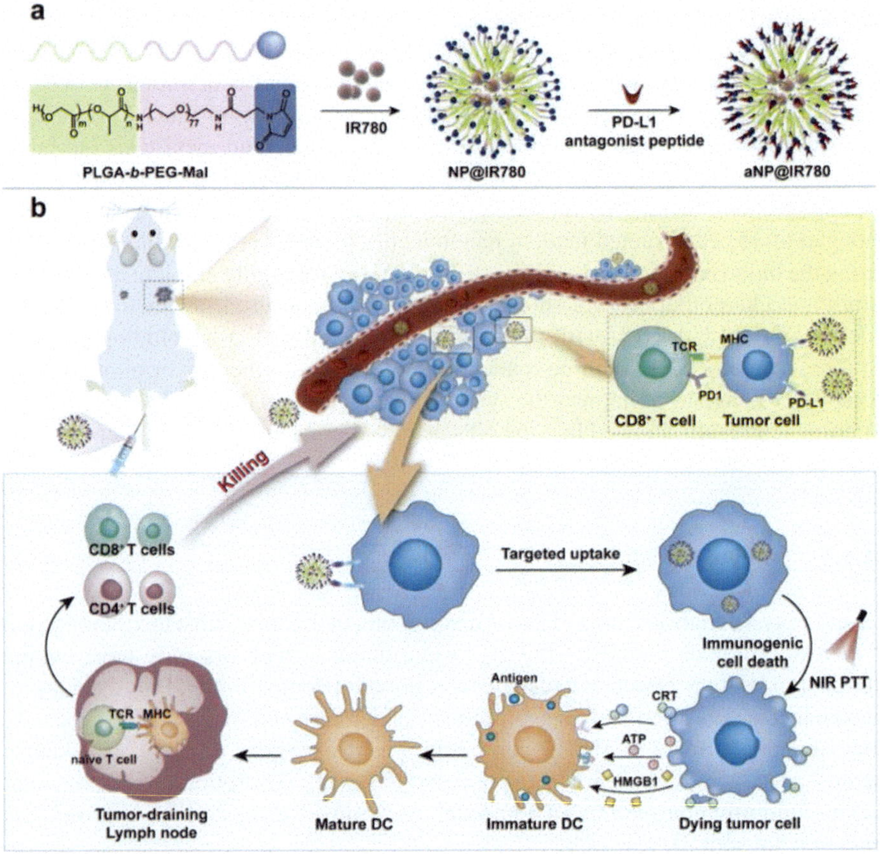

Fig. 7.2 Schematic illustration of the design and mechanism of the IR780 loaded polymeric PD-1 blocked NPs

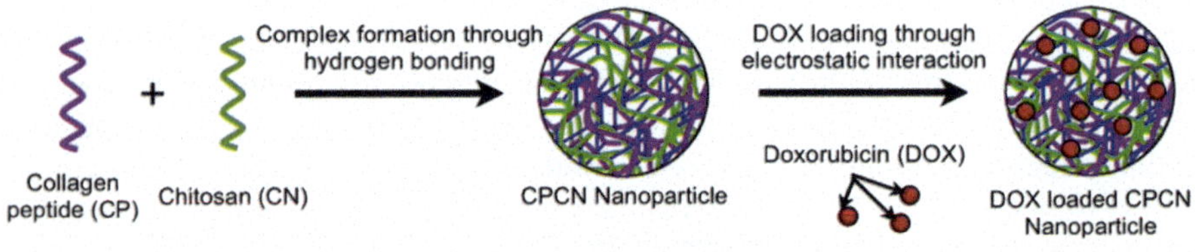

Fig. 7.3 Schematic diagram of CPCN NPs for the encapsulation of DOX

7.1.1.4 Liposomes

Liposomes are vesicular nanocarriers composed of bilayer phospholipids, which are both hydrophilic and hydrophobic and possess the properties such as better biocompatibility, low toxicity, as well as the ability of drug encapsulation and release.

Their applications on drug carriers have made significant progress. Liposomes can improve antitumor efficacy by appropriately modifying specific receptors that highly expressed in TEMs directly with phospholipids and/or polymer-specific ligands, thereby improving their aggregation in tumor tissues.

Buschiazzo et al. [9] reported a peptide containing a tissue-penetrating (TP) consensus motif fused to a neuroblastoma-targeting peptide (pep), capable of binding the neuropilin-1. This TP-pep increased the tumor-binding ability and homing in vivo. After conjugated to DOX, the formed TP-pep-SL@DOX showed an improved antineuroblastoma efficacy in three neuroblastoma animal models. Santos et al. [10] designed tumor-targeted nanovesicles by attaching the LinTT1 peptide to the surface of liposomes. Sorafenib (SRF) and DOX were then co-loaded into the LinTT1-functionalized liposomes (LinTT1-Lipo) to provide the synergy of the two agents. The results showed that the LinTT1-Lipo enhanced the treatment effect of the two chemotherapeutic agents DOX and SRF in estrogen receptor-positive (MCF-7) and triple-negative breast cancer (MDA-MB-231) cells.

As shown in Fig. 7.4, Lo et al. [11] designed pH-sensitive and peptide-modified liposomal and solid lipid NPs for the encapsulation of irinotecan and miR-200. These peptides included a cell-penetrating peptide (CPP) motif, a ligand targeting tumor neovascularization, and a mitochondria-targeting peptide motif. The peptide-modified NPs were further coated with a pH-sensitive PEG-lipid branch with the imine linkages. These NPs exhibited good pH-responsive release, internalization, and intracellular distribution at acidic pH of colon cancer HCT116 cells. Effective treatment was then achieved in colon cancer mice.

7.1.2 Peptide-Modified Proteins Nanomaterials

Proteins are extremely effective drug delivery systems because they are inherently biocompatible, biodegradable, and bioabsorbable. When genetically engineered, proteins can be easily tailored to perform specific activities with great homogeneity.

Mie et al. [12] established protein NPs for drug delivery systems using genetically modified elastin-like peptides with a fused polyaspartic acid tail. The iRGD peptide stimulated the integrin-dependent adhesion to cancer cells along with NRP-1-dependent internalization into tumor tissues. The PTX-loaded protein NPs that included iRGD moieties were found to be efficient at carrying the drug to cancer cells, and promoting the cancer cell death. Park et al. [13] generated a theranostic agent made of human serum albumin nanoparticles (HSA NPs), which include 5-fluorouracil (5-FU) and coated with LRP-1 targeting peptide. The formed HAS NPs could be used for tumor localization after discovering an LRP-1 targeting peptide in a radioresistant CRC PDX through in vivo peptide screening, as shown in Fig. 7.5. This strategy enhanced the ratio of complete tumor response in radio-resistant CRC and was clinically useful to improve the efficacy of neo-adjuvant electromagnetic radiation.

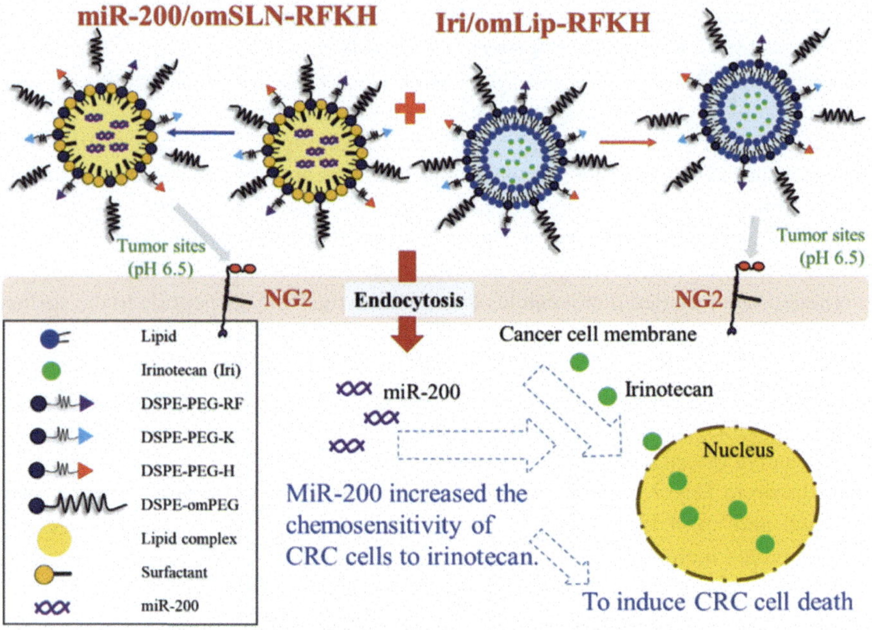

Fig. 7.4 Schematic diagram for the design of two kinds of NPs

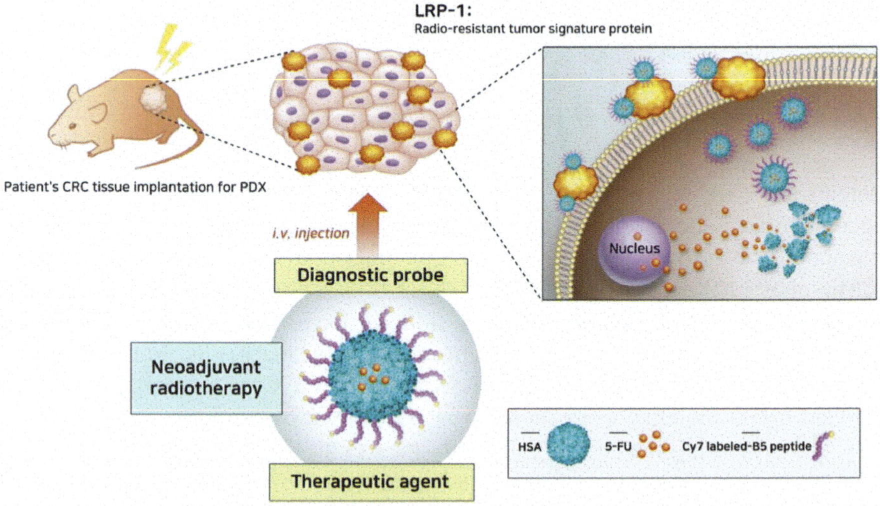

Fig. 7.5 Schematic illustration of the strategy for neoadjuvant radiotherapy in colorectal cancer

7.1.3 Peptide-Modified Membrane Nanomaterials

In the last decade, there have been an increased attention on the study of biomimetics as nanocarriers or nanoagents for disease treatment. Consequently, the investigations on the cell membrane-based NPs have illustrated anticancer therapeutic effect appears strategic. As glycosyls and protein molecule on the membrane surface autograph the NPs with features consisting a minimal reticuloendothelial system (RES), prolonged retention time, partial immunorecognition, accelerating drug delivery efficiency toward tumor sites prior to EPR effects. In addition, the use of peptide ligands as target molecules modified on cell membrane may boost the drug delivery efficiency.

Specific drug delivery system consisting extracellular vehicles (EVs), structure encapsulated-cell-derived membrane of diameter ranging 50–1000 nm which delivers their therapeutic payloads to the brain tissue have been explored. Ren et al. [14] developed the methotrexate (MTX)-loaded EV injection that functionalized with therapeutic Lys-Leu-Ala (LLA) for the molecular recognition between EV lipid bilayer membrane and ApoA-I mimetic peptides. More so, the functionalized peptide onto EV surfaces made available an advancement in the use of EV for specific target binding and therapeutic abilities for the treatment of glioma.

Lu and co-workers reported the synthesis of drug nanocrystals (NCs) upon a modified cell membrane encapsulation approach as targeted drug delivery system (Fig. 7.6), in which a facile insertion approach was adopted constituting avidin–biotin interactions illustrating how red blood cell (RBC) membrane was modified with targeting peptides [15]. Upon the modification with the cRGD, a tumor-targeting peptide, the derived nanocomposite RGD-RBC-NCs showed excellent features in smart therapeutic ability and tumor accumulation in orthotropic glioma and subcutaneous tumor type, respectively.

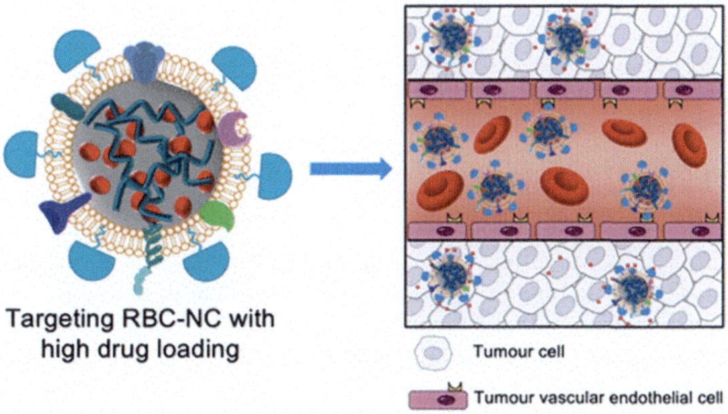

Fig. 7.6 Schematic illustration of the RBC membrane-coated drug NCs

7.1.4 Peptide-Modified Covalent Organic Frameworks (COFs) Nanomaterials

COFs are amorphous porous materials and composed primarily of light-weight elements (H, B, C, N, and O) formed by dynamic covalent connections to form two- or three-dimensional periodic structures. COFs have enormous potential across multiple sectors, including storing energy and separateness, catalysis, sense, and biomedicine, due to their low density, wide surface area, and excellent changeable pore size. However, when compared to metal–organic frameworks (MOFs), comparatively the large size and irregular architecture of COFs hinder their in vivo biocompatibility and bioavailability, limiting their biological applications [16].

Huang et al. [17] constructed multifunctional hollow COF nanospheres (named DCPT-2) to achieve interference-free cascade-targeting glioma medication delivery across the blood–brain barrier. The treatment with DCPT-2 resulted in an improved treatment outcome, with a significantly greater median survival time and handful effects. As shown in Fig. 7.7, they established a multifunctional core–shell NP (OPCPT) to aid in enhanced tumor therapy by uniformly attaching the PEG-COF with tumor-specific TKD peptide alteration to PTT mesoporous carbon nanospheres, using the polyethyleneimine-mediated interface polymerization [18]. In the acidic TME, COFs were degraded. Upon that, the NP charge was flipped, and the COF monomers could create $^1O_2/O_2$. Therefore, the OPCPT reached tumors deeply via positive charge/lysosome rupture transcytosis and restored its NIR-heating capacity for tumor-specific PTT in TME. Furthermore, TME-triggered $^1O_2$ dramatically reduced the lysosome autophagy via the membrane degradation and preferentially damaged mitochondria to enhance the cytochrome C release-activated apoptosis and tumor metastasis inhibition. This unique O_2 production pathway, in particular, alleviated tumor hypoxia following ROS therapy and downregulated hypoxia-inducible factor and its downstream proteins (Table 7.1).

7.2 Peptide-Modified Inorganic Nanomaterials

7.2.1 Peptide-Modified Metal and Metal Oxide Nanomaterials

Metal NPs have a wide range of prospects in biomedical field due to their special optical, magnetic, electrical, and catalytic properties. For example, AuNPs can kill cancer cells by releasing local heat, and both Fe^{2+} and Fe^{3+} can catalyze the production of ROS and promote apoptosis by the Fenton reaction. The combination of peptides and metal NPs can produce synergies effect and lead to a positive behavior

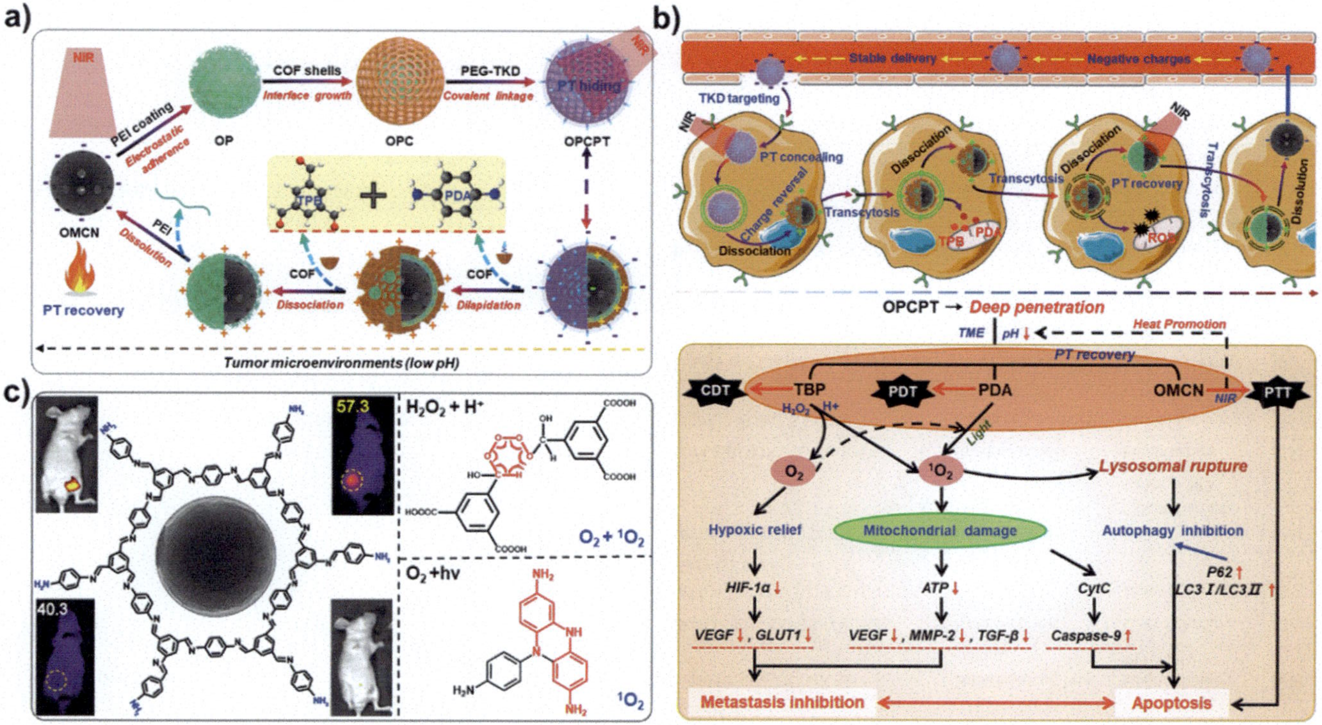

Fig. 7.7 Unveilings of the multifunctional NPs for augmented tumor combined therapy. **a** The synthesis and dissociation process of OPCPT. **b** The tumor-targeted deep penetration for metastasis-inhibited combination therapies. **c** Illustration of the molecular COF composition and mechanisms for $^1O_2$ and O_2 generated tumor combination therapy

Table 7.1 Peptide-modified organic nanomaterials

Type	Peptide	Sequence	Nanoparticle	Cancer	Therapy	References
Organic polymers	RGD	/	RGD-PC7A-POEG-PssCPT	Glioma	Chemotherapy	[19]
	cRGD	c (RGDfK)	cRGD-CH@Rlx	Breast cancer	Chemotherapy	[7]
	cRGD LTX-315	c (RGDfK) KKWWKK-Dip-K-NH2	cRGD-DSPE-PEG@LTX/siR	Breast cancer	Immunotherapy	[20]
	cRGD TAT	/	cRGD@TAT-DOX/ICG	Breast cancer, lung cancer	Chemotherapy, PDT, PTT	[21]
	iRGD	c (CRGDKGPDC)	iRGD-PEG-Azo-PLA@DOX	Breast cancer	Chemotherapy	[22]
			iRGD + CA-PLL-TK@siFGL1/siPD-L1	Lung cancer	Immunotherapy	[23]
			iRGD + NO-HA-DGL@DOX/ICG	Breast cancer	PTT, Chemotherapy, Release NO	[24]
	RGD-like	E16G6RGDK	E16G6RGDK-AD@siRNA	Prostate cancer	Genetherapy	[25]
	DPPA-1	NYSKPTDRQYHF	DPPA-1-CDM-SA@DOX	Colon cancer, breast cancer	Immunotherapy	[26]
	DPPA-1	NYSKPTDRQYHF	Dpeptide-PEG-PLGA@IR780	Colon cancer	Immunotherapy	[6]
	DPPA-1 peptide substrate of MMP-2	NYSKPTDRQYHF KLLGGPLGLAG	DEAP-KLLGGPLGLAG-DPPA-1 @NLG919	Cutaneous melanoma	Immunotherapy	[27]
	DPPA-1 MMP-cleavable peptide	NYSKPTDRQYHF PVGLIG	PEG/DPPA-PVGLIG@DOX	Cutaneous melanoma	Chemotherapy, Immunotherapy	[28]
	MMP2 inhibitor peptide	CCKIGLFRWR	MPL@pep-DOX	Hepatocarcinoma	Chemotherapy	[29]
	Ang	TFFYGGSRGKRNNFKTEEYC	Ang-PS-SAP	Glioma	Immunotherapy	[30]
	Ang TAT	TFFYGGSRGKRNNFKTEEYC YGRKKRRQRRRC	Ang/TAT-PEG-P (DTC-CL)@DTX	Glioma	Chemotherapy	[5]
	CPP	RLWMRWYSPRTRAYGC	SCPP-PS@MTX	Lung tumor	Chemotherapy	[31]

(continued)

Table 7.1 (continued)

Type	Peptide	Sequence	Nanoparticle	Cancer	Therapy	References
	RACPP	C (HE)10G5R6	CPP-HE-PEG-PLA@PTX	Breast cancer	Chemotherapy	[32]
	T4	NLLMAAS	mPEG1000-K (DEAP)-AAN-T4	Leukemia	Immunotherapy	[33]
	DMM (Melittin)	/	DMMPEG-VIPER	Colon cancer, breast cancer	Chemotherapy	[34]
	Peptide 1 Peptide 2	IPVSLRSG-C10-H10-L10-GFLG RTRKNGC-C10-H10-L10-GFLG	mPEG-Peptide1@DOX tLyP1-Peptide2@DOX	Breast cancer	Chemotherapy	[35]
	PNOC	/	PNOC20-PDA/DOX	Breast cancer	PTT, Chemotherapy, Release NO	[36]
	p28	LSTAADMQGVVTDGMASGLDKDYLKPDDC	p28-PLGA@GEF	Lung cancer	Chemotherapy	[2]
	NY-ESO-1 (85–111) NY-ESO-1 (117–143) NY-ESO-1 (157–165)	SRLLEFYLAMPFATPMEAELARRSLAQ PVPGVLLKEFTVSGNILTIRLTAADHR SLLMWITQC	PLGA@NY-ESO-1/IMM60	Sarcoma	Immunotherapy	[1]
	G12	DHLASLWWGTEL	G12-Lip@SF/IR780	Hepatocarcinoma	Chemotherapy, PTT	[37]
	E-selectin binding peptide	IELLQAR	PEG-Pep-SN38	Colon cancer	Chemotherapy	[4]
	KLN (KALA-2LMWP-NLS)	WEAKLAKALAKALAKHLAKALAKALKASEA-(VSRRRRRRGGRRRR)2-KPKKKRKV	KLN/pDNA	Breast cancer	Genetherapy	[38]
	/	WG	WG@porphyrin	Breast cancer	PDT	[39]
	anti-PD-L1 peptide Cathepsin B-cleavable peptide	CVRARTR FRRG	CVRARTR-FRRG-DOX	Breast cancer	Immunotherapy	[40]
	PCP	CNYSKPTDRQYHFK (C18)GGPAK (C18)GADYKPITVKVN	PCP@R848/DOX	Colon cancer	Immunotherapy	[41]
	PTP-7	FLGALFKALL	mPEG-AAN-PTP-7-CDM-PTX	Breast cancer	Chemotherapy	[42]
	FER-8	FEFEFRFK	FER-8 HG@PTX	Hepatocarcinoma	Chemotherapy	[43]

(continued)

Table 7.1 (continued)

Type	Peptide	Sequence	Nanoparticle	Cancer	Therapy	References
	pHK-pKV	MIASHLLAYFFTELN-KVLKQRAKKK	Pal-pHK-pKV	Lung cancer	Regulating suborganelles by self-assembly	[44]
	CF27	CGNKRTRGCPLGVRGNYSKPTDRGYHF	DTX-IR820-CF27	Cutaneous melanoma, colon cancer, lung cancer, breast cancer	PTT, Immunotherapy	[45]
Liposomes	CPP	RLWMRWYSPRTRAYGC	CPP33-CLP@siPLK1	Lung cancer	Genetherapy	[46]
	TAT	/	TAT-Lipo@Narc/siULK1	Hepatocarcinoma	Chemotherapy, Genetherapy	[47]
	LinTT1 (p32 protein targrted)	/	LinTT1-Lipo@DOX/SRF	Breast cancer	Chemotherapy	[10]
	pep (NB targeting) TP	YSHSHSYWLRSGGGC CRALKYSHSHSWLRSGGG	TP-pep-SL@DOX	Glioma	Chemotherapy	[9]
	RFKH	/	omSLN-RFKH omLip-RFKH	Colon cancer	Chemotherapy, Genetherapy	[11]
	RA-XII	/	HA-Lip@TPPP/RA-XII	Colon cancer	Chemotherapy	[48]
Membranes	cRGD	c (RGDyK)	RGD-RBC-NC@DTX	Glioma	Chemotherapy	[15]
	KLA-LDL	Ac-DWFKAFYDKVAEKFKEAF-GG-d (KLAKLAKKLAKLAK)-GG-RLTRKRGLKLA	KLA-LDL-Evs@MTX	Glioma	Chemotherapy	[14]
Proteins	iRGD	c (CRGDKGPDC)	ELP@PTX-D-iRGD	Lung cancer	Chemotherapy	[12]
	tLyP-1	CGNKRTR	tLyP-1-rHDL@PTX/GANT61	Breast cancer	Chemotherapy	[49]
	B5 peptide (LRP-1 targeting)	/	Cy7-B5-HAS@5-FU	Colorectal cancer	Chemotherapy	[13]
COF	TKD	CTKDNNLLGRFELSG	OMCN-PEI-COF-PEG-TKD	Breast cancer	CDT, PDT	[18]
	T10	CGGGHKYLRW	DOX@COF-PEG-T10	Glioma	Chemotherapy	[17]

on cancer. The combination of peptides and metal NPs has high potential for the treatment of cancer, while the toxic side effects and safety issues need further research and evaluation.

7.2.1.1 Gold

Gold is one of the least reactive chemical elements, contrary to colloidal gold, which provides optoelectronic, catalytic, antioxidant properties, and the possibility of the introduction of the surface modifications [50]. The unique optical properties made AuNPs widely used in PTT. When a specific wavelength of light interacted with gold atoms, colloidal AuNPs exhibited a unique localized surface plasmon resonance (LSPR) [51]. The emission frequencies, distinct wavelengths, and emission wavelengths of AuNPs are highly dependent on their surface properties, shape, size, and aggregation state, leading to a specific light-to-heat conversion efficiency range [52]. Modifying the surface of AuNPs with a cationic polymer or functional group can increase the surface area or the surface-to-volume ratio of the AuNPs, enhance their delivery-drugs ability, and protect them from enzymatic degradation by serum proteins in the blood stream. In addition, targeting peptides can also be grafted onto AuNPs to deliver the drug to specific targets [53].

For instance, Lu et al. functionalized gold nanostars (GNSs) by pHLIPs for enhancing the efficacy of tumor targeting and PTT, while minimizing the off-target effects on normal tissues [54]. GE11 can be specifically bound to the epidermal growth factor receptor (EGFR) and possesses low immunogenicity. Meneghetti et al. [55] achieved superior targeting activity by modifying PEGs with lysines and functionalizing with GE11 on AuNPs, compared to NPs decorated with the specific anti-EGFR antibody Cetuximab (C225).

Zhao et al. [56] synthesized a novel peptide-Au cluster ($Au_{10}Peptide_5$), which specifically targeted to EGFR. As illustrated in Fig. 7.8, when EGFR was active, the $Au_{10}Peptide_5$ replaced the natural ligand during the endocytosis and inhibited the cancer-related enzyme activity such as thioredoxin reductase1 (TrxR1), leading to oxidative stress-mediated apoptosis in tumor cells. For inactive EGFR, the binding of $Au_{10}Peptide_5$ inhibited the dimerization of EGFR, thereby retaining it in an inactive state for anticancer purposes.

7.2.1.2 Silver

Silver complexes have broad-spectrum anticancer activity, which can interfere with DNA replication and repair of tumor cells, promote the apoptosis of tumor cells, and exert the cytotoxic effects [57]. As shown in Fig. 7.9, Mostafavi et al. [58] fabricated silver NPs (AgNPs) from the staphylococcus aureus cell-free extract, and then executed the surface modification by conjugating TAT peptide. Therefore, the surface functionalization of AgNPs by employing potential in-house human-derived peptides could be a new and promising approach in cancer therapy.

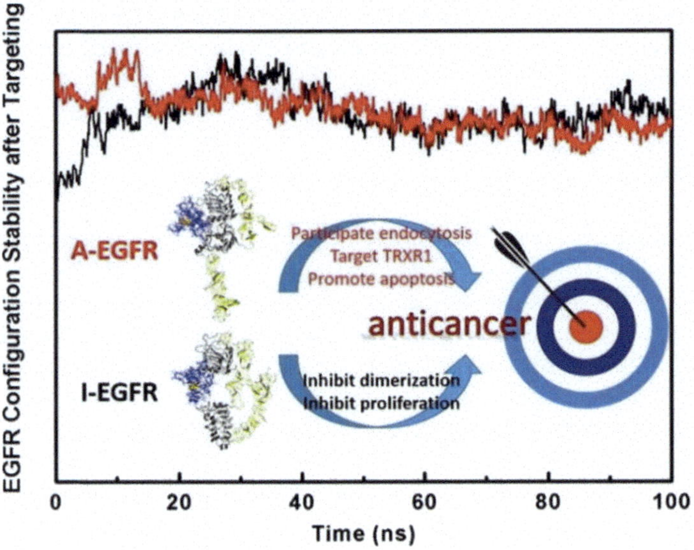

Fig. 7.8 Two antitumor pathways of $Au_{10}Peptide_5$

7.2.1.3 Iron Oxide

Iron plays an important role in the growth and metabolism of tumor cells, and the change of its content may have an impact on the development and treatment of tumors. Iron oxides are used in the imaging and treatment in the form of Fe_2O_3 and Fe_3O_4. The properties of the MRI, PTT, and ferroptosis of Iron oxide can bring a wide range of prospects for it.

Zhai et al. [59] designed a biomimetic nanoplatform for the delivery of chlorin e6 (Ce6), and PEP20 (CD47 inhibitory peptide), integrating oxygen-boosted photodynamic therapy (PDT), ferroptosis, and CD47-SIRPα blockade. The nanoplatform activated the ferroptosis via two modes, by down-regulating the GSH peroxidase 4 pathway and inducing the Fe^{2+} overload. Combining PEP20-mediated CD47-SIRPα blockade and PDT-mediated T-cell activation efficiently suppressed primary tumor growth and metastasis. Wu et al. [60] screened three targeting phages (HPC1, HPC2, and HPC4) and their respective displayed peptides (HSP1, HSP2, and HSP4) that targeting at SCLC and NSCLC cell lines. They synthesized the HSP-Dex-Fe_3O_4 and found that the T_2-weighted MR signal of HSP1-SPION could be decreased by up to 42%. Compared with the tight binding of HSP1 to cancer cell surfaces, HSP4 was preferentially endocytosed and intracellular drug delivery was improved.

As shown in Fig. 7.10, Yan et al. [61] constructed smart nanodrugs with high loading capacities. Multicomponent assembled nanodrugs had robust colloidal stability and ultrasensitive responses to pH and redox stimuli, of which the properties prolonged the blood circulation, increased the tumor accumulation, and enhanced the PDT efficacy of tumor.

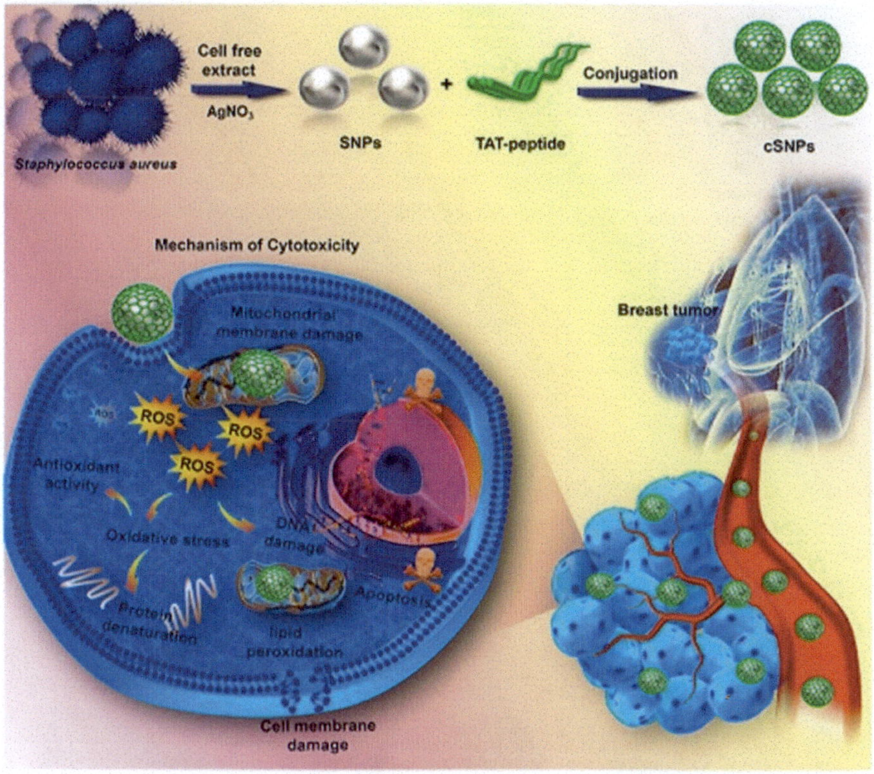

Fig. 7.9 Regulation mechanism of TAT@AgNPs

7.2.2 Peptide-Modified Nonmetal Nanomaterials

Inorganic nonmetallic materials have extensive application potential in tumor treatment, mainly including Se and carbon quantum dots (CQDs), which have excellent biocompatibility and chemical stability.

7.2.2.1 Se

Selenium (Se) is a trace element that must be implemented for life. It is integrated into selenoproteins as selenocysteine (Sec), which acts as the active center of its enzymatic activity. Many selenoproteins contain oxidoreductase activity and thus impact physiological redox equilibrium [62].

Cao et al. [63] synthesized a biocompatible siRNA carrier (RGDfC-SeNPs) by modifying the surface of selenium NPs (SeNPs) with the tumor-targeted molecular RGDfC peptide. Following that, the KLK12-siRNA was loaded onto the surface

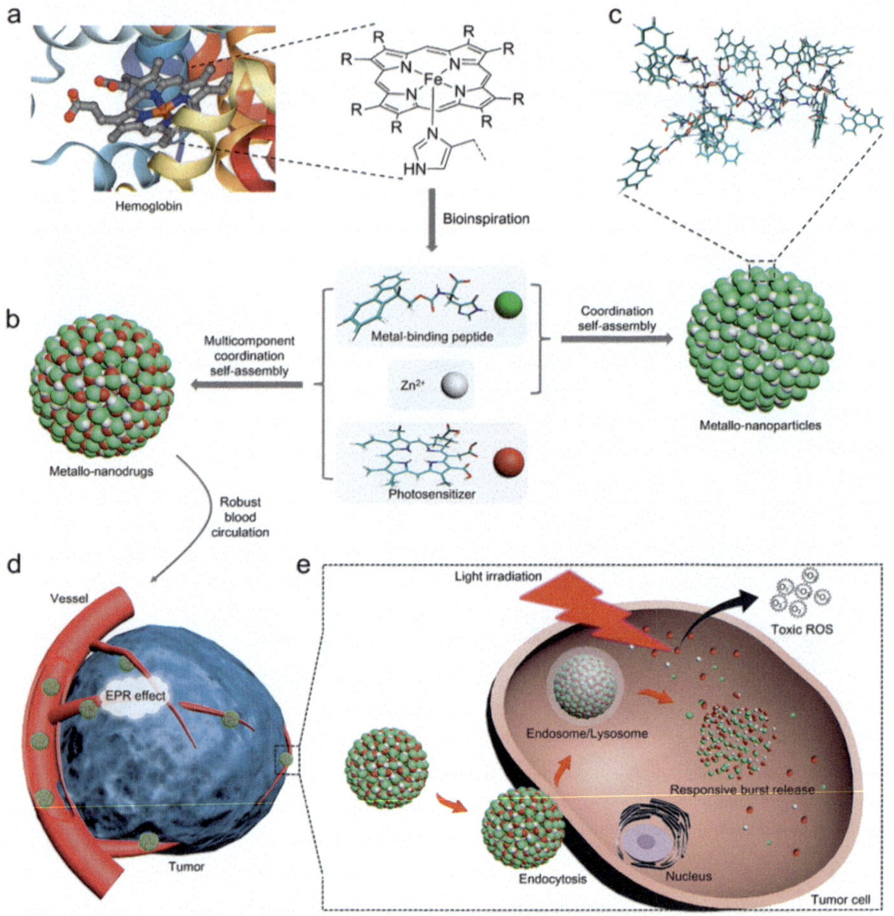

Fig. 7.10 Schematic illustration of supramolecular metallo-nanodrugs for efficient antitumor PDT

of RGDfC-SeNPs to form functionalized NPs. In HT-29 cells, the particle uptake was substantially greater. In an acidic TME, these particles generated siRNA more quickly. The KLK12 target gene was repressed in HT-29 cells, and the mitochondrial membrane potential was induced breakdown both in vitro and in vivo. Zhu et al. [64] constructed a NP of RGDfC-Se@siRNA that was more efficient in HeLa cells than in human umbilical vein endothelial cells (HUVECs). It caused the apoptosis of HeLa cells and demonstrated clear gene silencing efficacy in the HeLa cells, caused potential breakdown of mitochondrial membrane, and hindered the adhesion, migration, and proliferation of HeLa cells. Meanwhile, it raised the production of ROS in HeLa cells, indicating that mitochondrial dysfunction caused by ROS may have a role. Chen et al. [65] designed and synthesized exceptionally uniform and stable SeNPs (Se-5-Fu-Gd-P (Cet/YI-12)) with EGFR targeted and TME-responsive

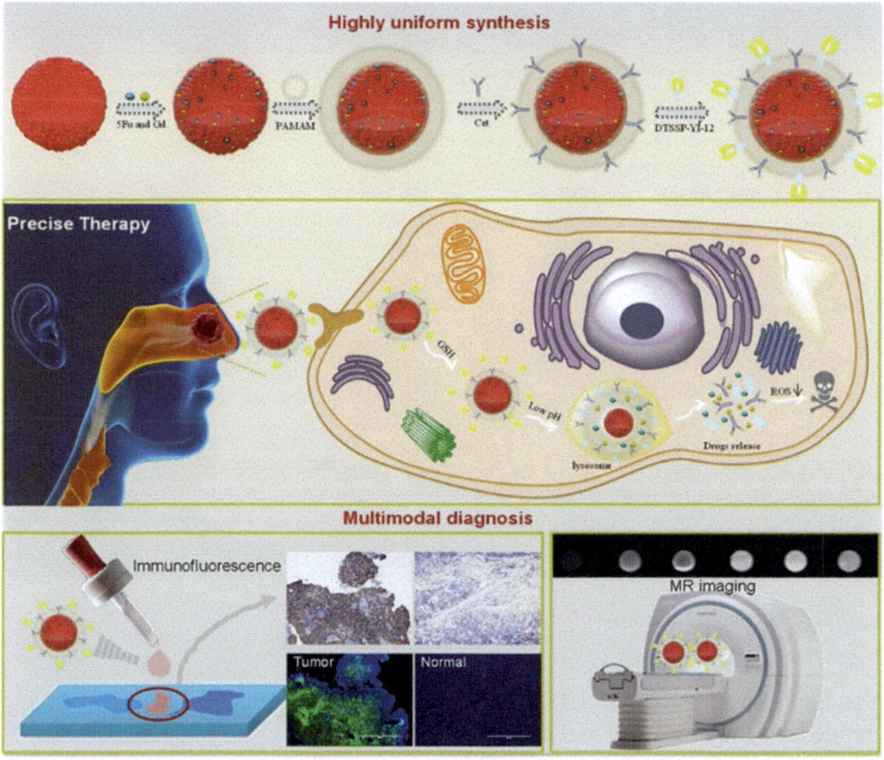

Fig. 7.11 Schematic illustration of SeNPs with EGFR

capabilities for the treatment and evaluation of nasopharyngeal carcinoma (NPC), as shown in Fig. 7.11.

7.2.2.2 CQDs

Sarkar et al. [66] designed a nonviral vector that could efficiently deliver the gene of interest while also detecting TNBC. They generated the CD-PAMAM conjugates (CDPs) by synthesizing CQDs from sweet lemon peel and conjugating them with different generations of polyamidoamine (PAMAM) dendrimers. The RGDS peptide was then attached to CDP in order to target the v3 integrin, which was overexpressed in TNBC. Comprehensive testing revealed CDP3 to be a promising gene carrier system for the TNBC gene therapy. Liu et al. [67] constructed a biomimetic nanodelivery system (named as GTDC@M-R NPs) for TNBC therapy through erythrocyte membrane (M)-camouflaged graphene oxide quantum dots (GOQDs, G). Through precise cell and nuclear targeting, immunological evasion, and increased DOX sensitivity, these GTDC@M-R NPs demonstrated synergistic treatment against TNBC

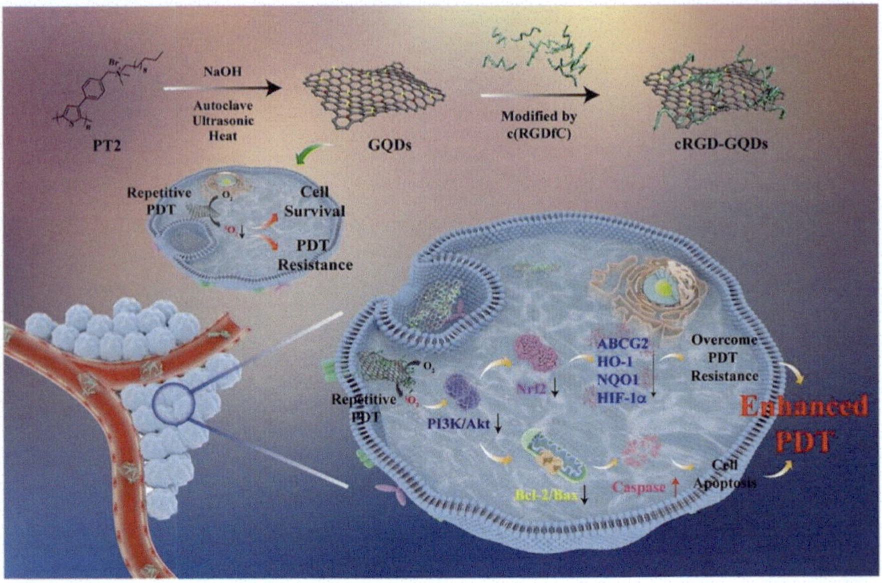

Fig. 7.12 Schematic illustration of cRGD-GQDs

cells. Inhibiting Bcl-2/BAX, COX-2, and VEGF-related signal pathways effectively inhibited the tumor growth and metastasis.

As shown in Fig. 7.12, Lin et al. [68] presented a high-performance nanosystem based on N, S-codoped graphene quantum dots (GQDs) to overcome the PDT resistance in cancer by uploading cyclic (RGDfC) peptide. The as-prepared cRGD-GQDs were biocompatible, had a high singlet oxygen quantum yield (0.95), and were pH-stable. Because of the unique binding ability of c(RGDfC) and integrin v3, the cRGD-GQDs successfully aggregated in tumor cells and generated large amounts of ROS on a continuous basis. The cRGD-GQDs demonstrated potent anticancer PDT effects in vitro and in vivo by blocking the survival pathways of cancer cells.

7.2.3 *Peptide-Modified Composite Nanomaterials*

7.2.3.1 Cisplatin

Cisplatin or (SP-4-2)-diamminedichloridoplatinum (II) can cause the necrosis or apoptosis by binding with genomic DNA (gDNA) or mitochondrial DNA (mtDNA) to create DNA lesions, block the production of DNA, mRNA and proteins, and activate several transduction pathways [69]. Lawler et al. [70] showed the Pt (IV)-M13 conjugate released active cisplatin upon intracellular reduction, and the effect potent in the glioblastoma (GBM) therapy. Compared with cisplatin, the Pt (IV)-M13 significantly increased the platinum uptake in intravenous administration, increasing the platinum levels in brain tissue and intratumorally. Qian et al. [71] constructed a nanosystem (PTX-SS-HPPH/Pt@RGD-NP) with RGD peptide, platinum NPs (PtNPs), and GSH. It could release drugs and enhance the accumulation time and permeability in TME, and alleviate hypoxia in tumor tissues, which enhanced the CDT-PDT performance for bladder cancer. Tang et al. [72] designed fluorescence-labeled magnetic NPs that linked with the CXCR1 targeting peptide (Cis@MFPPC), which relied on the pH response to release cisplatin to competitively inhibit the IL-8/CXCR1 pathway and to enhance the cisplatin sensitivity of circulating tumor cells (CTCs), as indicated in Fig. 7.13. The formed Cis@MFPPC could inhibit the OS growth and pulmonary metastasis by CDT.

MOFs, also known as secondary building units (SBUs), are porous hybrid crystalline, constructed from the metal-containing nodes, bridged by organic linkers,

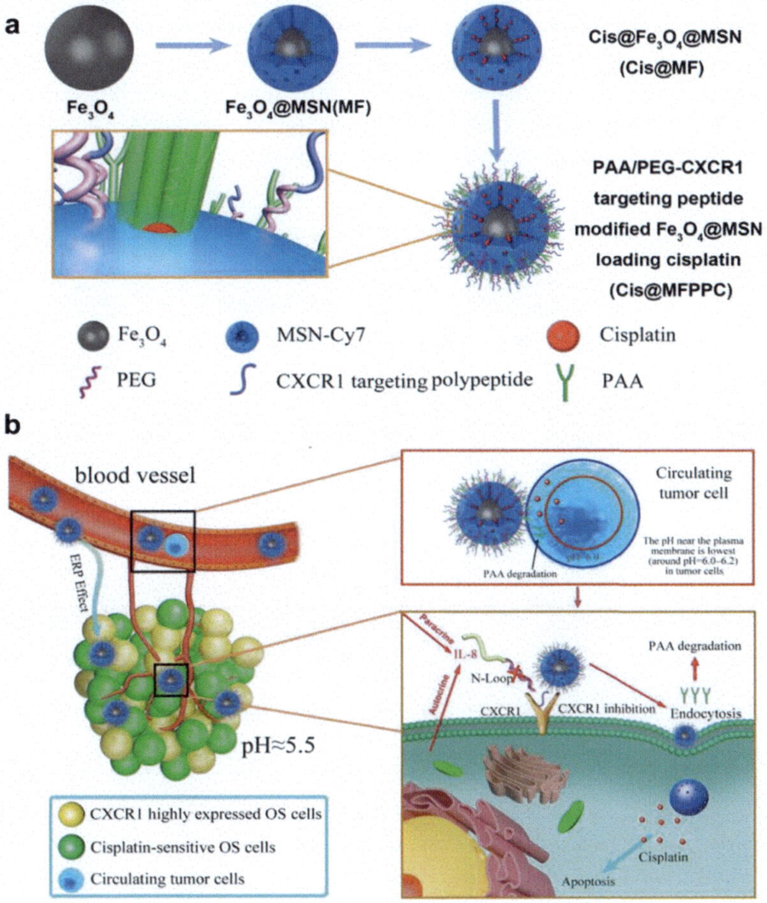

Fig. 7.13 Schematic representation and function mechanism of Cis@MFPPC

which can provide an effective drug loading platform, and achieve efficient loading and directional release of drugs by adjusting their pore size, surface chemistry, and other parameters [73]. In addition, MOFs can also realize the regulation of the drug release rate and drug stability by controlling their structure and composition, which are effective for the therapy and safety of drugs.

Lin et al. [74] designed a nanosystem (TM@P/DOX) with chemotherapy and PTT for bone tumor cell-targeting. The nanosystem was constructed by the hybridization of a core Mn-Co MOF and polydopamine as shell (TM@P), with a bone tumor cell-targeting peptide (BTTP) fixed to the surface of MOF. BTTP could be cut off by MMPs which was overexpressed in bone tumors. In addition, the nanosystem effectively inhibited the bone tumor growth and osteolysis.

In Fig. 7.14, Wu et al. [75] designed a NP of AP-ZIF-90@DOX, which could quickly release DOX with dual-responsive to high adenosine triphosphate (ATP) and low pH condition. Neuropeptide Y (NPY) could target at Y_1 reporter (Y_1R). Therefore, nano ZIF-90 combined with Y_1 ligand (Y_1L) with favorable biocompatibility and dual-responsiveness could be used as a promising nanosystem for the therapy of breast cancer. Further, they developed an Mn-ZIF-90 nanosystem [76] that modified with Y_1L, which exhibited high drug loading, great blood circulation stability, and high targetability of cell membrane and mitochondria. Therefore, these nanosystems showed great potential for the therapy of breast tumor.

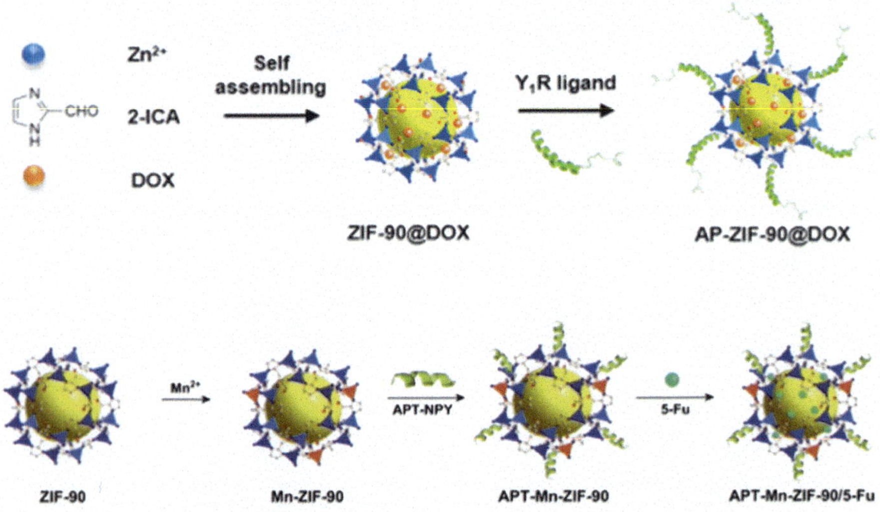

Fig. 7.14 Schematic representation of the nanosystems **a** AP-ZIF-90@DOX **b** APT-Mn-ZIF-90/5-Fu

7.2.3.2 Multimaterial Composites

Multimaterial composites combine different types of nanomaterials, possessing better physical and chemical properties compared with the single material, as shown in Table 7.2.

Shi et al. [85] reported hollow mesoporous silica nanocapsules (HMSs) with the Au nanostar (NS) coating and the modification of RGD peptide (HMSs@Au-PEG-RGD NSs/DOX) to encapsulate DOX for chemotherapy and PTT in vitro and in vivo, as shown in Fig. 7.15. The created nanocapsules released DOX in a pH/NIR dual-responsive manner, and specifically targeted cancer cells, which overexpressing $\alpha_v\beta_3$ integrin.

Teesalu et al. [95] designed nanoscale payloads (iron oxide nanoworms and metallic SNPs) functionalized by a novel targeting peptide (PL1) that interacted with both FN-EDB and TNC-C. This model was effective for GBM, prostate carcinoma xenografts, and non-malignant angiogenic neovessels induced by VEGF overexpression, resulted in reduced tumor growth and increased survival, compared with untargeted particles.

Wu et al. [103] synthesized multifunctional Er@nLiGdF$_4$@mSiO$_2$ nanocomposites with a core–shell structure and ultrasmall size, providing great potential for the breast cancer diagnosis and therapy. The photosensitizer MC540 loaded more highly through SiO_2 shell DSPE-PEG thin layer coating on the surface of nanocomposites, which reduced the acute toxicity of the nanocomposites. In addition, the NPY modification enhanced the targetability and PDT efficacy of the nanocomposites to Y_1R-overexpressed breast tumors. Furthermore, Li et al. [105] developed a combined cell membrane and bioresponsive ZIF system (Y_1L-RBC@ZIF-90@Ce6) for targeted PDT of breast cancer. The system enhanced the tumor accumulation with the NPY$_1$R-targeted guidance and long-term stability. With the dual-response of the ZIF-90 to pH and ATP, the system effectively suppressed tumors in vivo.

7.3 Conclusion

In this chapter, we divided various nanomaterials into organic, inorganic, and organic–inorganic hybrid nanomaterials, and discussed the roles of peptides in the materials, respectively, demonstrated the unique functions and advantages of peptides in the tumor diagnosis and treatment. From the above cases, it can be concluded that peptide-based nanomaterials can be connected to integrate multifunctional nanomaterials, target and integrate into tumor cells to develop multifunctional nanomaterials for the tumor diagnosis and treatment. Therefore, we believe that peptide-based multifunctional nanomaterials will greatly promote the development of antitumor research, advance clinical diagnosis, and ultimately overcome and surpass the challenges of the cancer treatment.

Table 7.2 Peptide-modified composite nanomaterials

Type	Peptide	Sequence	Nanoparticle	Cancer	Therapy	References
Au	RGD	Cyclo (-RGDfK)-YCC	Au4 (RGD)3	Lung cancer	CDT	[77]
		Arg-Gly-Asp	Au@CPL NCs	Breast cancer	Genetherapy	[78]
	GX1	CGNSNPKSC	Au NR@PAMAM-GX1	Colon cancer	Genetherapy, PTT	[79]
	MMP-9 substrate peptide	XLGLAG	pep@Au (I)-NHC	Breast cancer	CDT	[80]
	RGD-like	CKKKKKKGGRGDMFG	GNP-RGD-BLM	Breast cancer	Radiation therapy, Chemotherapy	[81]
	LK	CLKKLLKLLKKLLKLGGLKKLLKLLKKLLKLAGC	LK-PEG-BATrism	Breast cancer	PTT	[82]
	/	FFVLK	Au@DOX@PDA-FFVLK	Breast cancer	PTT	[83]
	GE11	YHWYGYTPQNVI	AuNP-GE11	Colon cancer	PTT	[55]
			Au10Peptide5	/	/	[56]
Au, Se	MSN	ESGRSAN	Au-Se@MSN (RES)	Breast cancer	Chemotherapy	[84]
Au, HMS	RGD	RGD	HMSs@Au-PEG-RGD NSs/DOX	Glioma	Chemotherapy, PTT	[85]
Au, Fe	Ang	TFFYGGSRGKRNNFKTEEYC	Ang-Fe-Au NPs	Glioma	Magnetothermal therapy	[86]
FeCD			Fe-CDs@Ang	Glioma	Chemotherapy	[87]
Fe	iRGD	/	iRGD + MIL-101 (Fe)@sor NPs	Hepatocellular carcinoma	Ferroptosis	[88]
	RGD	/	NapGFF-DOPA-RGD@Fe3 + /DOX	Hepatocarcinoma	CDT, PTT	[89]
	/	KKGFLGCREKA	Pep-SQ@USPIO	Breast cancer	PDT	[90]
	endoglin binding peptide	CAHKHVHHVPVRL	IONP-DOX-Poly IC-EBP	Breast cancer	Chemo-immunotherapy	[91]

(continued)

Table 7.2 (continued)

Type	Peptide	Sequence	Nanoparticle	Cancer	Therapy	References
	HSP1 HSP2 HSP4	GAMHLPWHMGTL NPWEEQGYRYSM NNPWREMMYIEI	HSP-Dex-Fe3O4	Lung cancer	Chemotherapy	[60]
	M2pep	YEQDPWGVKWWY	SPIO-M2pep	Breast cancer	Magnetothermal therapy	[61]
	TPP1 MMP2 substrate peptide	SGQYASHCWCWRDPGRSGGSK PLGLLG	SPIO NP@M	Lung cancer	Immunotherapy	[92]
Fe, PEG	A54	/	A54-PEG-gp (AAm-co-AN)@DOX	Hepatocarcinoma	PTT, Photochemotherapy	[93]
Fe, Protein	Pep20	Mal-PLGLAG-AWSATWSNYWRH	MP@CH/BSA NP	Breast cancer	PDT, Genetherapy, Immunotherapy, ferroptosis	[59]
Fe, P, Se	APP	(SNTSESF)2KFRVTQLAPKQIKE-COOH)	FePSe3@APP@CCM	Colon cancer	Immunotherapy	[94]
Fe, MSN, Pt, Organic polymers	CXCR1 targeting peptide	SAKELRCQCIKTYSKPFHPK	Cis@Fe_3O_4@MSNs-PEG/PAA-pep	Osteosarcoma	CDT	[72]
Ag, Fe	PL1	PPRRGLIKLKTS	PL1-SNP, PL1-NWs	Glioma, prostate cancer	Chemotherapy	[95]
Ag	TAT	/	TAT-SNP	Breast cancer	Chemotherapy	[58]
Pt, Organic polymers	cRGD	c (RGDfK)	DSPE-PEG-cRGD@PAM/Pt	Breast cancer	PTT	[96]
			cRGD-PLGA-PEG@Pt (IV) NP	Ovarian cancer	Chemotherapy	[97]
	TAT	/	DATATPt@mPEG-b-P (3-MeBEP)	Lung cancer	Chemotherapy	[98]
	/	GFLG	PEG-b-P (Pt–co-GFLG)-b-PEG@ICG	Lung cancer	PDT	[99]
	Wpep	TRLYTYMEVAMET	Wpep-PEG-DGL@Pt	Breast cancer	Immunotherapy	[100]
PCPDTBT	PT	PPSHTPT	DSPE-PEG-PT@PCPDTBT	Osteosarcoma	PDT, PTT	[101]

(continued)

Table 7.2 (continued)

Type	Peptide	Sequence	Nanoparticle	Cancer	Therapy	References
MoS_2QD	RGD	/	RGD-PEG-MoS_2QD-DOX	Hepatocarcinoma, Cervical cancer	Chemotherapy, PDT	[102]
MOF, PEG	NPY	INP-Nle-Bpa-RLRYX	Yb, Er@nLiGdF4@mSiO_2(MC540)/ DSPE-PEG-NPY	Breast cancer	PDT	[103]
MOF, mesoporous silica (MS)	dYNH	/	Au@MOF@MS-ICG-dYNH-PAA	Lung cancer	PTT	[104]
MOF (ZIF-90)	AP	YPSKPNNPGEDAPAEDLARYYSALRHYINLITRPRY-NH2	AP-ZIF-90@DOX	Breast cancer	Chemotherapy	[75]
	Y1L	YPSKPNNPGEDAPAEDLARYYSALRHYINLITRPRY-NH2	Y1L-RBC@ZIF-90@Ce6	Breast cancer	PDT	[105]
MOF (Mn-ZIF-90)	APT	ARHITYASASPRILTRARGLARTYR	APT-Mn-ZIF-90/5-Fu	Breast cancer	CDT	[76]
MOF (Mn, Co)	BTTP	D8-KCPGGWQQIGCK-R8-C	BTTP-MnCo MOF@PDA/DOX	Bone tumor	Chemotherapy, PTT	[74]
MOF	ZD2	CYTHVAARTHSEALASCYS	AuNS@MOF-ZD2	Breast cancer	PTT	[106]
Se	cRGD	c (RGDfC)	RGDfC-Se@siRNA	Cervical cancer	Genetherapy, CDT	[64]
	cRGD	c (RGDfC)	RGDfC-SeNPs-siKLK12	Colorectal cancer	Genetherapy	[63]
GOQD	/	GILGFVFTL	QD-Pep	Colorectal cancer	Immunotherapy	[107]
GQDs	cRGD	c (RGDfC)	cRGD-GQDs	Melanoma	PDT	[68]
CQDs	RGDS	RGDS	CD-PAMAM-RGDS	Breast cancer	Genetherapy	[66]
GOQDs, erythrocyte membrane	RGD TAT	RGD YGRKKRRQRRRR	GOQDs-PEG-TAT@DOX/CS-6@ MR-RGD	Breast cancer	Chemotherapy	[67]
Cu, Ga	PDGFB	SRNLIDC	PDGFB-CGO	Breast cancer	CDT	[108]
SiO_2, Yb, Tm	MTP	LTVSPWY	UCNCs@EPI-CD-MTP	Breast cancer	Chemotherapy	[109]

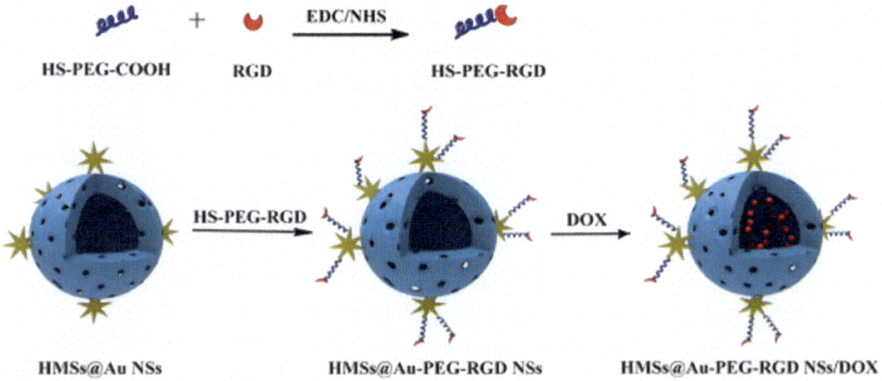

Fig. 7.15 Schematic representation of the nanosystem (HMSs@Au-PEG-RGD NSs/DOX)

References

1. Y. Dolen, U. Gileadi, J.L. Chen, M. Valente, J.H.A. Creemers et al., PLGA nanoparticles Co-encapsulating NY-ESO-1 peptides and IMM60 induce robust CD8 and CD4 T cell and B cell responses. Front. Immunol. **12**, 641703 (2021)
2. A.R. Garizo, F. Castro, C. Martins, A. Almeida, T.P. Dias et al., P28-functionalized PLGA nanoparticles loaded with gefitinib reduce tumor burden and metastases formation on lung cancer. J. Control. Release **337**, 329–342 (2021)
3. V. Yakati, S. Vangala, V.S. Madamsetty, R. Banerjee, G. Moku, Enhancing the anticancer effect of paclitaxel by using polymeric nanoparticles decorated with colorectal cancer targeting CPKSNNGVC-peptide. J. Drug Deliv. Sci. Tec. **68**, 103125 (2022)
4. T. Hao, Y. Fu, Y. Yang, S. Yang, J. Liu et al., Tumor vasculature-targeting PEGylated peptide-drug conjugate prodrug nanoparticles improve chemotherapy and prevent tumor metastasis. Eur. J. Med. Chem. **219**, 113430 (2021)
5. Y. Zhu, Y. Jiang, F. Meng, C. Deng, R. Cheng et al., Highly efficacious and specific anti-glioma chemotherapy by tandem nanomicelles co-functionalized with brain tumor-targeting and cell-penetrating peptides. J. Control. Release **278**, 1–8 (2018)
6. Y. Yu, J. Li, B. Song, Z. Ma, Y. Zhang et al., Polymeric PD-L1 blockade nanoparticles for cancer photothermal-immunotherapy. Biomaterials **280**, 121312 (2022)
7. A.S. Yadav, N.N.V. Radharani, M. Gorain, A. Bulbule, D. Shetti et al., RGD functionalized chitosan nanoparticle mediated targeted delivery of raloxifene selectively suppresses angiogenesis and tumor growth in breast cancer. Nanoscale **12**, 10664–10684 (2020)
8. S. Anandhakumar, G. Krishnamoorthy, K. Ramkumar, A. Raichur, Preparation of collagen peptide functionalized chitosan nanoparticles by ionic gelation method: an effective carrier system for encapsulation and release of doxorubicin for cancer drug delivery. Mater. Sci. Eng. C **70**, 378–385 (2017)
9. M. Ponzoni, F. Curnis, C. Brignole, S. Bruno, D. Guarnieri et al., Enhancement of tumor homing by chemotherapy-loaded nanoparticles. Small **14**, 1802886 (2018)
10. N. d'Avanzo, G. Torrieri, P. Figueiredo, C. Celia, D. Paolino et al., LinTT1 peptide-functionalized liposomes for targeted breast cancer therapy. Int. J. pharmaceut. **597**, 120346 (2021)
11. V. Juang, C.H. Chang, C.S. Wang, H.E. Wang, Y.L. Lo, PH-responsive PEG-shedding and targeting peptide-modified nanoparticles for dual-delivery of irinotecan and microRNA to enhance tumor-specific therapy. Small **15**, 1903296 (2019)

12. E. Kobatake, Y. Ikeda, M. Mie, Construction of protein nanoparticles for targeted delivery of drugs to cancer cells. Mater. Adv. **3**, 6262–6269 (2022)
13. K.J. Lee, E.J. Ko, Y.-Y. Park, S.S. Park, E.J. Ju et al., A novel nanoparticle-based theranostic agent targeting LRP-1 enhances the efficacy of neoadjuvant radiotherapy in colorectal cancer. Biomaterials **255**, 120151 (2020)
14. Z. Ye, T. Zhang, W. He, H. Jin, C. Liu et al., Methotrexate-loaded extracellular vesicles functionalized with therapeutic and targeted peptides for the treatment of glioblastoma multiforme. ACS Appl. Mater. Interfaces **10**, 12341–12350 (2018)
15. Z. Chai, D. Ran, L. Lu, C. Zhan, H. Ruan et al., Ligand-modified cell membrane enables the targeted delivery of drug nanocrystals to glioma. ACS Nano **13**, 5591–5601 (2019)
16. X. He, Z. Jiang, O.U. Akakuru, J. Li, A. Wu, Nanoscale covalent organic frameworks: from controlled synthesis to cancer therapy. Chem. Commun. **57**, 12417–12435 (2021)
17. T. Huo, Y. Yang, M. Qian, H. Jiang, Y. Du et al., Versatile hollow COF nanospheres via manipulating transferrin corona for precise glioma-targeted drug delivery. Biomaterials **260**, 120305 (2020)
18. Y. Qin, X. Zhang, U. Lächelt, R. Huang, Unveiling the multiple tumor-targeted impairments of covalent-organic framework-switched photothermal shielding nanoparticles. Adv. Funct. Mater. **33**, 2211075 (2023)
19. S. Wang, G. Yu, Z. Wang, O. Jacobson, R. Tian et al., Hierarchical tumor microenvironment-responsive nanomedicine for programmed delivery of chemotherapeutics. Adv. Mater. **30**, e1803926 (2018)
20. P. Cao Dai, N. Bao Loc, J. Jee-Heon, C. Jae-Hoon, J. Sung Giu, et al., Shaping the “hot” immunogenic tumor microenvironment by nanoparticles co-delivering oncolytic peptide and TGF-β1 siRNA for boosting checkpoint blockade therapy, Bioeng. Transl. Med. (2022)
21. H. Wu, C. You, F. Chen, J. Jiao, Z. Gao et al., Enhanced cellular uptake of near-infrared triggered targeted nanoparticles by cell-penetrating peptide TAT for combined chemo/photothermal/photodynamic therapy. Mater. Sci. Eng. C **103**, 109738 (2019)
22. B. Mamnoon, J. Loganathan, M.I. Confeld, N. De Fonseka, L. Feng et al., Targeted polymeric nanoparticles for drug delivery to hypoxic, triple-negative breast tumors. ACS Appl. Bio Mater. **4**, 1450–1460 (2021)
23. W.J. Wan, G. Huang, Y. Wang, Y. Tang, H. Li et al., Coadministration of iRGD peptide with ROS-sensitive nanoparticles co-delivering siFGL1 and siPD-L1 enhanced tumor immunotherapy. Acta Biomater. **136**, 473–484 (2021)
24. C. Hu, X. Yang, R. Liu, S. Ruan, Y. Zhou et al., Coadministration of iRGD with multistage responsive nanoparticles enhanced tumor targeting and penetration abilities for breast cancer therapy. ACS Appl. Mater. Interfaces **10**, 22571–22579 (2018)
25. Y. Dong, T. Yu, L. Ding, E. Laurini, Y. Huang et al., A dual targeting dendrimer-mediated siRNA delivery system for effective gene silencing in cancer therapy. J. Am. Chem. Soc. **140**, 16264–16274 (2018)
26. W. Zhu, Y. Bai, N. Zhang, J. Yan, J. Chen et al., A tumor extracellular pH-sensitive PD-L1 binding peptide nanoparticle for chemo-immunotherapy of cancer. J. Mater. Chem. B **9**, 4201–4210 (2021)
27. K. Cheng, Y. Ding, Y. Zhao, S. Ye, X. Zhao et al., Sequentially responsive therapeutic peptide assembling nanoparticles for dual-targeted cancer immunotherapy. Nano Lett. **18**, 3250–3258 (2018)
28. X. Zhu, C. Li, Y. Lu, Y. Liu, D. Wan et al., Tumor microenvironment-activated therapeutic peptide-conjugated prodrug nanoparticles for enhanced tumor penetration and local T cell activation in the tumor microenvironment. Acta Biomater. **119**, 337–348 (2021)
29. C. Qian, J. Wang, Y. Qian, R. Hu, J. Zou et al., Tumor-cell-surface adherable peptide-drug conjugate prodrug nanoparticles inhibit tumor metastasis and augment treatment efficacy. Nano Lett. **20**, 4153–4161 (2020)
30. Y. Jiang, W. Yang, J. Zhang, F. Meng, Z. Zhong, Protein toxin chaperoned by LRP-1-targeted virus-mimicking vesicles induces high-efficiency glioblastoma therapy in vivo. Adv. Mater. **30**, e1800316 (2018)

31. W. Yang, Y. Xia, Y. Fang, F. Meng, J. Zhang et al., Selective cell penetrating peptide-functionalized polymersomes mediate efficient and targeted delivery of methotrexate disodium to human lung cancer in vivo. Adv. Healthc. Mater. **7**, e1701135 (2018)
32. B. Tang, J.L. Zaro, Y. Shen, Q. Chen, Y. Yu et al., Acid-sensitive hybrid polymeric micelles containing a reversibly activatable cell-penetrating peptide for tumor-specific cytoplasm targeting. J. Control. Release **279**, 147–156 (2018)
33. L. Zhang, Y. Qi, H. Min, C. Ni, F. Wang et al., Cooperatively responsive peptide nanotherapeutic that regulates angiopoietin receptor Tie2 activity in tumor microenvironment to prevent breast tumor relapse after chemotherapy. ACS Nano **13**, 5091–5102 (2019)
34. S. Lv, M. Sylvestre, K. Song, S.H. Pun, Development of D-melittin polymeric nanoparticles for anti-cancer treatment. Biomaterials **277**, 121076 (2021)
35. P.H. Hsieh, W.Y. Huang, H.C. Wang, C.H. Kuan, T.Y. Shiue et al., Dual-responsive polypeptide nanoparticles attenuate tumor-associated stromal desmoplasia and anticancer through programmable dissociation. Biomaterials **284**, 121469 (2022)
36. Y. Ding, C. Du, J. Qian, C.-M. Dong, NIR-responsive polypeptide nanocomposite generates NO gas, mild photothermia, and chemotherapy to reverse multidrug-resistant cancer. Nano Lett. **19**, 4362–4370 (2019)
37. W. Mu, D. Jiang, S. Mu, S. Liang, Y. Liu, N. Zhang, Promoting early diagnosis and precise therapy of hepatocellular carcinoma by Glypican-3-targeted synergistic chemo-photothermal theranostics. ACS Appl. Mater. Interfaces **11**, 23591–23604 (2019)
38. S. Dehghani, M. Alibolandi, Z.A. Tehranizadeh, R.K. Oskuee, R. Nosrati et al., Self-assembly of an aptamer-decorated chimeric peptide nanocarrier for targeted cancer gene delivery. Coll. Surf. B **208**, 112047 (2021)
39. B. Sun, R. Chang, S. Cao, C. Yuan, L. Zhao et al., Acid-activatable transmorphic peptide-based nanomaterials for photodynamic therapy. Angew. Chem. Int. Ed. **59**, 20582–20588 (2020)
40. Y. Moon, M.K. Shim, J. Choi, S. Yang, J. Kim et al., Anti-PD-L1 peptide-conjugated prodrug nanoparticles for targeted cancer immunotherapy combining PD-L1 blockade with immunogenic cell death. Theranostics **12**, 1999–2014 (2022)
41. M. Sun, S. Yao, L. Fan, Z. Fang, W. Miao et al., Fibroblast activation protein-alpha responsive peptide assembling prodrug nanoparticles for remodeling the immunosuppressive microenvironment and boosting cancer immunotherapy. Small **18**, e2106296 (2022)
42. S. Zheng, Y. Cai, Y. Hong, Y. Gong, L. Gao et al., Legumain/pH dual-responsive lytic peptide–paclitaxel conjugate for synergistic cancer therapy. Drug Deliv. **29**, 1764–1775 (2022)
43. F. Raza, Y. Zhu, L. Chen, X. You, J. Zhang et al., Paclitaxel-loaded pH responsive hydrogel based on self-assembled peptides for tumor targeting. Biomater. Sci. **7**, 2023–2036 (2019)
44. D. Liu, A. Angelova, J. Liu, V.M. Garamus, B. Angelov et al., Self-assembly of mitochondria-specific peptide amphiphiles amplifying lung cancer cell death through targeting the VDAC1-hexokinase-II complex. J. Mater. Chem. B **7**, 4706–4716 (2019)
45. J. Peng, Q. Yang, Y. Xiao, K. Shi, Q. Liu et al., Tumor microenvironment responsive drug-dye-peptide nanoassembly for enhanced tumor-targeting, penetration, and photo-chemo-immunotherapy. Adv. Funct. Mater. **29**, 1900004 (2019)
46. M. Qiu, J. Ouyang, Y. Wei, J. Zhang, Q. Lan et al., Selective cell penetrating peptide-functionalized envelope-type chimeric lipopepsomes boost systemic RNAi therapy for lung tumors. Adv. Healthc. Mater. **8**, e1900500 (2019)
47. X. Wang, F. Wu, G. Li, N. Zhang, X. Song et al., Lipid-modified cell-penetrating peptide-based self-assembly micelles for co-delivery of narciclasine and siULK1 in hepatocellular carcinoma therapy. Acta Biomater. **74**, 414–429 (2018)
48. Y. Xu, Y. Yao, L. Wang, H. Chen, N. Tan, Hyaluronic acid coated liposomes Co-delivery of natural cyclic peptide RA-XII and mitochondrial targeted photosensitizer for highly selective precise combined treatment of colon cancer. Int. J. Nanomed. **16**, 4929 (2021)
49. C. Jiang, X. Wang, B. Teng, Z. Wang, F. Li et al., Peptide-targeted high-density lipoprotein nanoparticles for combinatorial treatment against metastatic breast cancer. ACS Appl. Mater. Interfaces **13**, 35248–35265 (2021)

50. K. Sztandera, M. Gorzkiewicz, B. Klajnert-Maculewicz, Gold nanoparticles in cancer treatment. Mol. Pharm. **16**, 1–23 (2019)
51. J.B. Vines, J.H. Yoon, N.E. Ryu, D.J. Lim, H. Park, Gold nanoparticles for photothermal cancer therapy. Front. Chem. **7**, 167 (2019)
52. M.A. Dheyab, A.A. Aziz, P.M. Khaniabadi, M.S. Jameel, N. Oladzadabbasabadi et al., Gold nanoparticles-based photothermal therapy for breast cancer. Photodiagn. Photodyn. **42**, 103312–103312 (2023)
53. S. Siddique, J.C.L. Chow, Gold nanoparticles for drug delivery and cancer therapy. Appl. Sci. **10**, 3824 (2020)
54. Y. Tian, Y. Zhang, Z. Teng, W. Tian, S. Luo et al., PH-dependent transmembrane activity of peptide-functionalized gold nanostars for computed tomography/photoacoustic imaging and photothermal therapy. ACS Appl. Mater. Interfaces **9**, 2114–2122 (2017)
55. F. Biscaglia, S. Rajendran, P. Conflitti, C. Benna, R. Sommaggio, et al., Enhanced EGFR targeting activity of plasmonic nanostructures with engineered GE11 peptide. Adv. Healthc. Mater. **6** (2017)
56. P. Zhang, J. Zhai, X. Gao, H. Zhao, W. Su, L. Zhao, Targeted peptide-Au cluster binds to epidermal growth factor receptor (EGFR) in both active and inactive states: a clue for cancer inhibition through dual pathways. Sci. Bull. **63**, 349–355 (2018)
57. Y. Tao, T. Aparicio, M. Li, K.W. Leong, S. Zha, J. Gautier, Inhibition of DNA replication initiation by silver nanoclusters. Nucleic Acids Res. **49**, 5074–5083 (2021)
58. S. Majeed, M. Saravanan, M. Danish, N.A. Zakariya, M.N.M. Ibrahim et al., Bioengineering of green-synthesized TAT peptide-functionalized silver nanoparticles for apoptotic cell-death mediated therapy of breast adenocarcinoma. Talanta **253**, 124026 (2023)
59. Z. He, H. Zhou, Y. Zhang, X. Du, S. Liu et al., Oxygen-boosted biomimetic nanoplatform for synergetic phototherapy/ferroptosis activation and reversal of immune-suppressed tumor microenvironment. Biomaterials **290**, 121832 (2022)
60. Y.H. Chi, J.K. Hsiao, M.H. Lin, C. Chang, C.H. Lan, H.C. Wu, Lung cancer-targeting peptides with multi-subtype indication for combinational drug delivery and molecular imaging. Theranostics **7**, 1612–1632 (2017)
61. W. Wang, F. Li, S. Li, Y. Hu, M. Xu et al., M2 macrophage-targeted iron oxide nanoparticles for magnetic resonance image-guided magnetic hyperthermia therapy. J. Mater. Sci. Technol. **81**, 77–8777 (2021)
62. A. Khurana, S. Tekula, M.A. Saifi, P. Venkatesh, C. Godugu, Therapeutic applications of selenium nanoparticles. Biomed. Pharmacother. **111**, 802–812 (2019)
63. Y. Xia, G. Tang, M. Guo, T. Xu, H. Chen, et al., Silencing KLK12 expression via RGDfC-decorated selenium nanoparticles for the treatment of colorectal cancer in vitro and in vivo, Mate.r Sci. Eng. C. **110**, 110594 (2020)
64. Y. Xia, G. Tang, C. Wang, J. Zhong, Y. Chen et al., Functionalized selenium nanoparticles for targeted siRNA delivery silence Derlin1 and promote antitumor efficacy against cervical cancer. Drug Deliv. **27**, 15–25 (2020)
65. J. Huang, W. Huang, Z. Zhang, X. Lin, H. Lin et al., Highly uniform synthesis of selenium nanoparticles with EGFR targeting and tumor microenvironment-responsive ability for simultaneous diagnosis and therapy of nasopharyngeal carcinoma. ACS Appl. Mater. Interfaces **11**, 11177–11193 (2019)
66. S. Ghosh, K. Ghosal, S.A. Mohammad, K. Sarkar, Dendrimer functionalized carbon quantum dot for selective detection of breast cancer and gene therapy. Chem. Eng. J. **373**, 468–484468 (2019)
67. J. Fan, B. Liu, Y. Long, Z. Wang, C. Tong et al., Sequentially-targeted biomimetic nano drug system for triple-negative breast cancer ablation and lung metastasis inhibition. Acta Biomater. **113**, 554–569 (2020)
68. Y. Li, R. Zhou, D. Xiao, S. Shi, S. Peng et al., Polypeptide uploaded efficient nanophotosensitizers to overcome photodynamic resistance for enhanced anticancer therapy. Chem. Eng. J. **403**, 126344 (2021)
69. S. Ghosh, Cisplatin: the first metal based anticancer drug. Bioorg. Chem. **88**, 102925 (2019)

70. J.L. Jimenez-Macias, Y.C. Lee, E. Miller, T. Finkelberg, M. Zdioruk et al., A Pt(IV)-conjugated brain penetrant macrocyclic peptide shows pre-clinical efficacy in glioblastoma. J. Control. Release **352**, 623–636 (2022)
71. Y. Hao, Y. Chen, X. He, R. Han, C. Yang et al., RGD peptide modified platinum nanozyme Co-loaded glutathione-responsive prodrug nanoparticles for enhanced chemo-photodynamic bladder cancer therapy. Biomaterials **293**, 121975 (2023)
72. X.G. Han, S.B. Yang, H.M. Mo, M.Q. Wang, F. Zhou, et al., Targeting of CXCR1 on osteosarcoma circulating tumor cells and precise treatment via cisplatin nanodelivery. Adv. Funct. Mater. **29**, 1902246 (2019)
73. G. Lan, K. Ni, W. Lin, Nanoscale metal-organic frameworks for phototherapy of cancer. Coordin. Chem. Rev. **379**, 65–81 (2019)
74. Y. Wang, J. Cui, J. Chen, J. Wan, Y. Liang et al., Novel bone tumor cell targeting nanosystem for chemo-photothermal therapy of malignant bone tumors. Chem. Eng. J. **446**, 136905 (2022)
75. Z. Jiang, Y. Wang, L. Sun, B. Yuan, Y. Tian et al., Dual ATP and pH responsive ZIF-90 nanosystem with favorable biocompatibility and facile post-modification improves therapeutic outcomes of triple negative breast cancer in vivo. Biomaterials **197**, 41–50 (2019)
76. Z. Jiang, B. Yuan, N. Qiu, Y. Wang, L. Sun et al., Manganese-Zeolitic Imidazolate frameworks-90 with high blood circulation stability for MRI-guided tumor therapy. Nanomicro. Lett. **11**, 61 (2019)
77. Z. Du, Z. He, J. Fan, Y. Huo, B. He et al., Au4 cluster inhibits human thioredoxin reductase activity via specifically binding of Au to Cys189. Nano Today **47**, 101686 (2022)
78. S. Yu, R. Wen, H. Wang, Y. Zha, L. Qiu et al., Chitosan-graft-Poly(l-lysine) dendron-assisted facile self-assembly of Au nanoclusters for enhanced x-ray computer tomography imaging and precise MMP-9 plasmid shRNA delivery. Chem. Mater. **31**, 3992–40073992 (2019)
79. L. Ye, Y. Chen, J. Mao, X. Lei, Q. Yang, C. Cui, Dendrimer-modified gold nanorods as a platform for combinational gene therapy and photothermal therapy of tumors. J. Exp. Clin. Cancer Res. **40**, 303 (2021)
80. Y. Marciano, V. Del Solar, N. Nayeem, D. Dave, J. Son et al., Encapsulation of gold-based anticancer agents in protease-degradable peptide nanofilaments enhances their potency. J. Am. Chem. Soc. **145**, 234–246 (2023)
81. C. Yang, K. Bromma, C. Di Ciano-Oliveira, G. Zafarana, M. van Prooijen, D. B. Chithrani, Gold nanoparticle mediated combined cancer therapy. Cancer Nanotechnol. **9** (2018)
82. M. Ha, S.H. Nam, K. Sim, S.E. Chong, J. Kim et al., Highly efficient photothermal therapy with cell-penetrating peptide-modified bumpy Au triangular nanoprisms using low laser power and low probe dose. Nano Lett. **21**, 731–739 (2021)
83. B. Xie, H. Zhao, M. Shui, Y.F. Ding, C. Sun et al., Spermine-responsive intracellular self-aggregation of gold nanocages for enhanced chemotherapy and photothermal therapy of breast cancer. Small **18**, e2201971 (2022)
84. X. Liu, J. Liu, S. Xu, X. Li, Z. Wang et al., Gold nanoparticles functionalized with Au-Se-bonded peptides used as gatekeepers for the off-target release of resveratrol in the treatment of triple-negative breast cancer. ACS Appl. Mater. Interfaces **15**, 2529–2537 (2023)
85. X. Li, L. Xing, Y. Hu, Z. Xiong, R. Wang et al., An RGD-modified hollow silica@Au core/shell nanoplatform for tumor combination therapy. Acta Biomater. **62**, 273–283 (2017)
86. S.P.C. Hsu, U. Dhawan, Y.-Y. Tseng, C.-P. Lin, C.-Y. Kuo et al., Glioma-sensitive delivery of Angiopep-2 conjugated iron gold alloy nanoparticles ensuring simultaneous tumor imaging and hyperthermia mediated cancer theranostics. Appl. Mater. Today **18**, 100510 (2020)
87. P. Muhammad, S. Hanif, J. Li, A. Guller, F.U. Rehman et al., Carbon dots supported single Fe atom nanozyme for drug-resistant glioblastoma therapy by activating autophagy-lysosome pathway. Nano Today **45**, 101530 (2022)
88. X. Liu, X. Zhu, X. Qi, X. Meng, K. Xu, Co-administration of iRGD with sorafenib-loaded iron-based metal-organic framework as a targeted ferroptosis agent for liver cancer therapy. Int. J. Nanomed. **16**, 1037–1050 (2021)
89. J. Li, S. Wu, X. Tian, X. Li, Fabrication of a multifunctional nanomaterial from a mussel-derived peptide for multimodal synergistic cancer therapy. Chem. Eng. J. **446**, 136837 (2022)

90. Y. Wang, L. Jiang, Y. Zhang, Y. Lu, J. Li et al., Fibronectin-targeting and Cathepsin B-activatable theranostic nanoprobe for MR/fluorescence imaging and enhanced photodynamic therapy for triple negative breast cancer. ACS Appl. Mater. Interfaces **12**, 33564–33574 (2020)
91. Q.G. Mu, G. Lin, M. Jeon, H. Wang, F.C. Chang et al., Iron oxide nanoparticle targeted chemo-immunotherapy for triple negative breast cancer. Mater. Today **50**, 149–169 (2021)
92. X. Meng, J. Wang, J. Zhou, Q. Tian, B. Qie, et al., Tumor cell membrane-based peptide delivery system targeting the tumor microenvironment for cancer immunotherapy and diagnosis. Acta Biomater. **127**, 266–275 (2021)
93. W.S. Li, X.J. Wang, S. Zhang, J.B. Hu, Y.L. Du et al., Mild microwave activated, chemo-thermal combinational tumor therapy based on a targeted, thermal-sensitive and magnetic micelle. Biomaterials **131**, 36–46 (2017)
94. X. Fang, X. Wu, Z. Li, L. Jiang, W.S. Lo et al., Biomimetic anti-PD-1 peptide-loaded 2D FePSe(3) nanosheets for efficient photothermal and enhanced immune therapy with multimodal MR/PA/thermal imaging. Adv. Sci. (Weinh) **8**, 2003041 (2021)
95. P. Lingasamy, A. Tobi, M. Haugas, H. Hunt, P. Paiste et al., Bi-specific tenascin-C and fibronectin targeted peptide for solid tumor delivery. Biomaterials **219**, 119373 (2019)
96. X. Xiong, Z. Xu, H. Huang, Y. Wang, J. Zhao et al., A NIR light triggered disintegratable nanoplatform for enhanced penetration and chemotherapy in deep tumor tissues. Biomaterials **245**, 119840 (2020)
97. H. Huang, Y. Dong, Y. Zhang, D. Ru, Z. Wu et al., GSH-sensitive Pt(IV) prodrug-loaded phase-transitional nanoparticles with a hybrid lipid-polymer shell for precise theranostics against ovarian cancer. Theranostics **9**, 1047–1065 (2019)
98. W. Jiang, J. Wang, J. Yang, Z. He, Z. Hou et al., Acidity-triggered TAT-presenting nanocarriers augment tumor retention and nuclear translocation of drugs. Nano Res. **11**, 5716–57345716 (2018)
99. W. Wang, G. Liang, W. Zhang, D. Xing, X. Hu, Cascade-Promoted photo-chemotherapy against resistant cancers by enzyme-responsive polyprodrug nanoplatforms. Chem. Mater. **30**, 3486–34983486 (2018)
100. L. Liu, Q. Chen, C. Ruan, X. Chen, Y. Zhang et al., Platinum-based nanovectors engineered with immuno-modulating adjuvant for inhibiting tumor growth and promoting immunity. Theranostics **8**, 2974–2987 (2018)
101. Y. Yuan, S. Diao, X. Ni, D. Zhang, W. Yi et al., Peptide-based semiconducting polymer nanoparticles for osteosarcoma-targeted NIR-II fluorescence/NIR-I photoacoustic dual-model imaging and photothermal/photodynamic therapies. J. Nanobiotechnology **20**, 44 (2022)
102. Y. Li, G. Dang, M. Rizwan Younis, Y. Cao, K. Wang, et al., Peptide functionalized actively targeted MoS(2) nanospheres for fluorescence imaging-guided controllable pH-responsive drug delivery and collaborative chemo/photodynamic therapy. J. Colloid Interface Sci. **639**, 302–313 (2023)
103. Z. Yu, Y. Xia, J. Xing, Z. Li, J. Zhen et al., Y1-receptor–ligand-functionalized ultrasmall upconversion nanoparticles for tumor-targeted trimodality imaging and photodynamic therapy with low toxicity. Nanoscale **10**, 17038–17052 (2018)
104. Y. Ma, L. Chen, X. Li, A. Hu, H. Wang et al., Rationally integrating peptide-induced targeting and multimodal therapies in a dual-shell theranostic platform for orthotopic metastatic spinal tumors. Biomaterials **275**, 120917 (2021)
105. Y. Wang, Y. Cao, Z. Jiang, Y. Li, B. Yuan et al., The neuropeptide Y(1) receptor ligand-modified cell membrane promotes targeted photodynamic therapy of zeolitic imidazolate frameworks for breast cancer. J. Phys. Chem. Lett. **12**, 11280–11287 (2021)
106. L. Zhang, C. Liu, Y. Gao, Z. Li, J. Xing et al., ZD2-engineered gold Nanostar@Metal-Organic framework nanoprobes for T(1)-weighted magnetic resonance imaging and photothermal therapy specifically toward triple-negative breast cancer. Adv. Healthc. Mater. **7**, e1801144 (2018)
107. M. Haider, R. Cagliani, J. Jagal, M.N. Jayakumar, B. Fayed et al., Peptide-functionalized graphene oxide quantum dots as colorectal cancer theranostics. J. Colloid Interface Sci. **630**, 698–713 (2023)

108. M. Yan, X. Qi, W. Xie, J. Xiao, Y. Gan et al., PDGF receptor-β-targeted copper-gadolinium-oxide self-assembled nanoclusters suppress tumor growth and metastasis via copper overload-mediated apoptosis and anti-angiogenesis. Chem. Eng. J. **457**, 141227 (2023)
109. K. Zhang, Z. Wan, H. Jiang, X. Xiao, Q. Feng et al., Dimer targeting peptide mediated precise and controllable drug delivery by upconversion nanocarriers for breast cancer therapy. Mater. Des. **203**, 109597 (2021)

Chapter 8
Peptide Assembling Nanomaterials for Cancer Therapy

Yi Cao, Xiaojiao Ge, Xiaolin Duan, Li Sun, Aiguo Wu, and Juan Li

Abstract Molecular self-assembly is ubiquitous in nature lives. As peptides are the basic building block of proteins to create fascinating functions in nature, the pharmacological activity of their self-assembly has ushered in a new era in the drug design and delivery. As a bottom-up method, the self-assembly is a powerful way to create higher-order structures from monomer at the nanoscale to supramolecular at the microscale, creating diverse materials that integrate different functional properties. By harnessing controllable self-assembly processes, peptide-modified nanomedicines can achieve site-specific responsiveness within tumor regions. The nanostructural modifications can also enhance the drug retention for the tumor cell elimination or enable size-variable delivery strategies. This synergistic approach to the peptide self-assembly enables tumor-targeted therapy with reduced side effects and has great potential for clinical application.

Keywords Peptide assembling · Hybrid materials · Cancer therapy · Nanomedicine

8.1 Peptide-Modified Assembling Nanomaterials

There have been many methods to selectively induce the peptide self-assembly and the formation of supramolecular monomer under physiological conditions of abnormal tissues, including the pH adjustment, extra-produced reactive oxygen species (ROS) or enzyme, ionic strength change, specific chemical interaction, and

Y. Cao · X. Ge · X. Duan · L. Sun · A. Wu (✉) · J. Li (✉)
Ningbo Institute of Materials Technology and Engineering, Chinese Academy of Sciences, Ningbo, China
e-mail: aiguo@nimte.ac.cn

J. Li
e-mail: lijuan@nimte.ac.cn

Ningbo Cixi Institute of Biomedical Engineering, Ningbo, China

G. Wei (ed.), *Peptide Nano-Chemistry and Nanotechnology*,
https://doi.org/10.1007/978-981-96-2387-7_8

controllable strategies in vitro, such as the temperature change, light irradiation, magnetic stimulation, and chemical reaction control [1].

8.1.1 TME-Triggered Assembling Peptide Nanomaterials

8.1.1.1 pH-Triggered Assembling Peptide Nanomaterials

The pH of TME was slightly acidic (less than pH 7), whereas normal cells have a pH of around 7.4. The intracellular pH can be slightly decreased by some viral and bacterial infections [2]. Zhang et al. developed an environment-responsive rotor molecule, which morphologically adapted in response to the pH and viscosity changes during the Golgi-endosomal transport [3]. These undefined nano-bio interactions suppressed regional tubulin polymerization and induced atypical prometaphase and metaphase oscillations, thereby inhibiting the proliferation of various cancer cells without causing overt neurotoxicity. In the subcutaneous carcinoma of the uterine cervix xenograft tumor model, the strategy showed better potent antiproliferative effects than the classical antimitotic drug taxol.

The reversible self-assembly method is an attractive way. TME-adapted self-assembly (TMAS) of pentapeptides has been reported by Yu et al. [4], as shown in Fig. 8.1, the reversible self-assembly of pentapeptide FF-Amp-FF (AmpF) regulated by the pH-sensitive cis → trans isomerization of 4-amino-proline (Amp) amide

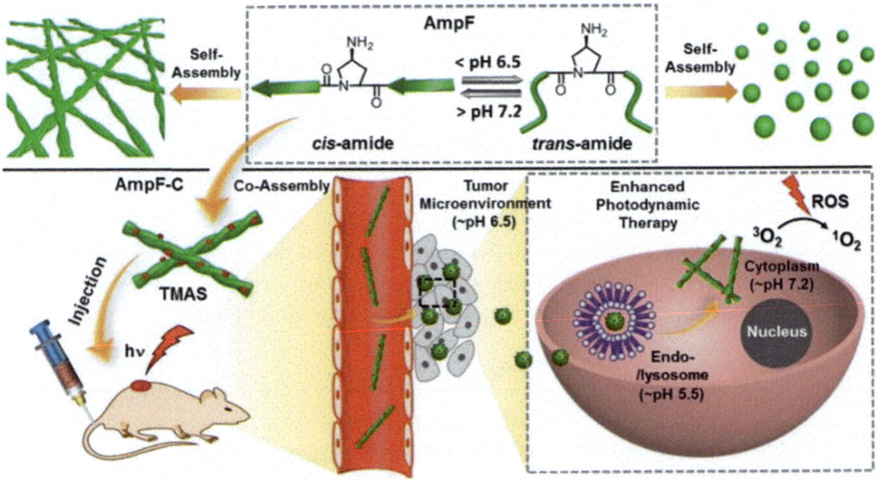

Fig. 8.1 Formation of TMAS via assembly of AmpF increase intracellular retention and induce cell death

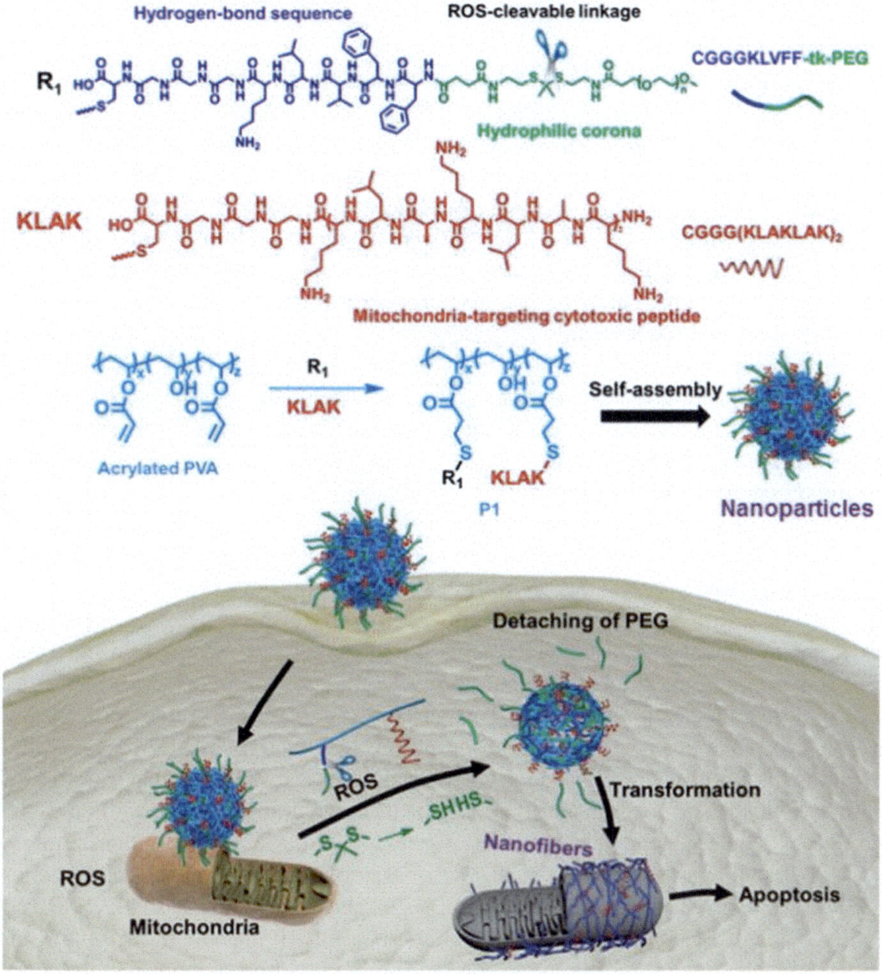

Fig. 8.2 Synthesis and mitochondria locational transformation of ROS-sensitive PPCs

bonds, into superhelices and NPs under alternating neutral and mildly acidic conditions. The reversible self-assembly method showed improved accumulation and optimized cellular uptake than the morphology-persistent nanomedicines at tumor sites, resulting in enhanced PDT and drug delivery efficiency, and enlightening the specific delivery of nanomedicines.

8.1.1.2 ROS-Triggered Assembling Peptide Nanomaterials

In most cancer cells, ROS is over-generated around mitochondria [5]. ROS-responsive chemical bonds have been taken for a useful tool in drug delivery systems to form organelle-located morphology transformation platform in this case.

As shown in Fig. 8.2, Wang et al. presented endogenous ROS-triggered morphology transformation of polymer-peptide conjugates (PPCs) for the cooperative interaction with mitochondria [6]. The functional fractions of PPCs are composed of a mitochondria-targeting cytotoxic peptide, and a β-sheet-forming peptide conjugated with PEGs through ROS-cleavable thioketal offers the ROS-triggered self-assembly ability. The thioketal linker can be cleaved by over-generated ROS around mitochondria in most cancer cells, leading to the transformation of PPCs from NPs to fibrous nanostructures. Consequently, the nanofibers showed enhanced multivalent cooperative interactions with mitochondria, causing selective cytotoxicity. This strategy provided a novel perspective for the development of nanomedicines that selectively enhance the drug efficacy.

8.1.1.3 Enzyme-Triggered Assembling Peptide Nanomaterials

Under physiological conditions, enzymes efficiently catalyze the formation and transformation of many biologically important molecules. Thus, the use of enzymes to form building blocks for molecular self-assembly is a powerful and mild preparation method for the construction of supramolecular materials [1]. Moreover, the enzyme expression is closely associated with the disease. The in situ formation of supramolecular nanomaterials at the disease sites or within abnormal cells may be facilitated by the high expression of certain enzymes at disease sites, such as tumors.

Xu et al. demonstrated the enzymatic self-assembly to selectively target the mitochondria of liver cancer cells [7]. In Fig. 8.3a, perimitochondrial ENTK cleaved the Flag-tag on the conjugate to selectively deliver CLRP into the mitochondria of cancer cells, thereby inhibiting the synthesis of mitochondrial proteins and inducing the release of cytochrome C into the cytosol, resulting in the death of the cancer cells. Liang et al. designed a pyrene-peptide conjugate (Fig. 8.3b), which was efficiently dephosphorylated by ALP into Py-Y-Lyso [8]. The excimer fluorescence of the Py-Yp-Lyso assemblies turned on and the assemblies specifically targeted the lysosomes once being uptake into HeLa cells, which enhanced the permeability of the lysosomal membrane and induced ultimate cancer cell death. Du et al. developed a novel tripeptide (FFY) that could be oxidized intracellularly and then self-assembled in situ into NPs with outstanding interference capability with microtubules and eventually reversed the drug resistance of melanoma [9]. With the catalysis of tyrosinase, FFY was oxidized to a melanin-like FFY dimer (mFFY) with a diquinone structure firstly and then self-assembled into the mFFY assemblies further. Effective intrinsic apoptosis of drug-resistant melanoma has been demonstrated using this tyrosinase-induced tripeptide assembly.

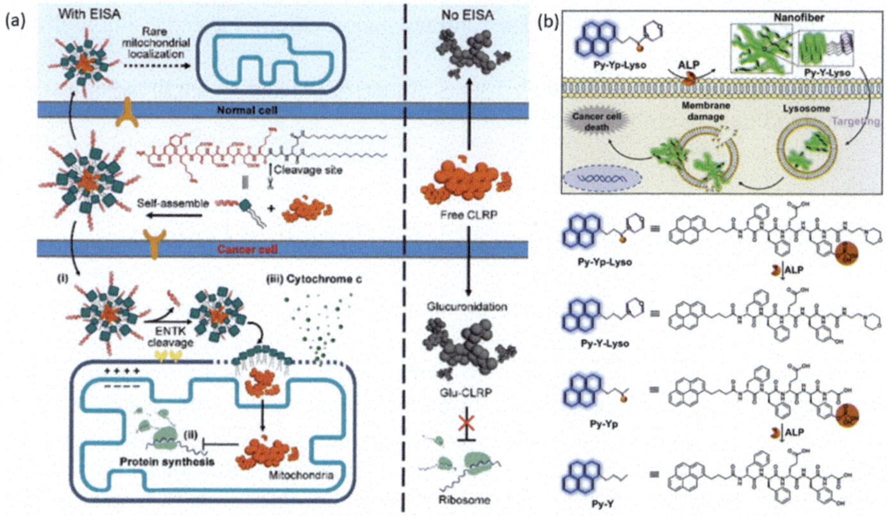

Fig. 8.3 **a** Illustration of perimitochondrial accumulation, mitochondria-targeting drug delivery, and mitochondrial outer membrane permeabilization of EISA, **b** schematic illustration of integration of enzyme-induced self-assembly and lysosome targeting for selective killing of cancer cells

8.2 GSH-Triggered Assembling Peptide Nanomaterials

As a common biomarker, GSH is overexpressed in many cancer cells. The use of GSH as a stimulus for the in situ self-assembly potentially allows the reduction of thiol groups by GSH.

Jeff et al. reported the spontaneous assembly of a peptide, which consisted of the cell-penetrating peptide RVRR and the anticancer agent olsalazine (Olsa) [10]. In Fig. 8.4, after entering the cytoplasm of high furin-expressing cells (HCT116 cells), the Olsa-RVRR underwent the reduction by GSH and furin near the Golgi complex cleave the peptide. Subsequently, amphiphilic oligomers were formed from the click reaction between two cleaved Olsa-RVRR molecules. These oligomers then self-assembled into Olsa NPs by intermolecular π–π stacking. The intracellular accumulation of Olsa NPs then hold the position of a reservoir of Olsa molecules that enhance CEST contrast and inhibit the DNA methylation for tumor therapy. Hou et al. fabricated Luffin P1 as an ideal template for multi-targeted epitope transfer. Then, the Luffin P1-derived p53-MDM2/MDMX and beta-catenin-Bcl9 inhibitor were fabricated (termed Lupbin) [11]. The disassembly of the nanostructure into smaller monomers of amorphous form by rupturing the disulfide bridge was triggered by intracellular GSH, leading to widespread distribution in the cytoplasm and the nucleus, thereby concurrently blocking the interactions between p53 and MDM2/MDMX in the cytoplasm and between β-catenin and Bcl9 in the nucleus. This strategy provided a potential high-stability nanoplatform for multi-targeted cancer therapy and extended the application of self-assembled nanomedicines to target intercellular protein–protein interactions.

To make it more clear, in Table 8.1 we summarize the studies on the synthesis of peptide-modified assembling nanomaterials.

8.2.1 Self-assembly of Sub-organelles Targeting Peptides

Organelles are subsystems within cells, and their activity and chemical composition can well reflect the metabolic status of living cells. Organelles regulate a variety of processes physiologically [38]. For example, lysosomes are fusion proteolytic organelles, which give them multiple functions, such as the degradation of cellular material, regulation of autophagy, cell transport, plasma membrane repair, and exosome biogenesis [39]. The endoplasmic reticulum is the site of all protein-secreting biogenesis and folding, is also the key to unfolded protein reactions, and communicates with almost all organelles through inter-organelle contact. Mitochondria are the most important organelles after the nucleus, responsible for the production of adenosine triphosphate for cells, the control of cellular oxidation–reduction homeostasis, and the regulation of cytoplasmic calcium levels. Table 8.2 describes the classification of organelle targeted self-assembly in tumor therapy.

8.2.1.1 Mitochondria

Recently, more studies have shown that the malignant proliferation, easy invasion, and metastasis of tumor cells are all related to mitochondria. So far, many mitochondria-targeting nanosystems have been designed and built for the cancer therapy, mainly to develop anti-tumor drugs by disrupting mitochondrial metabolism, ROS production, and endogenous mitochondrial apoptosis.

Xu et al. designed negatively charged branching peptides that responded by enzymes to target mitochondria [54]. As shown in Fig. 8.5, by binding an established protein marker (FLAG) to a self-assembled motif, the enzymatic response of the hydrophilic FLAG pattern (DDDDK) by enterokinase (ENTK) converts micelles into nanofibers, which were mainly localized in a mitochondria [39]. The combination of the precursor with DOX could effectively inhibit the activity of cancer cells. Ryu et al. proposed a phenylalanine dipeptide (FF) with mitochondrial targeting component, which could realize the local self-assembly of organelles of both parent peptides [16]. As a control of cell fate, it preferentially accumulated in mitochondria and reached a critical aggregation concentration, forming a fibrous nanostructure. The Mito-FF fibrils could cause mitochondrial dysfunction through the membrane destruction, thus leading to the cell apoptosis. In another study, Ryu et al. designed a mitochondrial penetration tripeptide Mito-FF containing a diphenylalanine building unit [46]. The Mito-FF and its mirror image pair were assembled to form super

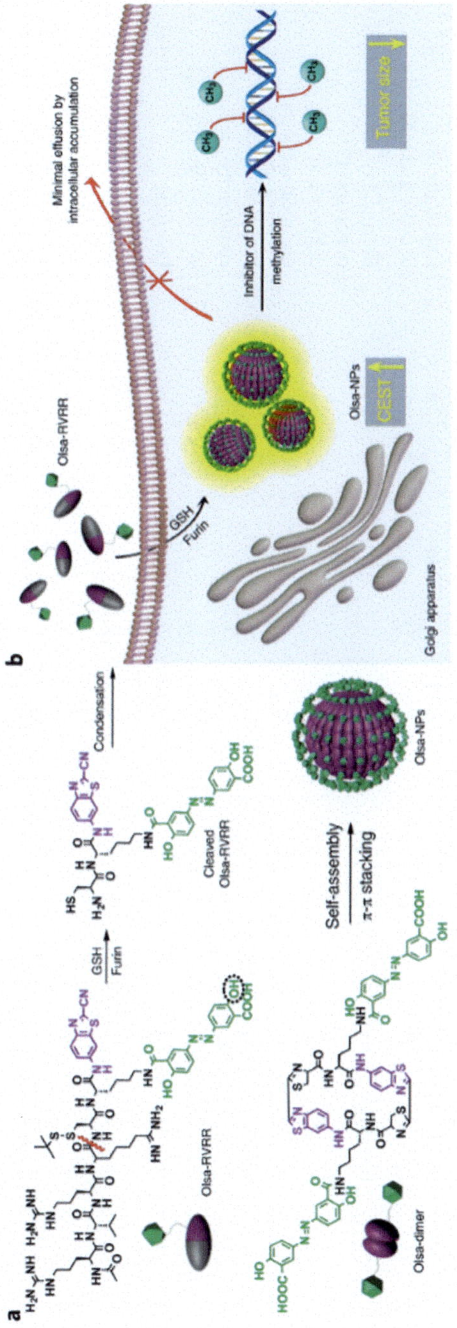

Fig. 8.4 Illustration of formation of Olsa NPs by intracellular furin-mediated reduction

Table 8.1 Peptide-modified assembling nanomaterials

Trigger	Sequence	Self-assembly part	Structure	Target/Action site	Application	References
pH and Viscosity	Nap-FF-BODIPY-aminosulfonic acid taurine	Nap-FF	Nanoscaffolds	Microtubule array	Suppressing regional tubulin polymerization	[3]
pH	N-ffl-ABS	N-ff	Fibers	Carbonic anhydrase (CA) IX	Provoking selective toxicities for hypoxic tumors	[12]
	FF-Amp-FF (AmpF)	FF-Amp-FF	Fibers	Tumor	Enhancing drug delivery and PDT	[4]
pH and Redox Stimuli	Fmoc-H/Zn^{2+}/Ce6 and Z-HF/Zn^{2+}/Ce6	Fmoc-H/Zn^{2+}	Spherical nanoparticles	Cells	Enhancing tumor-specific delivery and therapeutic effects	[13]
ROS	(CGGG-KLVFF-tk-PEG) + $(CGGGKLAKLAK)_2$	KLVFF	Fibers	Mitochondria	Provoking selective cytotoxicity against cancer cells	[6]
CES and ALP	Nap-ffy-OMe-OP	Nap-ffy-OMe	Fibers	Cell membrane	Providing chemotherapy by targeting downregulation in cancer cells	[14]
CES	NBD-ff-es-tau- (O)	FF	Fibers	Cells	Enhancing cellular uptake and intracellular delivery	[15]

(continued)

Table 8.1 (continued)

Trigger	Sequence	Self-assembly part	Structure	Target/Action site	Application	References
ENTK	Nap-ffk (GDYKDDDDK)	Nap-ffk	Fibers	Mitochondria	Targeting subcellular organelles for biomedicine	[16]
CtsB	CPT-LVFFGFLG-PEG-RGD	LVFFGF	Fibers	Tumor	Sustained drug release effectively preventing postsurgical local recurrences	[17]
MMP7	Nap-FFGPLGLARKRK	Nap-FF	Fibers	Cells	Efficient and targeted drug delivery and biomedical cancer treatment	[18]
Enzyme caspase-3/7	C_{16}-K (PpIX)-PEG_8-KDEVD-$_1$MT	/	Nanoparticles	Tumor	Producing ROS and trigger an intense immune response	[19]
	AVPIAQKDEVD (KLVFFAECG/ GCKLVFFAECG)-cyanine dyes	KLVFFAECG/ GCKLVFFAECG	Undefined structure and H-aggregates	XIAP	Modifying self-assembled peptides with functional molecules for bio-applications	[20]
	C_{16}-K (PpIX)GRRRR-AEEA-K (FAM)SDEVDSK (Dabcyl)PEG_8	/	Spherical nanoparticles	Cell membrane	Enhancing PDT efficiency and in situ therapeutic response	[21]

(continued)

Table 8.1 (continued)

Trigger	Sequence	Self-assembly part	Structure	Target/Action site	Application	References
GSH and furin	Olsalazine (Olsa)-RVRR	/	Nanoparticles	Cells	Tracking drug accumulation and therapeutic response	[10]
GSH	PMI、Bcl9p (43 amino)	/	Oligomeric nano ellipsoids	Nucleus	Multi-targeted cancer therapy	[11]
	Curcumin-FFE-CS-EE	Curcumin-FFE-SH	Fibers	Cells	Promoting the radiosensitivity of colorectal cancer cells	[22]
	PTX-buSS-pyridine-RGDRVVVGG	VVV	Filament hydrogel	ECM	Producing effective cytotoxicity against GBM cell lines and enhancing tumor penetration	[23]
	EEE-K (C_{16})-GG-C_{12}-AAGG-SS-GGEEEEEE、ER-EVM^{SeO}	/	Nanofibrils and nanoribbons	ER and Golgi apparatus	Inducing combinatorial organelle dysfunction	[24]
GSH and MMP-2	D$_{i}$CPT-PLGLAG-$_{i}$RGD	/	Filament hydrogel	/	Reducing tumor immunosuppression and enhancing cancer therapies efficacy	[25]
GSH and ALP	NBD-GFFpY-ss-ERGD	FF	Fibers	Cells	Cancer diagnostics and therapy	[26]

(continued)

Table 8.1 (continued)

Trigger	Sequence	Self-assembly part	Structure	Target/Action site	Application	References
ALP	Nap-fffy-Me	Nap- fffy -Me	Fibers	Cells	Translating regulatory circuits into therapeutic targets	[27]
	NBD-Nap-ffy$_p$e$_{Me2}$	Nap-ff	Fibers	Cells	Anticancer therapeutics	[28]
	Nap-ff$_p$YR	Nap-ff	Crescent-shaped assemblies	ER	Activating the caspase signaling cascade for cell death	[29]
	Nap-FFK (Dex)-Y(H_2PO_3)-OH (1-Dex-P)	/	Fibers	Cells	Enhancing anti-hepatic fibrosis effect	[30]
	Npx-ff$_p$y-hR + PTP1B and COX-2	Npx-ffpy-hR	Fibers	ER	Intracellular sequestration of enzymes on ER	[31]
Heating–cooling and ALP	NBD-βA-FFpYGTSFAEYWNLLSP	FFY	Fibers	Nuclear	Activating the p53 gene and enhancing cellular uptake and nuclear accumulation	[32]
P2-N3	SCYNTNHVPLSPKY-nitrogen heterocyclic-APIAQKDELEKLVFFAEC (Cy)	KLVFFAE	Nanorods	Cell membrane	Applications of diverse biomedical chemically engineering cell membranes	[34]

(continued)

Table 8.1 (continued)

Trigger	Sequence	Self-assembly part	Structure	Target/Action site	Application	References
Pillararene-based host–guest recognition	G_7CCERGDS	/	Sheets and nanoparticles	$\alpha_v\beta_3$integrin	Enhancing photodynamic therapy	[33]
RNA	thymine-kkfklkl (NBD)	kkfklkl	Membraneless molecular condensates	Nucleolus	Targeting subcellular organelles to induce toxicities	[34]
Click reaction	SC (DBCO)FPNWSLRPMNQM + MDEKAQKGPAKLVFFAC (Cy)EK (N3)G	KLVFF	Fibers	Cell membrane	Immune checkpoint blockade, tumor inhibition	[35]
ATG4B	P18-GKGSFGFTG-PAMAM	/	Fibers	Cells	Enhancing PA signals, autophagy-based treatment of diseases	[36]
ATP	thymine-ffkkfklkl	ffkkfklkl	Fibers	Cells	Disrupting intracellular ATP dynamics	[37]

Table 8.2 Classification of organelle targeted self-assembly in tumor therapy

Targeting site	Self-assembly peptide	Applied tumor	Structure	References
Lysosome targeting	FF	HeLa cell	Nanofiber	[36]
	FF	HeLa cell	Nanofiber	[40]
	VEALYL	SK-OV-3 cell	Nanofiber	[41]
Endoplasmic reticulum targeting	EIIYIIYE	Human cervical cancer cell	Nanofiber	[42]
	FFKY	OVSAHO cell	Nanofiber	[7]
	FF	HeLa cell	Nanofiber	[29]
Nuclear targeting	FFY	HepG2 cell	Nanoparticle	[43]
	GFFY	HepG2 cell	Nanoparticle	[44]
	FFY	HepG2 cell	Nanofiber	[45]
Mitochondria targeting	FF	MCF-7 cell	Nanofiber	[46]
	FF	SKBR-3 cell	Nanofiber	[47]
	FFK	HeLa cell	Nanofiber	[16]
	FF	Human cancer cell	Nanofiber	[48]
Golgi targeting	FF	HeLa cell	Nanofiber	[49, 50]
	FF	MDA-MB-231 cell	Micelle	[24]
Cell membrane targeting	GNNQQNY (αVβ3)	MCF-7	Nanofiber	[51]
	FFG (EphA2)	PC-3 cancer cell	Nanofiber	[52]
	KY	HeLa cell	Nanofiber	[53]

fibrils. The in vitro and in vivo intracellular co-administration of Mito-FF caused severe mitochondrial damage and cancer cell death.

8.2.1.2 Endoplasmic Reticulum (ER)

Self-assembled nanofibers had great promise in cancer therapy due to their in situ assembly, spatiotemporal responsiveness, and diverse biological activities. By delivering the drug in active form to the cell compartment, organelle targeting increases the concentration of the drug at the target of drug action, thereby improving the efficacy and reducing the side effects [55]. As the largest organelle, ER plays an important role in biosynthesis, sensing, and signaling in eukaryotic cells. In particular, ER is responsible for the synthesis, folding, and post-translational modification of proteins used in the secretory pathway, accounting for approximately 30% of the total proteome. Disrupting the protein-folding ability of the ER leads to ER stress, which ultimately activates the apoptotic signaling pathway and cell death [56].

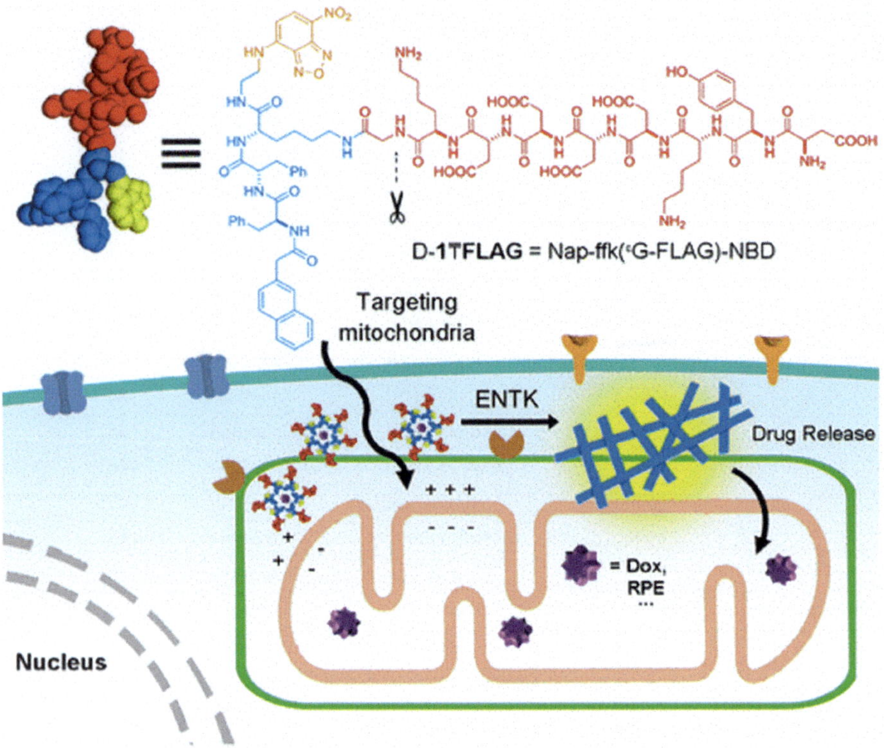

Fig. 8.5 ENTK triggers FLAG nanofibers to target mitochondria and induce cancer cell death

Xu and co-workers developed an ALP-induced peptide based on the high content of ALP in tumor cells, which could aggregate and self-assemble at the tumor site to achieve the effect of targeted therapy [29]. As shown in Fig. 8.6, after reaching the tumor tissue, the peptides generated supramolecular assembly on and inside the cancer cells through enzymatic reaction and interacted with the cell membrane. The crescent-shaped supramolecular assembly destroyed the integrity of the plasma membrane and then targeted the endoplasmic reticulum to induce the cancer cell death through the ER stress. Yu et al. prepared the precursor of a combination of peptides with the ER targeting and inhibition of chaperone glucose regulatory protein 78 (GRP78) [42]. The combination of peptides formed clear nanofibers that could be recycled by ER in human cervical cancer cells. The combination of peptides accumulated by ER also stimulated endoplasmic reticulum stress and inhibited the refolding activity of GRP78. Thus, endoplasmic reticulum stress could induce the cancer cell death.

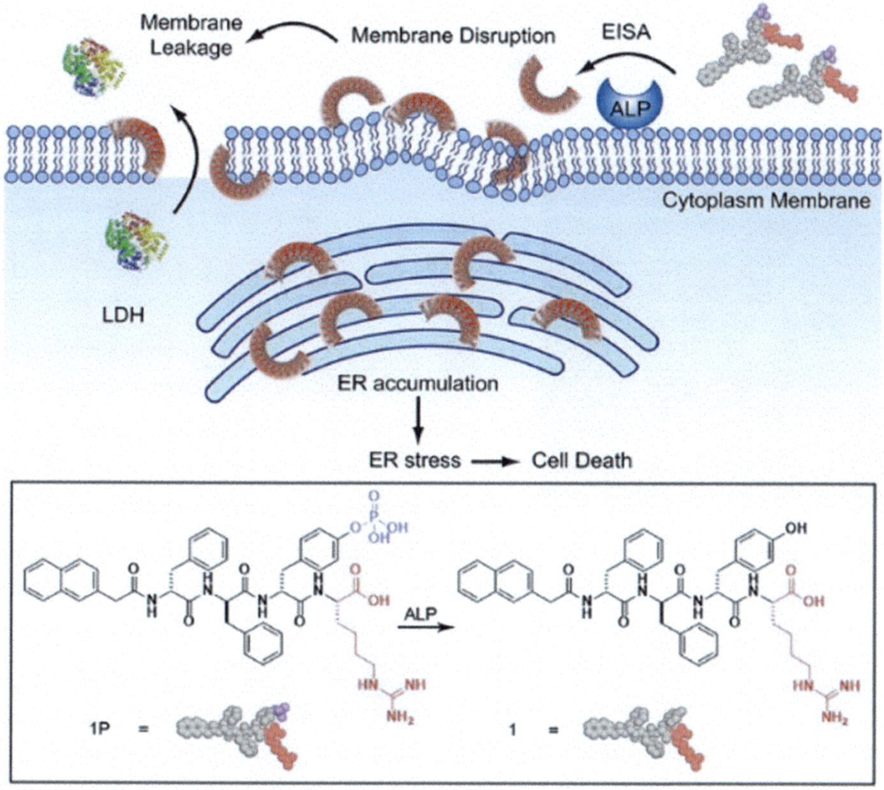

Fig. 8.6 Schematic diagram of ALP-triggered tripeptide self-assembly targeting endoplasmic reticulum to induce cell death

8.2.1.3 Golgi Apparatus

Golgi apparatus is a hub of intracellular transport, and an increasing number of studies have shown that the Golgi apparatus is a hub of different signaling pathways that drive the cancer cell survival and migration. As shown in Fig. 8.7, Xu et al. designed the phosphorylated short peptide pO1 and used sulfur atoms to change the oxygen atom of the phosphate ester bond in the short phosphate peptide to form pS1 [50]. Catalyzed by ALP, the short peptide pS1 assembled and rapidly targeted the cellular Golgi apparatus to selectively kill cancer cells through the self-assembly of enzymes.

At the same time, Golgi apparatus is also the main sub-organelle of drug resistance. The Tao group designed the convertible C_6RVRRF_4KY [49], which could self-assemble into non-toxic NPs in the water medium and transform into left spiral nanofibers in the Golgi apparatus of targeted cancer cells, mechanical destruction of the Golgi apparatus membrane, resulting in the collapse of the cell structure and eventually death of cancer cells. Yu et al. designed a GSH-triggered cross-linked

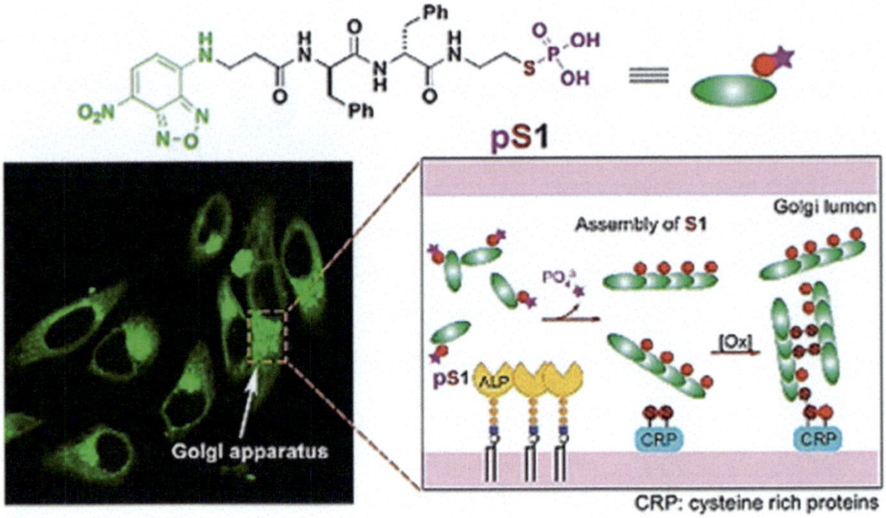

Fig. 8.7 Diagram of ALP triggering pS1 self-assembly targeting Golgi apparatus to induce cell death

amphiphilic peptide composed of derived peptides $E_3C_{16}E_6$ and EVMSeO [24]. The different organizational patterns of the two peptides promoted their GSH-induced self-aggregation into isolated nanofibers, where sulfhydryl and ER targeting groups were added to the cross-linked amphiphilic peptides to target both the endoplasmic reticulum and the Golgi apparatus, leading to combinational organelle dysfunction and cell death.

8.2.1.4 Lysosomes

It is found that the development of tumors is closely related to the lysosome and its enzymes [57]. In the process of cell canceration, the position and membrane permeability of lysosome will change, and a large amount of cathepsin will be released into the cytoplasm, which interacts with the lysogen-activating system in the blood system to participate in the degradation of extracellular matrix, thus enhancing the invasion and motility of tumor cells, accelerating the formation of tumor blood vessels and promoting the tumor growth [58].

Stu et al. successfully constructed cell-penetrating transformable peptide NPs, as shown in Fig. 8.8 [59]. The NPs composed of amphiphilic peptides and assembled into NPs. Due to the pH changes in lysosomes, the conformation of the NPs was changed into nanofibers, resulting in the lysosome destruction and cell death. Wang et al. designed LTP by using lysosome acidification and pH-responsive natural protein VEALYL and modifying aromatic groups at the N-terminal to improve the assembly capacity of the peptide. LTP maintained a soluble random oligomer state at neutral or alkaline pH and formed an amyloid fiber structure under acidic conditions [41]. The in vitro and in vivo experiments had shown that the cell death induced by the aggregation of artificial biomolecules formed by pH-responsive peptides in cancer cell lysosomes could significantly improve their efficacy in cancer cells and tumor-bearing mouse models.

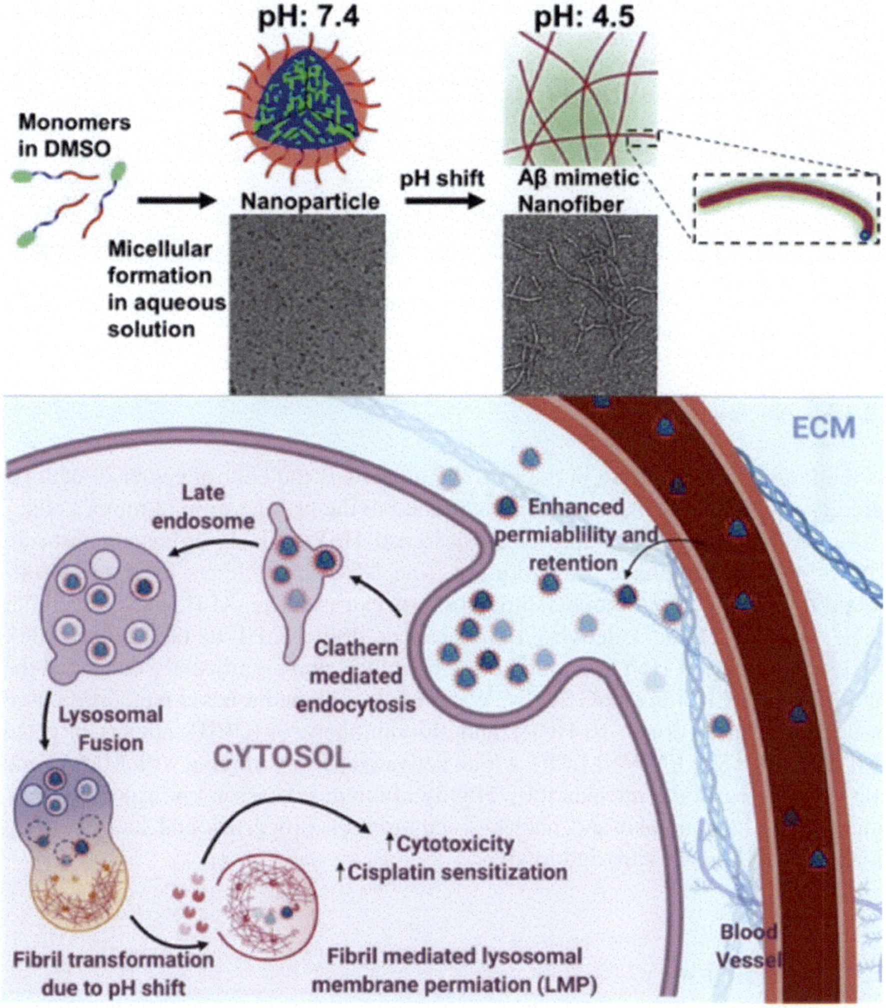

Fig. 8.8 Apoptosis induced by converting peptide nanomaterials

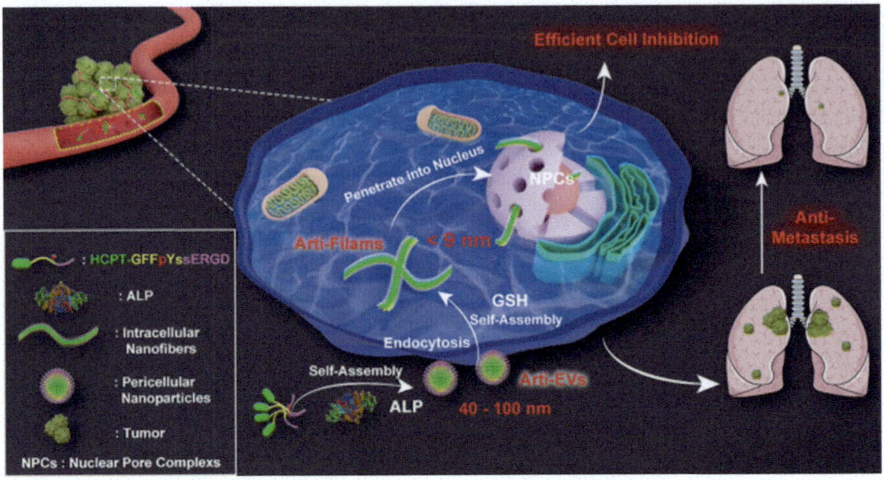

Fig. 8.9 Strategy diagram of nuclear targeting of organelle cascade delivery to effectively inhibit liver cancer and lung metastasis

8.2.1.5 Nucleus

As the largest cell structure in the cell, the nucleus is the control center of cellular heredity and metabolism, and is closely related to the proliferation of tumor cells.

As shown in Fig. 8.9, Yang et al. designed HpYss, 10-hydroxycamptothecin (HCPT)-peptide conjugate, transforming from NPs to nanofibers by a series self-assembly under the continuous stimulation of extracellular ALP and intracellular GSH, enabling efficient delivery of anticancer drug HCPT to the nucleus [38]. Nuclear targeting of HpYss amplified the apoptosis and significantly enhanced the inhibition of HCPT on HepG2 cells. Wang et al. designed a novel nanodrug based on dual anticancer drugs, 10-HCPT and chloramphenicol (CRB), nuclear targeted PMI peptide (TSFAEYWNLLSP), which activated p53 by binding with MDM2 and MDMX located in the nucleus [60]. Highly efficient self-assembly tripeptide FFY enhanced cellular uptake and nuclear accumulation properties and had excellent anticancer abilities in vitro and in vivo.

8.3 Conclusion

To summarize, we reviewed the advanced applications of peptide-based self-assembled nanosystems in cancer diagnosis and treatment in this chapter. The high biocompatibility and the self-assembly nature of peptides make them excellent candidates for the self-assembly strategies. Using endogenous triggers, such as pH, ROS, enzyme, and GSH, the nanosystems could undergo the changes in morphology and function to achieve specific killing and retention of tumors via the molecular self-assembly. This specific self-assembly strategy effectively reduced the retention in non-tumor regions, decreased the off-target effects, and enhanced the potential for clinical applications. Targeted delivery to subcellular organelles is a more precise and efficient targeting strategy that concentrates therapeutic agents in subcellular compartments, which is expected to reduce the administered dose and prevent recurrence, invasion, and metastasis, overcoming multidrug resistance and completely inducing the eradication of tumors.

References

1. J. Gao, J. Zhan, Z. Yang, Enzyme-instructed self-assembly (EISA) and hydrogelation of peptides. Adv. Mater. **32**, 1805798 (2020)
2. S. Yamamoto, K. Nishimura, K. Morita, S. Kanemitsu, Y. Nishida et al., Microenvironment pH-induced selective cell death for potential cancer therapy using nanofibrous self-assembly of a peptide amphiphile. Biomacromol **22**, 2524–2531 (2021)
3. G. Li, X. Hu, X. Wu, Y. Zhang, Microtubule-targeted self-assembly triggers prometaphase-metaphase oscillations suppressing tumor growth. Nano Lett. **21**, 3052–3059 (2021)
4. M. Li, Y. Ning, J. Chen, X. Duan, N. Song et al., Proline isomerization-regulated tumor microenvironment-adaptable self-assembly of peptides for enhanced therapeutic efficacy. Nano Lett. **19**, 7965–7976 (2019)
5. P.T. Schumacker, Reactive oxygen species in cancer cells: live by the sword, die by the sword. Cancer Cell **10**, 175–176 (2006)
6. D.B. Cheng, X.H. Zhang, Y.J. Gao, L. Ji, D. Hou et al., Endogenous reactive oxygen species-triggered morphology transformation for enhanced cooperative interaction with mitochondria. J. Am. Chem. Soc. **141**, 7235–7239 (2019)
7. H. He, X. Lin, J. Guo, J. Wang, B. Xu, Perimitochondrial enzymatic self-assembly for selective targeting the mitochondria of cancer cells. ACS Nano **14**, 6947–6955 (2020)
8. C. Wu, C. Wang, T. Zhang, G. Gao, M. Wei et al., Lysosome-targeted and fluorescence-turned "On" cytotoxicity induced by alkaline phosphatase-triggered self-assembly. Adv. Healthc. Mater. **11**, 2101346 (2022)
9. M. Sun, C. Wang, M. Lv, Z. Fan, J. Du, Intracellular self-assembly of peptides to induce apoptosis against drug-resistant melanoma. J. Am. Chem. Soc. **144**, 7337–7345 (2022)
10. Y. Yuan, J. Zhang, X. Qi, S. Li, G. Liu et al., Furin-mediated intracellular self-assembly of olsalazine nanoparticles for enhanced magnetic resonance imaging and tumour therapy. Nat. Mater. **18**, 1376–1383 (2019)
11. W. He, J. Yan, F. Sui, S. Wang, X. Su et al., Turning a luffa protein into a self-assembled biodegradable nanoplatform for multitargeted cancer therapy. ACS Nano **12**, 11664–11677 (2018)
12. J. Li, K. Shi, Z.F. Sabet, W. Fu, H. Zhou, et al., New power of self-assembling carbonic anhydrase inhibitor: short peptide; constructed nanofibers inspire hypoxic cancer therapy, Sci. Adv. **5**, eaax0937 (2019)
13. S. Li, Q. Zou, Y. Li, C. Yuan, R. Xing, X. Yan, Smart peptide-based supramolecular photodynamic metallo-nanodrugs designed by multicomponent coordination self-assembly. J. Am. Chem. Soc. **140**, 10794–10802 (2018)
14. Z. Feng, H. Wang, R. Zhou, J. Li, B. Xu, Enzyme-instructed assembly and disassembly processes for targeting downregulation in cancer cells. J. Am. Chem. Soc. **139**, 3950–3953 (2017)

15. J. Zhou, X. Du, C. Berciu, S.J. Del Signore, X. Chen et al., Cellular uptake of a taurine-modified, ester bond-decorated D-peptide derivative via dynamin-based endocytosis and macropinocytosis. Mol. Ther. **26**, 648–658 (2018)
16. H. He, J. Wang, H. Wang, N. Zhou, D. Yang et al., Enzymatic cleavage of branched peptides for targeting mitochondria. J. Am. Chem. Soc. **140**, 1215–1218 (2018)
17. D.B. Cheng, X.H. Zhang, Y.J. Gao, D. Wang, L. Wang et al., Site-specific construction of long-term drug depot for suppression of tumor recurrence. Small **15**, 1901813 (2019)
18. M. Cao, S. Lu, N. Wang, H. Xu, H. Cox et al., Enzyme-triggered morphological transition of peptide nanostructures for tumor-targeted drug delivery and enhanced cancer therapy. ACS Appl. Mater. Interfaces **11**, 16357–16366 (2019)
19. W. Song, J. Kuang, C.X. Li, M. Zhang, D. Zheng et al., Enhanced immunotherapy ased on photodynamic therapy for both primary and lung metastasis tumor eradication. ACS Nano **12**, 1978–1989 (2018)
20. R. Zheng, J. Yang, M. Mamuti, D.Y. Hou, H.W. An et al., Controllable self-assembly of peptide-cyanine conjugates in vivo as fine-tunable theranostics. Angew. Chem. Int. Ed. **60**, 7809–7819 (2021)
21. W. Ma, S.N. Sha, P.L. Chen, M. Yu, J.J. Chen et al., A cell membrane-targeting self-delivery chimeric peptide for enhanced photodynamic therapy and in situ therapeutic feedback. Adv. Healthc. Mater. **9**, 1901100 (2020)
22. H. Xu, T. Wang, C. Yang, X. Li, G. Liu et al., Supramolecular nanofibers of curcumin for highly amplified radiosensitization of colorectal cancers to ionizing radiation. Adv. Funct. Mater. **28**, 1707140 (2018)
23. R.W. Chakroun, F. Wang, R. Lin, Y. Wang, H. Su et al., Fine-tuning the linear release rate of paclitaxel-bearing supramolecular filament hydrogels through molecular engineering. ACS Nano **13**, 7780–7790 (2019)
24. X. Liu, M. Li, J. Liu, Y. Song, B. Hu, et al., In situ self-sorting peptide assemblies in living cells for simultaneous organelle targeting, J. Am. Chem. Soc., **144**, 9312–9323 (2022)
25. F. Wang, H. Su, D. Xu, W. Dai, W. Zhang et al., Tumour sensitization via the extended intra-tumoural release of a STING agonist and camptothecin from a self-assembled hydrogel. Nat. Biomed. Eng. **4**, 1090–1101 (2020)
26. J. Zhan, Y. Cai, S. He, L. Wang, Z. Yang, Tandem molecular self-assembly in liver cancer cells. Angew. Chem. Int. Ed. **57**, 1813–1816 (2018)
27. J. Zhou, X. Du, X. Chen, J. Wang, N. Zhou et al., Enzymatic self-assembly confers exceptionally strong synergism with NF-κB targeting for selective necroptosis of cancer cells. J. Am. Chem. Soc. **140**, 2301–2308 (2018)
28. Z. Feng, X. Han, H. Wang, T. Tang, B. Xu, Enzyme-instructed peptide assemblies selectively inhibit bone tumors. Chem **5**, 2442–2449 (2019)
29. Z. Feng, H. Wang, S. Wang, Q. Zhang, X. Zhang et al., Enzymatic assemblies disrupt the membrane and target endoplasmic reticulum for selective cancer cell death. J. Am. Chem. Soc. **140**, 9566–9573 (2018)
30. W. Tang, Z. Zhao, Y. Chong, C. Wu, Q. Liu et al., Tandem enzymatic self-assembly and slow release of dexamethasone enhances its antihepatic fibrosis effect. ACS Nano **12**, 9966–9973 (2018)
31. Z. Feng, H. Wang, B. Xu, Instructed assembly of peptides for intracellular enzyme sequestration. J. Am. Chem. Soc. **140**, 16433–16437 (2018)
32. C. Liang, X. Yan, R. Zhang, T. Xu, D. Zheng et al., Enhanced cellular uptake and nuclear accumulation of drug-peptide nanomedicines prepared by enzyme-instructed self-assembly. J. Control. Release **317**, 109–117 (2020)
33. H. Zhu, H. Wang, B. Shi, L. Shangguan, W. Tong et al., Supramolecular peptide constructed by molecular Lego allowing programmable self-assembly for photodynamic therapy. Nat. Commun. **10**, 2412 (2019)
34. H. Wang, Z. Feng, W. Tan, B. Xu, Assemblies of d-peptides for targeting cell nucleolus. Bioconjugate Chem. **30**, 2528–2532 (2019)

35. Z.-W. Deng, C.-S. Yuan, T. Wang, X.-G. Chen, Y. Liu, Peptide-based assemblies as immune checkpoint inhibitor delivery systems for enhanced immunotherapy. Appl. Mater. Today **23**, 101063 (2021)
36. Y. Lin, Y. Wang, S. Qiao, H. An, J. Wang et al., "In vivo self-assembled" nanoprobes for optimizing autophagy-mediated chemotherapy. Biomaterials **141**, 199–209 (2017)
37. H. Wang, Z. Feng, Y. Qin, J. Wang, B. Xu, Nucleopeptide assemblies selectively sequester ATP in cancer cells to increase the efficacy of doxorubicin. Angew. Chem. Int. Ed. **57**, 4931–4935 (2018)
38. J. Zhan, Y. Wang, S. Ma, Q. Qin, L. Wang et al., Organelle-inspired supramolecular nanomedicine to precisely abolish liver tumor growth and metastasis. Bioact. Mater. **9**, 120–133 (2022)
39. H. He, J. Guo, X. Lin, B. Xu, Enzyme-instructed assemblies enable mitochondria localization of histone H2B in cancer cells. Angew. Chem. Int. Ed. **59**, 9330–9334 (2020)
40. X. Yang, H. Lu, Y. Tao, L. Zhou, H. Wang, Spatiotemporal control over chemical assembly in living cells by integration of acid-catalyzed hydrolysis and enzymatic reactions. Angew. Chem. Int. Ed. **60**, 23797–23804 (2021)
41. J. Wang, L. Hu, H. Zhang, Y. Fang, T. Wang, H. Wang, Intracellular condensates of oligopeptide for targeting lysosome and addressing multiple drug resistance of cancer. Adv. Mater. **34**, 2104704 (2021)
42. N. Song, Y. Song, B. Hu, X. Liu, X. Yu et al., Persistent endoplasmic reticulum stress stimulated by peptide assemblies for sensitizing cancer chemotherapy. Adv. Healthc. Mater. **12**, 2202039 (2022)
43. S. Liu, Q. Zhang, A.N. Shy, M. Yi, H. He et al., Enzymatically forming intranuclear peptide assemblies for selectively killing human induced pluripotent stem cells. J. Am. Chem. Soc. **143**, 15852–15862 (2021)
44. T. Xu, C. Liang, D. Zheng, X. Yan, Y. Chen et al., Nuclear delivery of dual anticancer drug-based nanomedicine constructed by cisplatinum-induced peptide self-assembly. Nanoscale **12**, 15275–15282 (2020)
45. J. Wu, W. Ding, G. Han, W. You, W. Gao et al., Nuclear delivery of dual anti-cancer drugs by molecular self-assembly. Biomater. Sci. **9**, 116–123 (2021)
46. M.T. Jeena, L. Palanikumar, E.M. Go, I. Kim, M.G. Kang et al., Mitochondria localization induced self-assembly of peptide amphiphiles for cellular dysfunction. Nat. Commun. **8**, 26 (2017)
47. S. Wang, Y. Ma, C. Ma, K. Liu, Z. Huo, Y. Shang, Supramolecular nanofibers formed by enzyme-instructed self-assembly for SKBR-3 cell selective inhibition. Chem. Asian J. **17**, e202200301 (2022)
48. L. Jiang, S. Zhou, X. Zhang, C. Li, S. Ji et al., Mitochondrion-specific dendritic lipopeptide liposomes for targeted sub-cellular delivery. Nat. Commun. **12**, 2390 (2021)
49. X. Li, C. Cao, P. Wei, M. Xu, Z. Liu et al., Self-assembly of amphiphilic peptides for recognizing high furin-expressing cancer cells. ACS Appl. Mater. Interfaces **11**, 12327–12334 (2019)
50. W. Tan, Q. Zhang, J. Wang, M. Yi, H. He, B. Xu, Enzymatic assemblies of thiophosphopeptides instantly target golgi apparatus and selectively kill cancer cells. Angew. Chem. Int. Ed. **60**, 12796–12801 (2021)
51. M.D. Wang, G.T. Lv, H.W. An, N.Y. Zhang, H. Wang, In situ self-assembly of bispecific peptide for cancer immunotherapy. Angew. Chem. Int. Ed. **61**, 202113649 (2022)
52. J. Li, Y. Fang, Y. Zhang, H. Wang, Z. Yang, D. Ding, Supramolecular self-assembly-facilitated aggregation of tumor-specific transmembrane receptors for signaling activation and converting immunologically cold to hot tumors, Adv Mater. **33**, 2008518 (2021)
53. R.C. Guo, X.H. Zhang, P.S. Fan, B.L. Song, Z.X. Li et al., In vivo self-assembly induced cell membrane phase separation for improved peptide drug internalization. Angew. Chem. Int. Ed. **60**, 25128–25134 (2021)
54. J. Qin, N. Gong, Z. Liao, S. Zhang, P. Timashev et al., Recent progress in mitochondria-targeting-based nanotechnology for cancer treatment. Nanoscale **13**, 7108–7118 (2021)

55. W. Li, J. Yang, L. Luo, M. Jiang, B. Qin et al., Targeting photodynamic and photothermal therapy to the endoplasmic reticulum enhances immunogenic cancer cell death. Nat. Commun. **10**, 3349 (2019)
56. X. Ma, N. Gong, L. Zhong, J. Sun, X.J. Liang, Future of nanotherapeutics: targeting the cellular sub-organelles. Biomaterials **97**, 10–21 (2016)
57. L. Hu, Y. Li, X. Lin, Y. Huo, H. Zhang, H. Wang, Structure-based programming of supramolecular assemblies in living cells for selective cancer cell inhibition. Angew. Chem. Int. Ed. **60**, 21807–21816 (2021)
58. M. Borkowska, M. Siek, D.V. Kolygina, Y.I. Sobolev, S. Lach et al., Targeted crystallization of mixed-charge nanoparticles in lysosomes induces selective death of cancer cells. Nat. Nanotechnol. **15**, 331–341 (2020)
59. C.M. Baehr, L. Zhang, Y. Wu, A. Domokos, W. Xiao et al., Transformable amyloid-beta mimetic peptide amphiphiles for lysosomal disruption in non-small cell lung cancer. Biomaterials **277**, 121078 (2021)
60. Y. Zhou, Y. Chen, Y. Tan, R. Hu, M.M. Niu, An NRP1/MDM2-targeted D-peptide supramolecular nanomedicine for high-efficacy and low-toxic liver cancer therapy. Adv. Healthc. Mater. **10**, e2002197 (2021)

Chapter 9
Peptide Hydrogels for Drug and Gene Delivery

Zixuan Xiao and Zhiqiang Su

Abstract Self-assembling peptide nanomaterials exhibit great potential in biomedical fields, especially for the delivery of various drugs and genes for treating disease therapy. This chapter introduces the design and synthesis of 3D peptide hydrogels for drug and gene delivery applications. The fabrication strategies of functional peptide hydrogels via both physical and chemical cross-linking are introduced, and functional tailoring of the self-assembling peptide hydrogels is discussed. In addition, the techniques on applying the synthesized peptide hydrogels for delivering different drugs and genes are demonstrated.

Keywords Peptide hydrogels · Material design · Drug delivery · Gene delivery

9.1 Introduction

Hydrogels have a unique 3D interconnectivity and network structure, high water-absorption capacity, and good biocompatibility. These properties are suitable for various biological applications, such as the tissue engineering, wound dressings, and controlled drug loading and release [1–5]. Since Wichterle et al. reported the contact lenses using hydrogels in 1960, hydrogels have been widely used in biomedical field [6]. The preparation of hydrogel materials is relatively extensive. Polymer molecules, such as natural polysaccharides, gelatin [7], peptides [8], and naturally modified biomacromolecules, have been used to prepare various hydrogels [9, 10]. Among these materials that can be synthesized into hydrogels, peptides are often used to synthesize biomedical hydrogels due to their design ability, self-assembly function, and other excellent biological properties. Previously, the synthesis principles of peptide hydrogels have been explored in various ways in order to investigate how they were synthesized, which can be divided into both physical and chemical cross-linking

Z. Xiao · Z. Su (✉)
State Key Laboratory of Chemical Resource Engineering, Beijing Key Laboratory of Advanced Functional Polymer Composites, Beijing University of Chemical Technology, Beijing 100029, China
e-mail: suzq@mail.buct.edu.cn

G. Wei (ed.), *Peptide Nano-Chemistry and Nanotechnology*,
https://doi.org/10.1007/978-981-96-2387-7_9

[11, 12]. Physical cross-linking consists of the hydrogen bonding, hydrophobic interactions, electrostatic interaction, and chain entanglement, and chemical cross-linking consists of the covalent bonding between polymer chains.

Peptide-based hydrogels have also gained the popularity in both nanotechnology and materials science due to their excellent biocompatibility, high biodegradability, ease of modification, and good bioactivity. In recent years. The use of peptide-based hydrogels as novel smart materials has revealed good applications in scientific research and clinical biomedicine [13, 14].

In this chapter, we introduce the direction of the peptide hydrogels as carriers for the drug and gene delivery. Peptide hydrogels can be used as biological carriers to load drugs or gene molecules and deliver functional factors to the lesion sites. Such an approach maintains the properties of peptides, releases the medicine for a long time, and reduces the drug irritation.

9.2 Peptide Hydrogels for Drug Delivery

So far, the impact caused by tumors is still a significant problem worldwide. Although researchers are constantly working to solve this problem, suitable therapeutic tools are not yet at the level of demand. Therefore, new therapeutic vectors should continue to be developed. One of the vectors is the nanomaterials as drug carriers, which are superior in terms of their physical and chemical properties, high drug loading capacity, and low rejection rate in vivo [15]. Based on that, peptide-based hydrogels have been considered as promising carriers for the cancer therapy. They can enhance the therapeutic effect of drugs without negative reactions. In addition, the outer layer of the nanomaterials can be adjusted to achieve specific drug delivery [16]. Many different anticancer drugs, including doxorubicin (DOX), curcumin, fluorouracil (FU), and paclitaxel (PTX), have been loaded into/onto self-assembled peptide nanostructures for preclinical and clinical cancer therapy.

9.2.1 Delivery of DOX

DOX has been widely used clinically for its therapeutic effects on a wide range of cancers. DOX has the ability of killing tumor cells of all growth cycles in the direction of biomedical use. The self-assembly of peptide hydrogels from an RGD-derived peptide (naphthylacetic acid-Gly-Phe-Phe-Tyr-Gly-Arg-Gly-Asp-His, abbreviated as 1-RGDH) has been demonstrated to generate electrostatic interactions with DOX, as shown in Fig. 9.1e [17]. The interactions were achieved using the negatively charged residues of amino acids in the peptide sequence, which enhanced the hydrogel network structure and delivered drugs by this system. In addition, this hydrogel mixed with DOX showed the pH-responsive release properties, which accelerated the drug release under weakly acidic conditions. The prepared hydrogel and DOX system were promising to inhibit the growth of A549, MCF-7, and HeLa cells better than the DOX solution alone. In another study, a synthetic peptide amphiphile (PA) with the sequence of $RADA_{32}$ self-assembled into peptide nanohydrogels (Fig. 9.1a) [18]. The formed nanohydrogel was further used to fabricate the melittin-RADA32-DOX (MRD) hydrogel for the melanoma cell therapy. Single dose injection of the MRD hydrogel resulted in over 95% delay in primary melanoma tumor growth due to the loading of toxin and DOX, accompanied by the recruitment of activated natural killer cells in the tumor (Fig. 9.1b).

In another study, an extremely potent cationic peptide anti-tumor agent consisting of 26 amino for loading DOX and melittin was developed [19]. A cationic peptide composed of 26 amino acids was used for the fabrication of a hydrogel scaffold. Dual delivery of melittin and DOX from the hydrogel induced an effective and sustained anti-tumor effect. The Ade-FFFF peptide hydrogels were prepared by the pH-switching method at the concentrations above a critical value. It was demonstrated that the Ade-FFFF hydrogels at a concentration of 25 mM were able to store 1 mM of DOX by recovering and maintaining their hydrogel form. In the Ade-FFFF molecule, approximately 55% of DOX was released by the hydrogel degradation within 12 days with a near-linear release curve ($R^2 = 0.9366$).

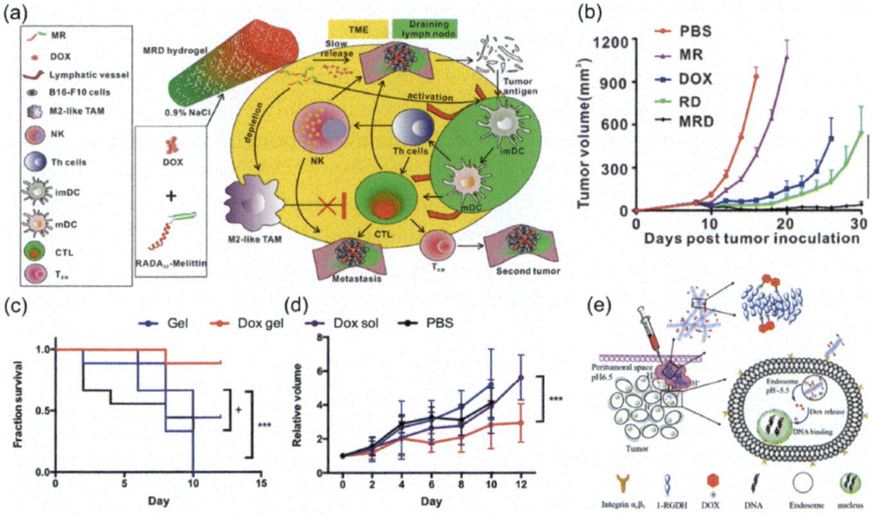

Fig. 9.1 (**a**) Preparation of MRD hydrogels and mechanism mediating anti-melanoma effect. (**b**) Tumor growth curves of mice with intra-tumor injection of PBS, DOX, RD, MR, or MRD hydrogels. Reproduced from Ref. [18], Copyright 2018, American Chemical Society. Survival curves (**c**) and tumor growth curves (**d**) of mice treated by DOX-loaded hydrogels. Reproduced from Ref. [19] Copyright 2020, Royal Society of Chemistry. (**e**) Injectable DOX-reinforced 1-RGDH hydrogels as drug delivery systems for anti-tumor treatment. Reproduced from Ref. [17] Copyright 2019, Royal Society of Chemistry

9.2.2 Delivery of Curcumin

Curcumin is a hydrophobic polyphenol drug that extracted from turmeric root. It has been shown to have antioxidant, anti-inflammatory, and anti-tumor properties. However, curcumin is a hydrophobic drug with a short half-life, so the choice of drug carrier is challenging for the therapeutic effect [20]. The using of peptide hydrogels is a good choice for the delivery of the curcumin drug.

Haines-Butterick et al. designed a curcumin-delivered peptide hydrogel (VKVKVKVDPLPTKVKVKV-NH_2) for the drug delivery [21]. This hydrogel responded to an increase in the ionic strength of the medium. The self-assembly of peptides was induced by triggering the sol–gel transition by increasing the ionic strength, and the formed hydrogel was successfully used as a vehicle for the sustained release of curcumin (4 mM) over two weeks. The peptide hydrogel exhibited shear-thinning properties, increased the stability of the drug under physiological conditions, and maintained its biological activity. In another case, Liu et al. encapsulated curcumin in the tumor-targeting self-assembled peptide nanofibers (PNFs) that formed by Nap-GFFYGRGD, and obtained a 32% encapsulation efficiency [22]. The curcumin encapsulated in 5 mg/mL of the peptide was 0.5 mM, and this delivery system released 60% of the drug over two days. It had high cellular uptake in HepG2 liver cancer cells. In addition, it was found to accumulate in tumor cells after the intravenous injection into a mouse model.

Panda and co-workers created nanoparticles that allow curcumin to be released continuously over 36 h [23]. The nanoparticles were mainly produced by the peptide self-assembly in methanol, in which the methanol evaporated promoted the drug loading rate to reach 30% and the entrapment rate reach 92%. They further tested the drug delivery in different cancer cell lines, including the human cervical cancer cell line (HeLa), MCF-7, and the human liver cancer cell line (HuH-7). The cytotoxic effect of the curcumin nanoparticles system was higher than that of native curcumin. In addition, this formulation inhibited the growth of B16F10 melanoma tumors in mice.

The synergistic effect of different drugs was utilized to achieve good therapeutic effects with lower administration doses and reduced side effects. For instance, Karavasili et al. evaluated the therapeutic efficacy of self-assembled peptide hydrogels to co-deliver adriamycin and curcumin on the head and neck cancer cells [24]. The dual-loaded peptide hydrogels were able to control the drug release rate based on the water solubility of drugs. Significant cell growth inhibition has been observed after the treatment with the combined drug-loaded hydrogel formulation compared to the corresponding pure drug solution.

9.2.3 Delivery of FU

5-Fluorouracil (5-FU) is the only chemotherapeutic agent with potent activity against colorectal cancer. Continuous delivery and release of this antineoplastic agent are essential for better cancer treatment. However, the widely used chemotherapeutic agent 5-FU shows adverse effects by destroying normal cells, which can be reduced when using the designed peptide hydrogels as the carriers.

Ischakov and co-workers demonstrated that Fmoc-FF aromatic dipeptide building blocks self-assembled into hydrogels in an aqueous solution [25]. They proposed an expansion scheme to assemble the peptides into hydrogel nanoparticles (HNPs) as potential drug carriers. The Fmoc-FF-based HNPs were prepared with vitamin E-TPGS as the stabilizer of the emulsifier. Next, the HNPs were used to continuously encapsulate and release two chemotherapeutic drugs, DOX and 5-Fu (Fig. 9.2a). It was found that 5-Fu was released 50% from the HNPs within 5 h and released momentum to the platform within 12 h (Fig. 9.2b). In contrast, DOX released from the HNP matrix was relatively slow, resulting in 50% of the drugs being released after 20 h and 80% within 55 h.

Emtiazi et al. assembled diphenylalanine peptides nanotubes (FNTs) and coupled them with folic acid/magnetic nanoparticles (FA/MNPs) for the synthesis of hybrid nanocarriers (Fig. 9.2c) [26]. Anti-tumor drug, 5-FU, and the anti-inflammatory drug, flufenamic acid (FFA), were loaded into FNTs during the peptide self-assembly in the liquid phase. The in vitro release tests were conducted using the FNT hydrogels that prepared in a mixture of 75% toluene and 25% ethanol. The results showed that 5-FU released from the FNTs within 4 h. In another study, Soumi et al. used low molecular weight peptide-drug couplings for the photocontrol transfer of 5-Fu [27]. The drug was covalently linked to a short peptide by a photolytic connector. By designing peptide sequences, the peptide-drug couplers formed the drug-carrying peptide-based hydrogels at a pH of 6.0–8.0. The cellular experiments on human cervical HeLa cell lines demonstrated that the system was biocompatible. In addition, the photolytic connector made the carrier capable of cutting and releasing drugs under 365 nm irradiation.

Previously, Xu et al. designed a peptide containing an RGD sequence, Phe-Phe-Arg-Gly-Asp-Phe [28], in which the Phe residue was designed to adjust the hydrophilicity/hydrophobicity of the peptide and assist the supramolecular self-assembly into a hydrogel. The formed hydrogel was used to load the anti-proliferation model drug, 5-FU. The 5-FU-loaded hydrogel effectively inhibited the scleral flap fibrosis in rabbit eye filtration, as shown in Fig. 9.2d and e.

9.2.4 Delivery of PTX

PTX has been considered the first approved herbal-derived chemotherapy drug. However, in the cancer therapy, the PTX resistance has become a major problem

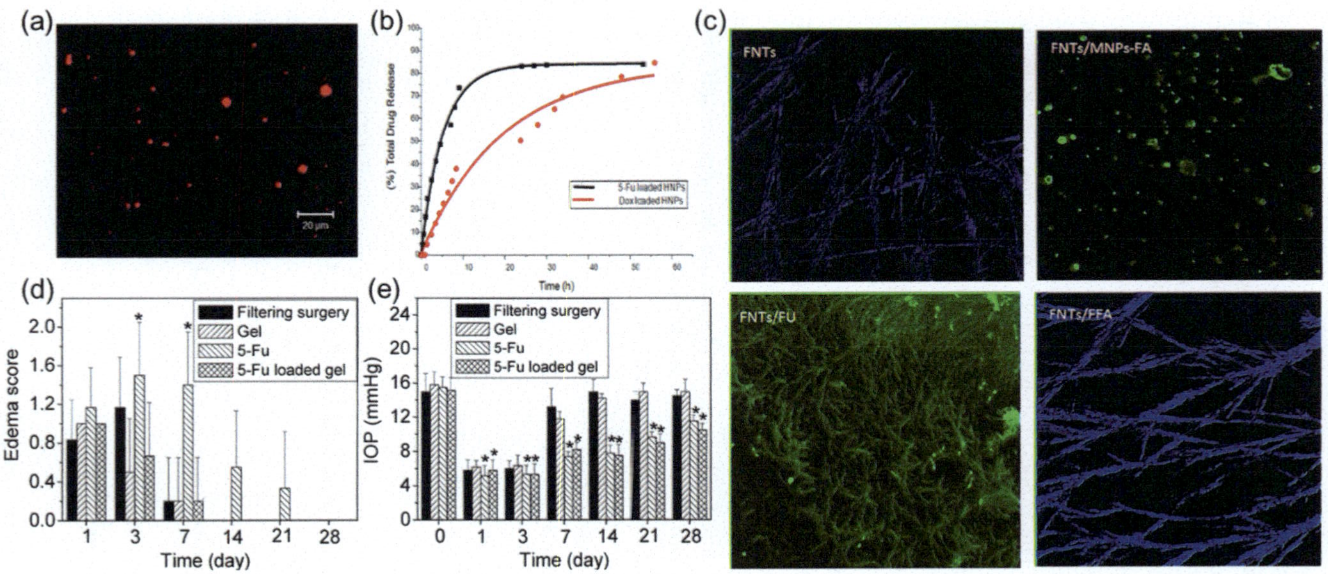

Fig. 9.2 Drug encapsulation and release of 5-FU. **a** Confocal microscopy observation of hydrogel particles encapsulated with drug. **b** Drug release profile at 37 °C. Reproduced from Ref. [25] Copyright 2013, Elsevier. **c** Confocal laser scanning microscopy of FNT, FNT/MNP-FA, FNT/5-FU, and FNT/FFA. Reproduced from Ref. [26] Copyright 2017, Elsevier. **d** Corneal edema in rabbit eyes treated with 5-FU-loaded hydrogel. **e** Mean intraocular pressure of 5-FU hydrogel-loaded treated rabbit eyes. Reproduced from Ref. [28] Copyright 2010, American Chemical Society

and it affects the survival rate of many tumor patients. To address this problem, Liu et al. used a self-assembled peptide (RADA16) to control its spontaneous and rapid formation of a hydrogel under physiological conditions, which was further used for the release of PTX [29]. It was found that RADA16 interacted with PTX and the amphiphilic peptides to stabilize hydrophobic drugs in aqueous solutions. The particle size of PTX in the RADA16 solution was significantly reduced. The higher the peptide concentration, the longer the half-life of the drug released, and the more evident the inhibitory effect on the proliferation of the MDA-MB-435S breast cancer cell [30].

Raza et al. studied the development of a PTX drug delivery system for tumor targeting using the pH-responsive FER-8 peptide hydrogel (Fig. 9.3a) [31]. It was found that at pH of 7.4, the FER-8 peptide self-assembled to form a stable hydrogel, while the degradation rate of the hydrogel was higher under acidic conditions. The in vitro cytotoxicity studies were also carried out. Compared to free medicines, the FER-8 peptide hydrogel increased the accumulation of medications in HepG2 cells and effectively inhibited the growth of HepG2 tumor cells. Therefore, the designed peptide hydrogel could enhance the tumor targeting, prolonged the stay time, reduced systemic side effects, and increased the drug accumulation at tumor sites.

By spontaneously binding the designed PTX-peptide couplers into supramolecular filamentous hydrogels, Chakroun et al. maintained the continued drug release over several months (Fig. 9.3b) [32]. Four pre-solutions of the hydrogels containing PTX with various peptide sequences were designed. By incorporating the opposite-charged amino acids and alkyl chains into the peptides to regulate the release kinetics of the supramolecular hydrogels, the efficacy of drug release could be regulated by modulating the dissociation. The binding of the RGDR-containing peptides to drugs promoted the tumor suppression while preserved cytotoxic effects.

Currently, it is not easy to achieve high anti-tumor efficacy using single-drug therapy. Therefore, Liu et al. designed an injectable pH-responsive OE peptide hydrogel as a carrier material for the anti-tumor drugs gemcitabine (GEM) and PTX (Fig. 9.3c) [33]. It released drugs simultaneously at the tumor site to achieve anti-tumor effects. The therapeutic effects of the drug-carrying peptide hydrogel were demonstrated by the in vitro and in vivo experiments. The results showed that the peptide hydrogel had different structures at different pH values and were highly sensitive to the pH.

9.3 Peptide Hydrogels for Gene Delivery

With the advancement of modern technology and a better understanding of the disease mechanisms, there has been a dramatic change in the diagnosis and treatment of many diseases. Gene therapy is a promising biological therapy that has attracted much attention in recent years. In this section, we summarize recent work on the peptide hydrogels as gene carriers. The focus is on the applications in the treatment of the cystic fibrosis, cancer, AIDs, hemophilia, cardiovascular lesions, and other diseases.

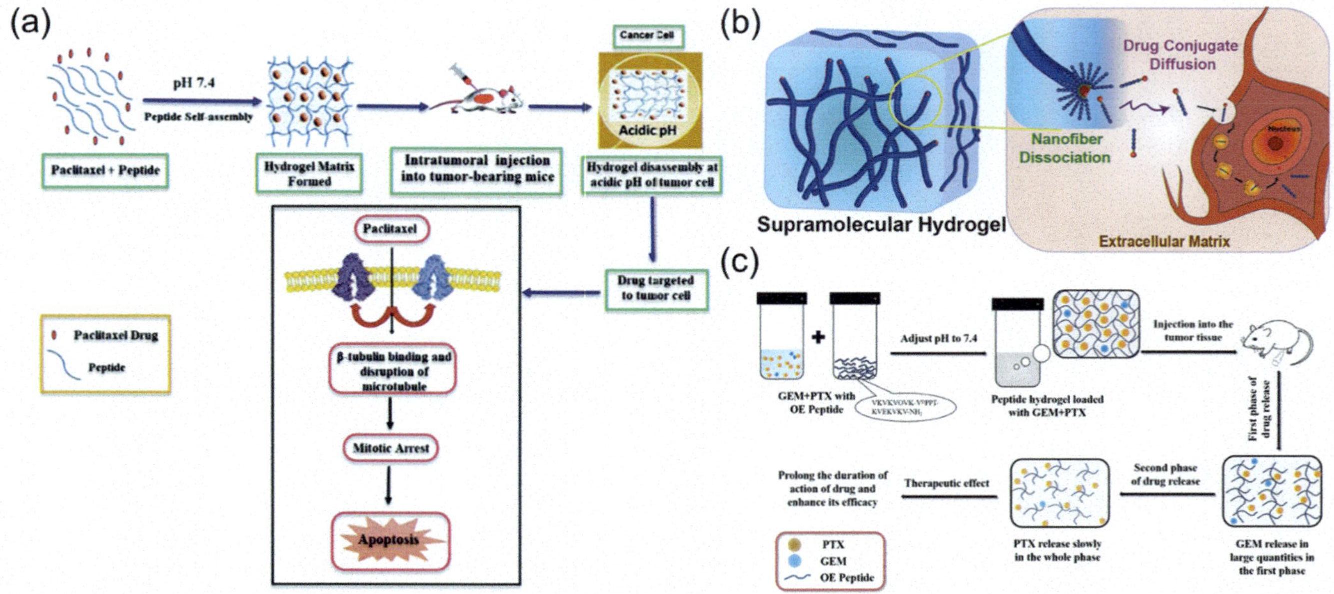

Fig. 9.3 **a** Schematic diagram of pH-responsive peptide hydrogels for delivery of PTX. Reproduced from Ref. [31] Copyright 2019, Royal Society of Chemistry. **b** Drug release process from gel of supramolecular hydrogel. Reproduced from Ref. [32] Copyright 2019, American Chemical Society. **c** Diagram of PTX/GEM loading and releasing. Reproduced from Ref. [33] Copyright 2022, MDPI

In addition to the commonly used deoxyribonucleic acid (DNA), nucleases, and several types of ribonucleic acid (RNA), other small interfering RNA (siRNA) and microRNA (miRNA) can also be used to regulate the gene expression [34, 35]. So far, non-viral vectors (e.g., gene guns, liposomes, and particle-mediated gene transfer) have been developed as safer alternatives [36–38].

9.3.1 Delivery of DNA

The delivery of DNA therapeutics via the self-assembled peptide hydrogels can substantially facilitate the tissue accumulation of genes through the high viscosity and low mobility of hydrogels.

Tian et al. prepared peptide-based nanofibrous hydrogels to deliver functional DNA vaccines to facilitate the efficacy of HIV vaccines (Fig. 9.4a) [39]. The structure of the end group of G-NMe allowed for better loading of DNA. The resulting hydrogel vector had a DNA-releasing effect, which was attributed to the strongly activated humoral and cellular immune responses against HIV by the intramuscular, intradermal, and subcutaneous injections. Cellular immune response by intramuscular, intradermal, and subcutaneous injection showed that this peptide-based hydrogel nanocarrier had great potential for delivering DNA vaccines (Fig. 9.4b).

Medina et al. investigated the in vivo immunostimulatory potential of DNA-encapsulated peptide hydrogels [40]. They demonstrated that DNA vectors maintained their rheological properties during the injection by controlling the gene encapsulation and released through the charge of DNA molecules. In the experiments, the carriers were prepared with peptides of different charges, where peptides with a net charge of + 5 retained a significant amount of DNA. Peptides with larger charges inhibited the dissociation of peptides, thus interfering with the cellular internalization.

Hydrogels with localized gene delivery functions are under the development. The duration, level, and localization of current gene expression are usually insufficient for the regeneration. Shepard and co-workers developed a method to modify hydrogels with affinity peptides to enhance the vector retention and increase the transfection within the matrix [41]. Hydrogels were modified with lysine-based repeat sequences (K4 and K8), resulting in 25% more loaded genes than unmodified hydrogels. These peptide-modified hydrogels were subsequently applied to an in vitro co-culture model to induce the NGF expression and promote the neurite growth.

Gojgini et al. cross-linked matrix metalloproteinase (MMP)-degradable peptides with hyaluronic acid (HA) to form a hydrogel material by the Michael reaction [42]. As shown in Fig. 9.4c, HA was cross-linked with MMP-degradable peptide to form the HA-AC/RGD hydrogel. After loading the DNA/polyethyleneimine (PEI) into the hybrid hydrogel, the hydrogel exhibited higher transgene expression than the stiffer hydrogel. In addition, the RGD sequence promoted precise targeting of the hydrogel to tumor cells. The in vitro model experiments enabled better design of scaffolds for local gene therapy.

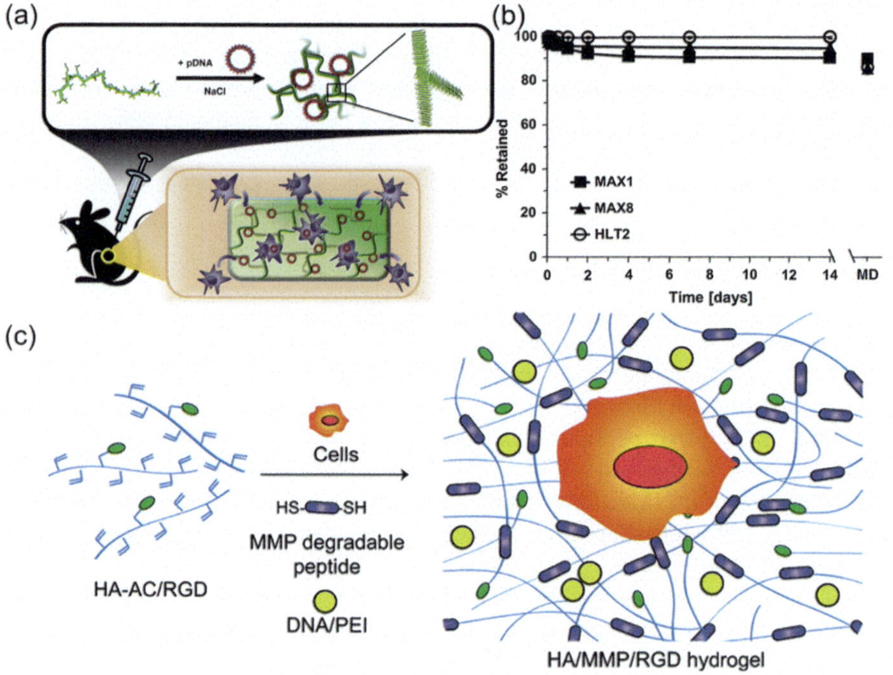

Fig. 9.4 DNA delivery with peptide hydrogels: **a** DNA encapsulated into peptide hydrogel for delivery of vector to cells. **b** Amount of DNA retained encapsulated in peptide hydrogel. Reproduced from Ref. [40] Copyright 2015, Elsevier. **c** Schematic diagram of HA/MMP/RGD hydrogel formation and loaded DNA particles. Reproduced from Ref. [42] Copyright 2011, American Chemical Society

9.3.2 Delivery of RNA

Kinpe et al. prepared peptide-based hydrogels for the delivery of siRNA by linking the enzyme-sensitive peptides to poly (methacrylic acid-co–N-vinyl-2-pyrrolidone) (P(MAA-co-NVP)) for the treatment of inflammatory bowel disease [43]. The P(MAA-co-NVP) backbone was designed to collapse and protect the encapsulated nanogels from the degradation at low pH in the stomach. Hydrogels were prepared using the enzyme-sensitive peptides. As shown in Fig. 9.5a, in the presence of enzymes (e.g., matrix metalloproteinase 2 (MMP-2), protease, trypsin, and lysozyme), the peptide junctions in the hydrogel network were disrupted, thereby releasing the encapsulated RNA. In a similar study, Cai and co-workers prepared degradable MMP-2sp hydrogels by cross-linking carboxymethyl chitosan with the MMP-2 substrate peptide CPLGLAGC, as indicated in Fig. 9.5b [44]. The MMP-2 degradable hydrogels loaded with siRNA multistranded bodies, which were further adsorbed onto electrospun fibers to inhibit the fibroblast proliferation. Wang et al. used siRNA to treat myocardial infarction (MI) [45]. The vector was prepared by

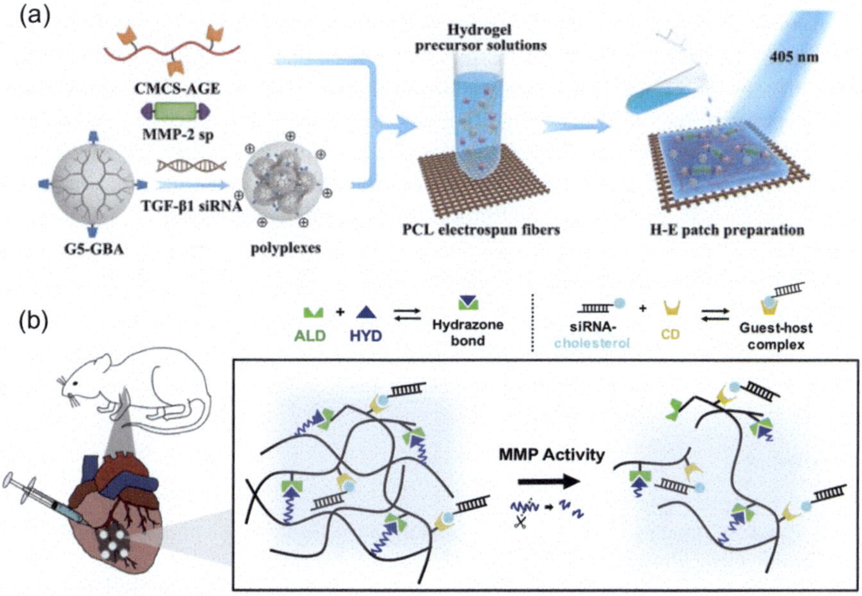

Fig. 9.5 **a** Preparation of hydrogel patches in which encapsulated TGF-β1 siRNA is encapsulated. Reproduced from Ref. [44] Copyright 2021, Wiley–VCH. **b** Schematic representation of preparation of peptide hydrogels and loading siRNA acting on MI. Reproduced from Ref. [45] Copyright 2018, Elsevier

HA that modified with hydrazide or aldehyde and mixed with peptides in response to the protease activity to form a self-healing hydrogel.

Kanazawa and co-workers developed a topical skin drug for atopic dermatitis (AD) and prepared a hydrogel containing anti-RelA siRNA and functional peptides [46]. Moreover, silk protein was selected as the hydrogel matrix for the hydrogel preparation. After topical application through the hydrogel containing functional peptide, siRNA was more easily delivered to the site of the AD-induced ear skin of mouse than through the preparation without functional peptide.

9.4 Conclusions

Peptide hydrogels have been widely accepted for delivering therapeutic agents due to their unique designability and good biocompatibility. In summary, this chapter presented advances in peptide hydrogels for the drug and gene delivery. Functional peptide hydrogels loaded with drugs exhibited better efficacy than the drug components alone. In the preparation of hydrogel carriers, different functional carriers were obtained from other raw materials of peptides. For the drug delivery, the current status of applying four commonly used therapeutic agents, DOX, curcumin, FU, and PTX, was presented and discussed. For the gene delivery, peptide hydrogels revealed very promising applications as the gene carriers by giving non-viral vectors and avoiding adverse effects. There are many modification methods to treat the damaged tissues in vivo, especially for the cancer gene therapy. Nowadays, the gene delivery for the disease treatment is a well-established tool with the commonly used gene fragments, DNA, RNA, etc. In addition, the loading of multiple drug components into functional peptide hydrogels is a promising technique for future cancer therapy.

Acknowledgements ZQS acknowledges the financial support from the National Natural Science Foundation of China (NSFC, Grant no. 51873016) and the Joint Project of BRC-BC (Biomedical Translational Engineering Research Center of BUCT-CJFH) (XK2020-11).

References

1. S.M. Pawde, K. Deshmukh, S. Parab, Preparation and characterization of poly(vinyl alcohol) and gelatin blend films. J. Appl. Poly. Sci. **109**, 1328–1337 (2008)
2. B.V. Slaughter, S.S. Khurshid, O.Z. Fisher, A. Khademhosseini, N.A. Peppas, Hydrogels in regenerative medicine. Adv. Mater. **21**, 3307–3329 (2009)
3. J. Wu, X. Zhao, D. Wu, C.C. Chu, Development of a biocompatible and biodegradable hybrid hydrogel platform for sustained release of ionic drugs. J. Mater. Chem. B **2**, 6660–6668 (2014)
4. J. Zhu, F. Li, X. Wang, J. Yu, D. Wu, Hyaluronic acid and polyethylene glycol hybrid hydrogel encapsulating nanogel with hemostasis and sustainable antibacterial property for wound healing. *ACS Appl. Mater. Interfaces* **10**, 13304–13316 (2018)
5. E. Caló, V.V. Khutoryanskiy, Biomedical applications of hydrogels: a review of patents and commercial products. Eur. Poly. J. **65**, 252–267 (2015)
6. O. Wichterle, D. Lím, Hydrophilic gels for biological use. *Nature*, **185**, 117–118 (1960)
7. T. Göckler, S. Haase, X. Kempter, R. Pfister, B.R. Maciel, A. Grimm, T. Molitor, N. Willenbacher, U. Schepers, Tuning superfast curing thiol-norbornene-functionalized gelatin hydrogels for 3D bioprinting. Adv. Health. Mater. **10**, 2100206 (2021)
8. C. Echalier, S. Jebors, G. Laconde, L. Brunel, P. Verdié, L. Causse, A. Bethry, B. Legrand, H. Van Den Berghe, X. Garric, D. Noël, J. Martinez, A. Mehdi, G. Subra, Sol–gel synthesis of collagen-inspired peptide hydrogel. Mater. Today **20**, 59–66 (2017)
9. S. Peers, A. Montembault, C. Ladaviere, Chitosan hydrogels for sustained drug delivery. J. Control. Release **326**, 150–163 (2020)
10. Z. Zou, L. Wang, Z. Zhou, Q. Sun, D. Liu, Y. Chen, H. Hu, Y. Cai, S. Lin, Z. Yu, B. Tan, W. Guo, Z. Ling, X. Zou, Simultaneous incorporation of PTH(1–34) and nano-hydroxyapatite into Chitosan/Alginate Hydrogels for efficient bone regeneration. Bioact. Mater. **6**, 1839–1851 (2021)
11. O. Qianqian, K. Songzhi, H. Yongmei, J. Xianghong, L. Sidong, L. Puwang, L. Hui, Preparation of nano-hydroxyapatite/chitosan/tilapia skin peptides hydrogels and its burn wound treatment. Int. J. Biol. Macromol. **181**, 369–377 (2021)
12. A. Bayón-Fernández, A. Méndez-Ardoy, C. Alvarez-Lorenzo, J.R. Granja, J. Montenegro, Self-healing cyclic peptide hydrogels. J. Mater. Chem. B **11**, 606–617 (2023)
13. S. Kuroyanagi, N. Shimada, S. Fujii, T. Furuta, A. Harada, K. Sakurai, A. Maruyama, Highly ordered polypeptide with UCST phase separation behavior. J. Am. Chem. Soc. **141**, 1261–1268 (2019)
14. T. Masuda, N. Shimada, A. Maruyama, A thermoresponsive cationic comb-type copolymer enhances membrane disruption activity of an amphiphilic peptide. Biomacromolecules **19**, 1333–1339 (2018)

15. J. Shi, Z. Xiao, N. Kamaly, O.C. Farokhzad, Self-assembled targeted nanoparticles: evolution of technologies and bench to bedside translation. Acc. Chem. Res. **44**, 1123–1134 (2011)
16. J. Boekhoven, R.H. Zha, F. Tantakitti, E. Zhuang, R. Zandi, C.J. Newcomb, S.I. Stupp, Alginate-peptide amphiphile core-shell microparticles as a targeted drug delivery system. RSC Adv. **5**, 8753–8756 (2015)
17. L. Mei, K. Xu, Z. Zhai, S. He, T. Zhu, W. Zhong, Doxorubicin-reinforced supramolecular hydrogels of RGD-derived peptide conjugates for pH-responsive drug delivery. Org. Biomol. Chem. **17**, 3853–3860 (2019)
18. H. Jin, C. Wan, Z. Zou, G. Zhao, L. Zhang, Y. Geng, T. Chen, A. Huang, F. Jiang, J.P. Feng, J.F. Lovell, J. Chen, G. Wu, K. Yang, Tumor ablation and therapeutic immunity induction by an injectable peptide hydrogel. ACS Nano **12**, 3295–3310 (2018)
19. K. Baek, A.D. Noblett, P. Ren, L.J. Suggs, Self-assembled nucleo-tripeptide hydrogels provide local and sustained doxorubicin release. Biomater. Sci. **8**, 3130–3137 (2020)
20. M. Salem, S. Rohani, E.R. Gillies, Curcumin, a promising anti-cancer therapeutic: a review of its chemical properties, bioactivity and approaches to cancer cell delivery. RSC Adv. **4**, 10815–10829 (2014)
21. A. Altunbas, S.J. Lee, S.A. Rajasekaran, J.P. Schneider, D.J. Pochan, Encapsulation of curcumin in self-assembling peptide hydrogels as injectable drug delivery vehicles. Biomaterials **32**, 5906–5914 (2011)
22. J. Liu, J. Liu, H. Xu, Y. Zhang, L. Chu, Q. Liu, N. Song, C. Yang, Novel tumor-targeting, self-assembling peptide nanofiber as a carrier for effective curcumin delivery. Int. J. Nanomedicine **9**, 197–207 (2014)
23. S. Alam, J.J. Panda, V.S. Chauhan, Novel dipeptide nanoparticles for effective curcumin delivery. Int. J. Nanomedicine **7**, 4207–4222 (2012)
24. C. Karavasili, D.A. Andreadis, O.L. Katsamenis, E. Panteris, P. Anastasiadou, Z. Kakazanis, V. Zoumpourlis, C.K. Markopoulou, S. Koutsopoulos, I.S. Vizirianakis, D.G. Fatouros, Synergistic antitumor potency of a self-assembling peptide hydrogel for the local co-delivery of doxorubicin and curcumin in the treatment of head and neck cancer. Mol. Pharmaceutics **16**, 2326–2341 (2019)
25. R. Ischakov, L. Adler-Abramovich, L. Buzhansky, T. Shekhter, E. Gazit, Peptide-based hydrogel nanoparticles as effective drug delivery agents. Bioorg. Med. Chem. **21**, 3517–3522 (2013)
26. G. Emtiazi, T. Zohrabi, L.Y. Lee, N. Habibi, A. Zarrabi, Covalent diphenylalanine peptide nanotube conjugated to folic acid/magnetic nanoparticles for anti-cancer drug delivery. *J. Drug Deliv. Sci. Technol.* **41**, 90–98 (2017)
27. S. Das, H. Horo, U. Goswami, L.M. Kundu, Synthesis of a peptide conjugated 5-fluorouracil gelator prodrug for photo-controlled release of the antitumor agent. ChemistrySelect **4**, 6778–6783 (2019)
28. X.-D. Xu, L. Liang, C.-S. Chen, B. Lu, N.-L. Wang, F.-G. Jiang, X.-Z. Zhang, R.-X. Zhuo, Peptide hydrogel as an intraocular drug delivery system for inhibition of postoperative scarring formation. ACS Appl. Mater. Interfaces **2**, 2663–2671 (2010)
29. J. Liu, L. Zhang, Z. Yang, X. Zhao, Controlled release of paclitaxel from a self-assembling peptide hydrogel formed in situ and antitumor study in vitro. Int. J. Nanomedicine **6**, 2143–2153 (2011)
30. J. Liu, Controlled release of paclitaxel from a self-assembling peptide hydrogel formed in situ and antitumor study in vitro. Int. J. Nanomedicine **6**, 2143–2153 (2022)
31. F. Raza, Y. Zhu, L. Chen, X. You, J. Zhang, A. Khan, M.W. Khan, M. Hasnat, H. Zafar, J. Wu, L. Ge, Paclitaxel-loaded pH responsive hydrogel based on self-assembled peptides for tumor targeting. Biomater. Sci. **7**, 2023–2036 (2019)
32. R.W. Chakroun, F. Wang, R. Lin, Y. Wang, H. Su, D. Pompa, H. Cui, Fine-tuning the linear release rate of paclitaxel-bearing supramolecular filament hydrogels through molecular engineering. ACS Nano **13**, 7780–7790 (2019)
33. Y. Liu, Y. Ran, Y. Ge, F. Raza, S. Li, H. Zafar, Y. Wu, A.C. Paiva-Santos, C. Yu, M. Sun, Y. Zhu, F. Li, PH-sensitive peptide hydrogels as a combination drug delivery system for cancer treatment. Pharmaceutics **14**, 652 (2022)

34. G.M. Rubanyi, The future of human gene therapy. Mol. Aspe. Med. **22**, 113–142 (2001)
35. N. Akhtar, M. Akram, H.M. Asif, K. Usmanghani, S.M.A. Shah, S.A. Rao, M. Uzair, G. Shaheen, K. Ahmad, Gene therapy: a review article. J. Med. Plant. Res. **5**, 1812–1817 (2011)
36. L. Aps, M.B. Tavares, J.H.K. Rozenfeld, M.T. Lamy, L.C.S. Ferreira, M.O. Diniz, Bacterial spores as particulate carriers for gene gun delivery of plasmid DNA. J. Biotechnol. **228**, 58–66 (2016)
37. P. Peking, U. Koller, S. Hainzl, S. Kitzmueller, T. Kocher, E. Mayr, A. Nystrom, T. Lener, J. Reichelt, J.W. Bauer, E.M. Murauer, A gene gun-mediated nonviral RNA trans-splicing strategy for Col7a1 repair. Mol. Ther. Nucleic Acids **5**, e287 (2016)
38. H. Yin, R.L. Kanasty, A.A. Eltoukhy, A.J. Vegas, J.R. Dorkin, D.G. Anderson, Non-viral vectors for gene-based therapy. Nat. Rev. Genet. **15**, 541–555 (2014)
39. Y. Tian, H. Wang, Y. Liu, L. Mao, W. Chen, Z. Zhu, W. Liu, W. Zheng, Y. Zhao, D. Kong, Z. Yang, W. Zhang, Y. Shao, X. Jiang, A peptide-based nanofibrous hydrogel as a promising DNA nanovector for optimizing the efficacy of HIV vaccine. Nano Lett. **14**, 1439–1445 (2014)
40. S.H. Medina, S. Li, O.M. Howard, M. Dunlap, A. Trivett, J.P. Schneider, J.J. Oppenheim, Enhanced immunostimulatory effects of DNA-encapsulated peptide hydrogels. Biomaterials **53**, 545–553 (2015)
41. J.A. Shepard, P.J. Wesson, C.E. Wang, A.C. Stevans, S.J. Holland, A. Shikanov, B.A. Grzybowski, L.D. Shea, Gene therapy vectors with enhanced transfection based on hydrogels modified with affinity peptides. Biomaterials **32**, 5092–5099 (2011)
42. S. Gojgini, T. Tokatlian, T. Segura, Utilizing cell-matrix interactions to modulate gene transfer to stem cells inside hyaluronic acid hydrogels. Mol. Pharmaceutics **8**, 1582–1591 (2011)
43. J.M. Knipe, L.E. Strong, N.A. Peppas, Enzyme- and pH-responsive microencapsulated nanogels for oral delivery of siRNA to induce TNF-α knockdown in the intestine. Biomacromolecules **17**, 788–797 (2016)
44. C. Cai, W. Wang, J. Liang, Y. Li, M. Lu, W. Cui, C. Fan, L. Deng, Y. Li, F. Wang, S. Liu, MMP-2 responsive unidirectional hydrogel-electrospun patch loading TGF-β1 siRNA polyplexes for peritendinous anti-adhesion. Adv. Funct. Mater. **31**, 2008364 (2021)
45. L.L. Wang, J.J. Chung, E.C. Li, S. Uman, P. Atluri, J.A. Burdick, Injectable and protease-degradable hydrogel for siRNA sequestration and triggered delivery to the heart. J. Contr. Relea. **285**, 152–161 (2018)
46. T. Kanazawa, Y. Shizawa, M. Takeuchi, K. Tamano, H. Ibaraki, Y. Seta, Y. Takashima, H. Okada, Topical anti-nuclear factor-kappa B small interfering RNA with functional peptides containing sericin-based hydrogel for atopic dermatitis. Pharmaceutics **7**, 294–304 (2015)

Chapter 10
Peptide Nanomaterials for Bioimaging Applications

Ziyu Zhou and Xin Zhou

Abstract Biomedical imaging is a powerful tool for the diagnosis and treatment of diseases. Various small molecules, polymers, and nanomaterials have been used as contrast-enhancing imaging agents to achieve multifunctional imaging applications. Among them, peptides and peptide-based materials have been attracted growth interesting due to their excellent biocompatibility and diverse biological functions. This article summarized recent advances in developing peptide-based nanomaterial for bioimaging applications, including fluorescence imaging (FLI), photoacoustic imaging (PAI), magnetic resonance imaging (MRI), positron emission tomography (PET), and X-ray computed tomography (CT). In addition to biomedical imaging, peptides and peptide-based materials with therapeutic functions used for the treatment of disease were also involved. Moreover, the perspective and challenge on the future of peptide nanomaterials for bioimaging were also discussed. We hope such a work would stimulate this fast-developing research area.

Keywords Peptide nanomaterials · Bioimaging · Fluorescent imaging · Disease diagnosis · Biomedicine

10.1 Introduction

Nanomaterials can be designed to possess unique physical, chemical, biological, and optical properties through careful manipulation of their size, structure, and function. This enables a diverse range of applications, including photothermal therapy, biosensing, drug delivery, and photodynamic therapy [1–3]. The rapid advancement of nanomaterials with exceptional optical properties, including nanoclusters [4, 5], luminescent nanoparticles [6, 7], and quantum dots [8, 9], has revolutionized the fields of biology and biomedicine [10]. Consequently, the unique physical and chemical properties of nanomaterials enable them to interact with biological systems at the molecular level, providing unprecedented opportunities for the development of

Z. Zhou · X. Zhou (✉)
College of Chemistry and Chemical Engineering, Qingdao University, Qingdao 266071, China
e-mail: zhouxin@qdu.edu.cn

G. Wei (ed.), *Peptide Nano-Chemistry and Nanotechnology*,
https://doi.org/10.1007/978-981-96-2387-7_10

innovative technologies that can significantly enhance our ability to understand and treat complex diseases. With the continued advancement of nanotechnology, we can expect further breakthroughs in the biomedical field [11, 12].

Peptides exhibit distinct biological functions, making them valuable constituents of biomaterials due to their bioactivity, biocompatibility, and biodegradability [13]. Peptides can also serve as stabilizers for hybrid nanomaterials, thereby enhancing their biocompatibility and bioactivity [14]. With the advancement of peptide synthesis technology, motif-designed peptides with precise amino acid sequences can now be readily obtained [15]. By coupling these peptides with nanomaterials, peptide-functionalized nanomaterials can be produced, which exhibit enhanced affinity, biocompatibility, and targeting, all of which are critical for biomedicine applications [16].

Peptide nanomaterials have garnered considerable interest in the cutting-edge bioimaging field owing to their exceptional biocompatibility and diverse physicochemical properties [17–19]. Peptides can be synthesized and self-assembled into nanoparticles with tailored shapes and physicochemical properties, which can produce imaging signals, such as fluorescence or be loaded with contrast agents [20, 21]. The utilization of these nanoparticles provides enhanced imaging efficiency and decreased cytotoxicity when compared to conventional contrast agents [22]. Peptides can also be employed to functionalize pre-existing contrast agents or imaging platforms, allowing them to target specific lesions with responsiveness [23].

Previous reviews on the synthesis, functionalization, and biological applications of bio-nanomaterials have been published. For example, Peng et al. provided an overview of soft fluorescent nanoparticles for biological and biomedical imaging. This review covered the primary synthetic methods, structures, optical properties, and surface modifications of these nanoparticles [24]. Qi et al. described a range of peptide-engineered nanomaterials for bioimaging and disease treatment, covering molecular design, self-assembly, functionalization, and biomedical applications [25]. Hu et al. emphasized protein and peptide-mediated fluorescent advancements in the synthesis of metal nanoclusters, with an emphasis on synthesis protocols and formation mechanisms [26]. Although these significant progress have been made, until now only few review on the peptide-based nanomaterials for bioimaging applications has been released. Therefore, this comprehensive review summarized recent advancements in various bioimaging modalities and explores the potential of peptide nanomaterials. It is hoped that this review will provide guidance for the optimal design of functional peptide nanomaterials and promote practical bioimaging applications.

10.2 The Applications of Peptide Nanomaterials for Bioimaging

Over the years, a range of sophisticated and precise bioimaging techniques have been developed with high spatial and temporal resolution [27]. These techniques have enabled researchers to capture sensitive and detailed information about biological processes and structures, contributing significantly to our understanding of complex biological phenomena [28].

Peptide nanomaterials are extensively utilized in bioimaging due to their exceptional physical and chemical properties, biological activities, and significantly enhanced biocompatibility [29]. Based on their diverse bioimaging applications, we have categorized peptide nanomaterials into five sections, including fluorescence imaging (FLI), magnetic resonance imaging (MRI), positron emission tomography (PET), photoacoustic imaging (PAI), and X-ray computed tomography (CT). A brief summary of their individual bioimaging applications is provided [19, 30, 31].

10.2.1 FLI

FLI is a noninvasive technique that enables the visualization of physiological activity in living organisms [31]. FLI utilizes fluorescent reporter groups, such as the inorganic or organic materials, to label and visualize biological structures [32]. In recent years, fluorescence techniques have been extensively employed in molecular biology and the in vivo metabolic studies [33].

Fluorescence resonance energy transfer (FRET) is a commonly used design method for activatable FLI systems [34]. To activate FRET signals, reactive peptides are incorporated into the design of FLI platforms. In a recent study, Gao et al. reported a protease-activated fluorescence platform based on FRET between a fluorophore and a Fe_3O_4 nanocrystal for the tumor pH mapping [35]. A reactive peptide substrate, Gly-Pro-Leu-Gly-Leu-Pro-Gly-Lys-Gly-Gly, was used as a linker between the Fe_3O_4 quencher and the pH-sensitive fluorophore. The responsive peptide substrate could respond to protease and activate the fluorescence signals. Furthermore, a gastric cancer-specific monoclonal antibody was covalently attached to the peptide to target the cancer cells. Based on the pH-response of the fluorophore to the tumor microenvironment (TME), the resulting probe successfully imaged subcutaneous tumors in vivo at different time intervals within 60 min of injection. The pH-mapping images of tumors with pH values ranging from 6.6 to 7.6 were obtained. The tumor size on the pH-mapping images at 60 min was comparable to that of the extracted tumor, indicating the accurate response of this responsive peptide substrate.

Fluorescent nanomaterials that have been modified with specific peptides have proven to be effective for the in vivo nontoxic imaging and nuclear targeting. One example of this is the work conducted by Chen et al., who synthesized a CdSe/ZnS nanocrystalline with the peptide coupling that contained the SV40 large T antigen nuclear localization signal (NLS) [36]. This coupling actively transported the complex to the nucleus. The selected peptide coupling had low cytotoxicity, high sensitivity, and biostability, as well as the resistance to intracellular aggregation and lysosomal degradation. The presence of NLS facilitated the passage of QDs through the nuclear pore complex activity into it. The prepared NLS-QDs complexes could be retained in living cells for up to 7 days with negligible cellular effects. Additionally, the migration of the CdSe/ZnS-peptide conjugates showed significant changes in the complex from the cytoplasm to the nucleus, and the accumulation of the conjugates in the nucleus could be observed over a long observation period.

Naturally excited peptide assembly remains an attractive method for preparing fluorescent NPs for bioimaging. In a recent study, Yan and co-workers synthesized dipeptide-engineered NPs (CDP NPs) with tunable fluorescence emission for bioimaging through the covalent cross-linking assembly [37]. The modulation of the fluorescence emission of NPs has been easily achieved by adjusting the process time of covalent binding. The cytotoxicity and cell imaging studies have demonstrated that the resulting CDP NPs were highly stable even after the intracellular uptake, enabling their use for sustained in vivo multicolor FLI.

The thermally activated delayed fluorescence NPs (TADF NPs), with long-lived luminescence and low toxicity, have been utilized for the time-resolved FLI. However, due to their poor biocompatibility, low cell membrane-permeable TADF NPs typically require several hours to enter cells for fluorescence imaging. Recently, Zhu et al. developed an amphiphilic cell-penetrating peptide (CPP), $F_6G_6(rR_3)R_2$, for transporting hydrophobic TADF fluorophores across the cellular barrier for time-resolved cellular imaging [38]. The amphiphilic CPP self-assembled with each of the three TADF fluorophores to synthesize TADF nanocomposites (Fig. 10.1a). The resulting NPs were almost non-cytotoxic and exhibited strong cell membrane permeability in both HeLa and 3T3 cells, indicating their biomarking ability. Moreover, HeLa cells emitted strong fluorescent signals after only 5 min of the incubation with different NPs. As the incubation time was extended to 15 min, the fluorescence emission of HeLa cells significantly increased, suggesting the accumulation of the TADF NPs in the cells (Fig. 10.1b). In contrast, the cells without CPP did not exhibit any fluorescent signal. Additionally, the fluorescence lifetime of the TADF NPs was estimated to be a few μs, allowing for their delayed fluorescence to be clearly observed in cells from 0.05 to 2 μs (Fig. 10.1c). Therefore, the long-lived TADF enabled the synthetic NPs to be utilized for the time-resolved FLI of living cells under the oxygen-containing conditions.

NPs that can receive and emit near-infrared (NIR, wavelength 700–1100 nm) light have unique properties that make them useful for in vivo imaging, including depth imaging and excellent spatial resolution. Gao et al. have demonstrated a bifunctional peptide fabrication strategy that enables precise organelle-targeted bioimaging by combining material-bound TATs with NPs [39]. Compared with the naked nucleus targeting peptide, the hybrid peptide enhanced the good surface coverage of AuNPs. The specific recognition of AuNPs by the Au-binding peptide mediated the efficient targeted assembly of the targeting peptide on AuNPs, enhancing the targeting ability and achieving accurate targeting and high efficiency of AuNPs on the cancer cell

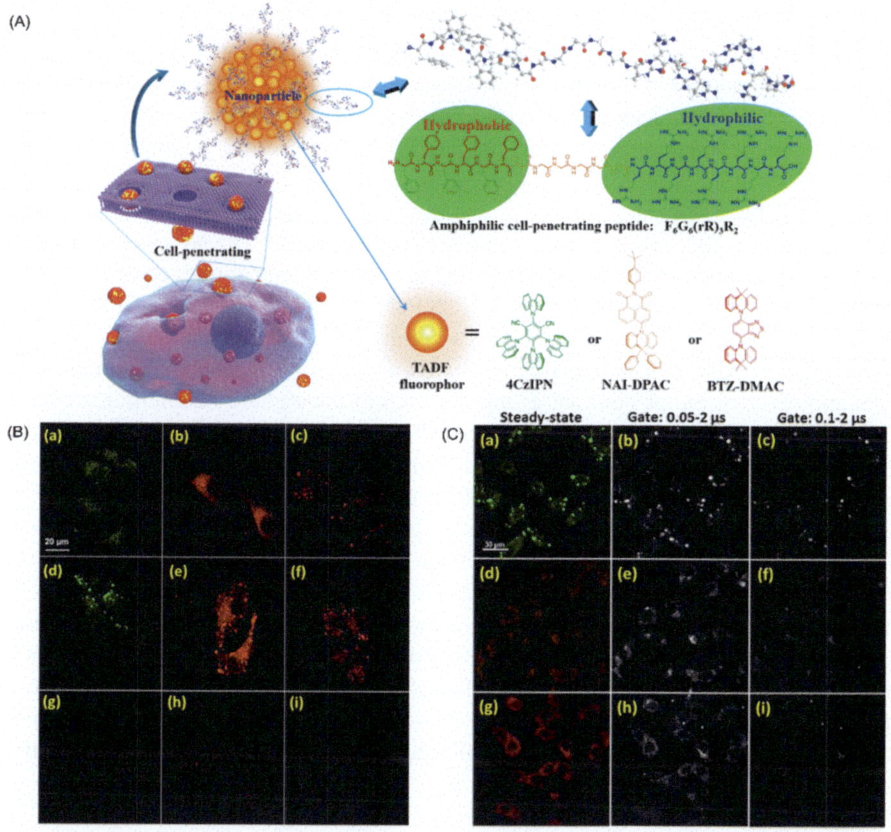

Fig. 10.1 **a** Synthesis of the cell-penetrating peptide NPs. **b** The comparison was made between HeLa cells incubated with peptide-functionalized NPs and those incubated with naked NPs. **c** Fluorescence images were captured of HeLa cells after the incubation with CPP-functionalized NPs. Reproduced with the permission from Ref. [38]. Copyright 2018, American Chemical Society

nuclei. Additionally, the AuNPs and bifunctional peptide complex exhibited good biocompatibility, enabling accurate and efficient bioimaging of the nucleus of a hepatocellular carcinoma cell line (HepG2). This preparation strategy of bifunctional peptide-coupled NPs provides a convenient nanoplatform for effective targeting the cell organelles.

To avoid the toxic accumulation of mitochondrial targeting molecules in the mitochondrial matrix, He et al. developed a new method for targeting mitochondria [40]. They designed a branched peptide (D-1TFLAG) for specific targeting and enhanced bioimaging of mitochondria (Fig. 10.2a), in which the hydrophilic branched peptide containing the FLAG motif (DDDK) can specifically deliver small molecule drugs into the mitochondria. Subsequently, intestinal kinase (ENTK) on cancer cell mitochondria recognized the hydrophilic FLAG motif and promoted the micelle self-assembly into nanofibers. In this way, the dye effectively accumulated on the mitochondria and produced an enhanced fluorescent signal in the tumor cells. Furthermore, this branched peptide revealed potential for enhanced bioimaging because the negatively charged peptide nanofibers had little toxicity to cells.

Li et al. developed a strategy to synthesize high upconversion luminescence (UCL)-emitting NPs in the infrared region (Fig. 10.2b) [41]. The NPs were encapsulated by a silica shell and combined with the tumor-targeting peptides containing PSP motifs (l-SP5-H and l-SP5-C) to form UCNP@SiO_2-l-SP5-H and UCNP@SiO_2-l-SP5-C, respectively. These designed NPs combined with tumor-targeting peptides were found to have higher targeting ability for human colon cancer cells than other NPs and could serve as contrast agents for UCL/T1-weighted magnetic resonance (MR) dual-mode bioimaging. The prepared UCNP@SiO_2-l-SP5-C was able to detect HCT116 tumors with a volume size of 13 mm^3, providing a promising method for early imaging and diagnosis of colon cancer using the designed functional NPs.

Peptides with fluorescent properties have great potential for bioimaging. The π-π stacking and the Zn^{2+} coordination have been shown to affect the fluorescence properties of peptides. For instance, Fu et al. designed a self-assembled fluorescent tripeptide NP that utilizes the π-π stacking effect and the Zn^{2+} coordination for bioimaging applications (Fig. 10.3a–c) [42]. The π-electrons of tryptophan (W) provided the off-domain electrons for fluorescence, and the self-assembled NPs (TPNPs) made from tryptophan-tryptophan-tryptophan (WWW) tripeptide have little toxicity to normal cells. Therefore, the formed TPNPs exhibited green visible fluorescence under the UV light irradiation and could resonate fluorescently with DOX, enabling them to deliver DOX and detect its release.

To detect the tumors that are rich in epidermal growth factor receptor (EGFR), Yu et al. developed an EGFR-binding peptide, GE11, for targeted PDT and bioimaging delivery of zinc(II) phalloidin (Fig. 10.4a) [43]. The binding peptide could target EGFR on tumors and facilitate the internalization of EGFR-expressing cancer cells. In comparison with the tumor-bearing mice that were injected with non-targeting compounds, the tumor-bearing mice that were injected with phthalocyanine peptide coupling (Pc-GE11) showed higher fluorescence intensity in the tumor region at 24–72 h (Fig. 10.4b). This GE11 peptide-conjugated photosensitizer could be used as a fluorescent probe for the detection and diagnosis of EGFR-rich tumors. Notably, Li et al. synthesized AuNCs with bright green fluorescence by combining ghrelin-6 (GHRP-6) with the fluorescent sensitizer NaBr via the aqueous synthesis [44]. The high selectivity for Fe^{3+} and high fluorescence intensity of the GHRP-6-gold nanocrystals could be used for cellular imaging, and this simple green aqueous phase strategy provided useful ideas for the synthesis of novel functional peptides for bioimaging by combining with metal fluorescent nanocrystals.

In a different approach, Ni et al. investigated the potential application of amyloid peptides as novel imaging probes for the second harmonic generation (SHG) [45].

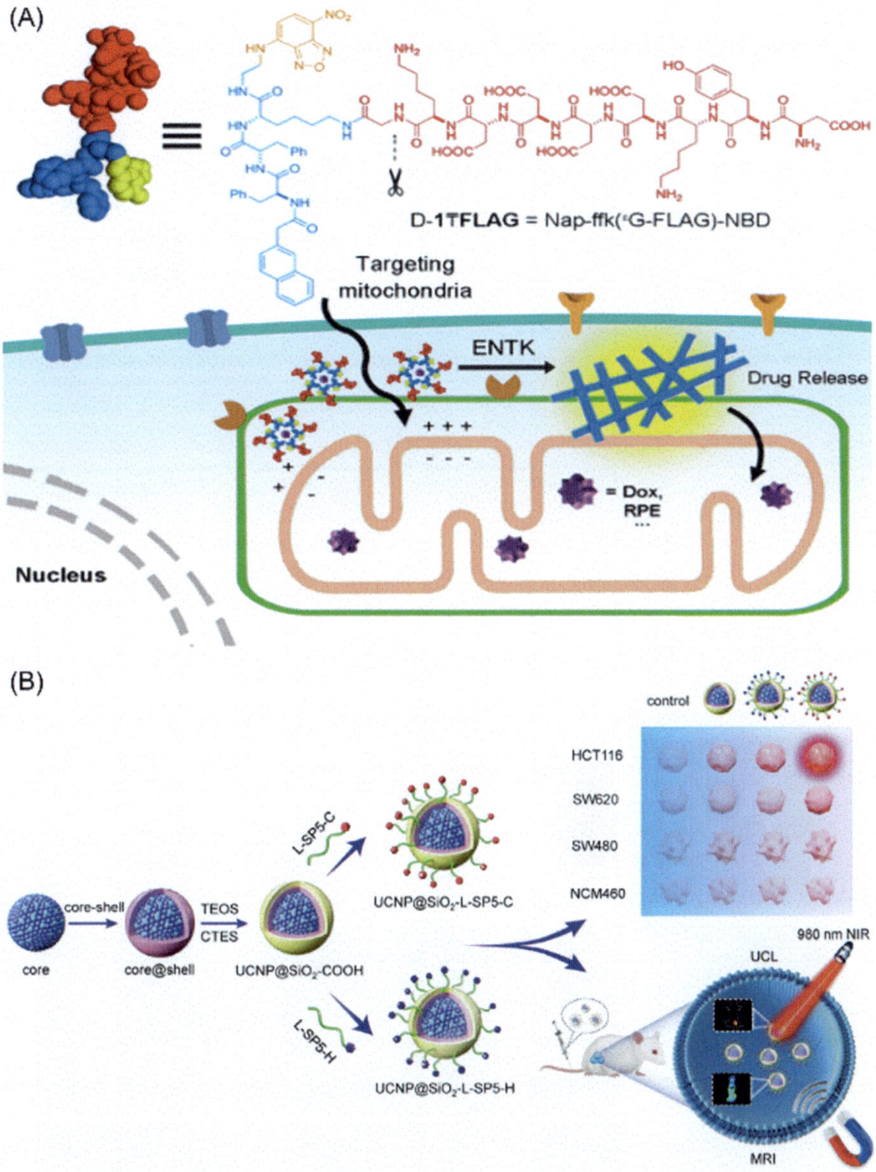

Fig. 10.2 **a** Diagram was created to illustrate formation process of branched peptides for enhanced mitochondrial imaging. Reproduced with permission from Ref. [40]. Copyright 2018, American Chemical Society. **b** Schematic illustration was created to demonstrate preparation process of peptide-modified UCNPs for UCL/MR imaging. Reproduced with permission from Ref. [41]. Copyright 2020, American Chemical Society

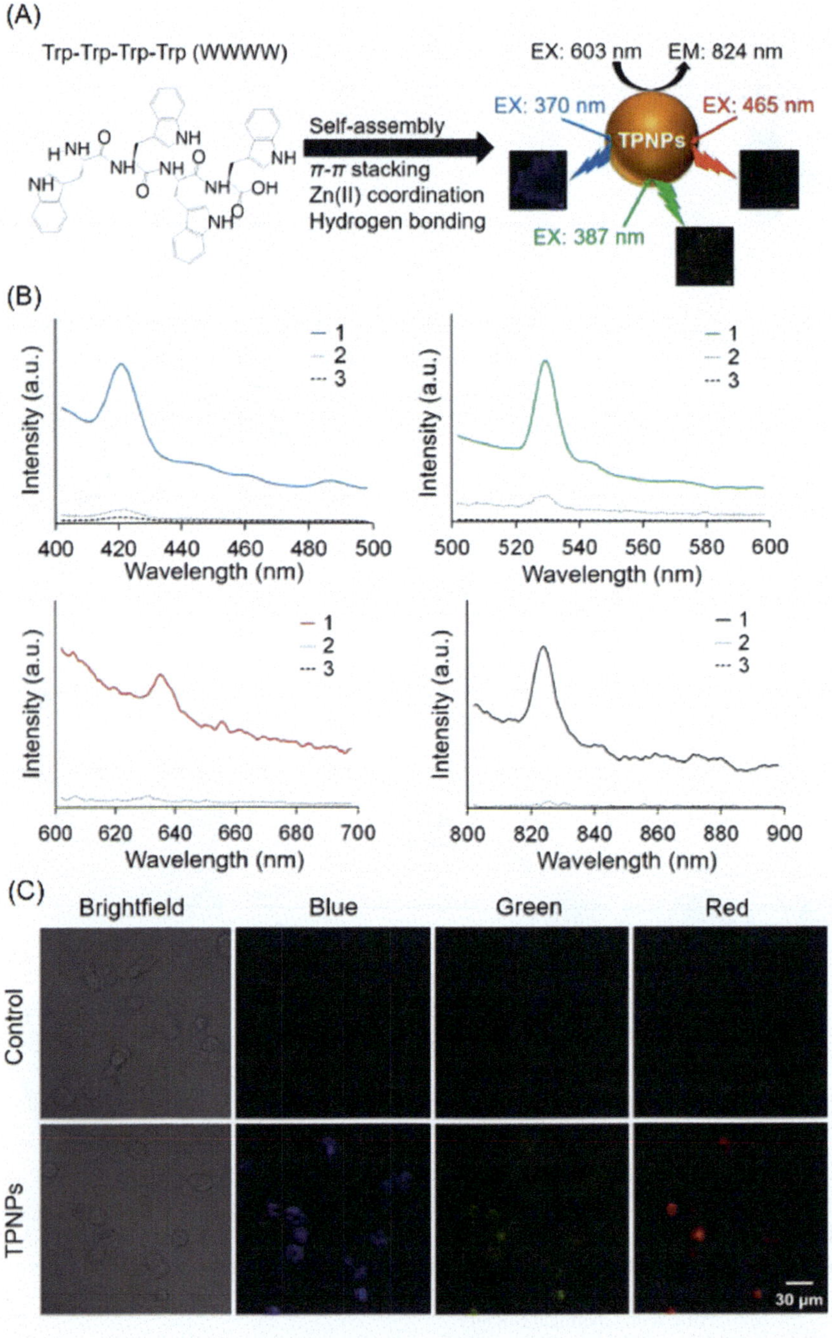

Fig. 10.3 **a** WWW peptide self-assembled with Zn(II) to construct TPNPs through a combination of π-π stacking, zinc coordination, and hydrogen bonding. **b** Optical characterizations of TPNPs **c** Confocal fluorescence images of MCF-7 cells incubated with TPNPs. Reproduced with the permission from Ref. [42]. Copyright 2023, Chinese Chemical Letters

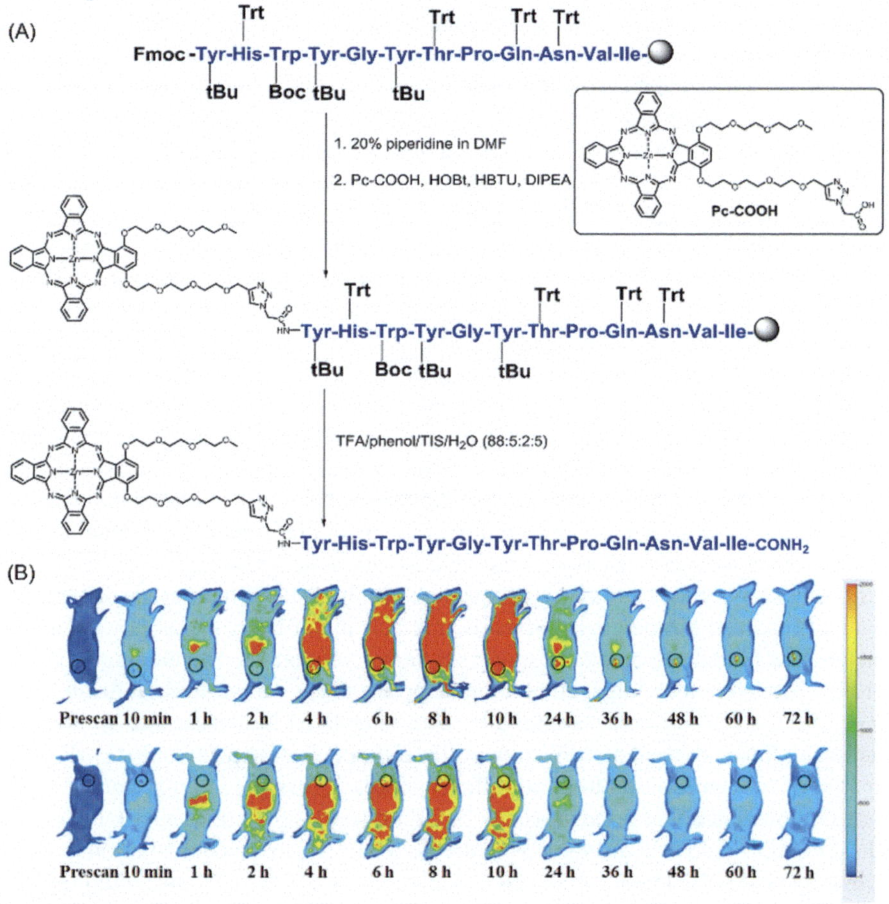

Fig. 10.4 **a** Synthetic route of zinc (II) phthalocyanine conjugated with GE11 peptide (Pc-GE11) was illustrated. **b** Fluorescence images were captured of A431 tumor-bearing nude mice before and after intravenous injection of Pc-GE11 (up) and Pc-rGE11 (down) over a period of 3 days. Reproduced with the permission from Ref. [43]. Copyright 2018, Dyes and Pigments

They found that self-assembled amyloid peptide nanomaterials have ultrashort structures and nonlinear optical properties, resulting in deeper penetration, lower photodamage, and better imaging period. This work provided a potential strategy for the application of peptide nanodrugs in bioimaging without any fluorescent labeling. To enhance the chemotherapeutic effect, small molecule drugs, DNA, and RNA were encapsulated to prepare novel SHG probes that combine the therapeutic power of bioimaging and drug encapsulation.

Real-time imaging of cellular organelles is an increasingly important aspect of modern biomedical analysis and disease diagnosis. Organelle-targeted NPs have emerged as one of the most commonly used techniques for subcellular sensing and imaging. In order to achieve efficient organelle-targeted bioimaging, the biocompatibility, specificity, and binding efficiency are particularly important. In this regard, Gao et al. designed and investigated AuNPs prepared from bifunctional peptides with superb binding affinity and nuclear-targeting ability for efficient nuclear-targeted bioimaging [46]. This design strategy enabled the targeted assembly of peptides through a medium of Au-binding peptides, specifically assembled on the surface of AuNPs, while leaving the nuclear-targeting ends open for the access. The bifunctional peptide exhibited a binding capacity 1.5-fold higher than that of the bare nuclear-targeting peptide, ensuring good surface coverage of the nanoparticles and enhanced nuclear-targeting ability. The prepared AuNPs were incubated with HepG2 cells for 24 h, resulting in a cell survival rate over 90% with good biocompatibility. Precise and efficient cell nuclear bioimaging of HepG2 cells was performed using the prepared AuNPs under confocal laser scanning microscopy, dark field/fluorescence microscopy, and transmission electron microscopy. The high surface coverage and targeted binding mode of the bifunctional peptide-AuNPs appeared to be a promising strategy for the construction of the organelle-targeted drugs. This study highlighted the potential of the organelle-targeted NPs for subcellular sensing and imaging and provided a valuable contribution to the field of biomedical research.

Luminescent probes and NPs with extended excited state lifetimes are critical for the time-resolved bioimaging. Typically, the cell membranes act as physiological barriers that hinder the uptake of many synthetic compounds. Developing biocompatible visualizers with high cell membrane permeability remains a significant challenge, particularly for non-metallic NPs with long-lived luminescence. To address this issue, Zhu et al. designed an amphiphilic cell-penetrating peptide, $F_6G_6(rR)_3R_2$, capable of transporting hydrophobic fluorophores across the cellular barrier [38]. They demonstrated that three classical TADF molecules, 4CzIPN, NAI-DPAC, and BTZ-DMAC, could self-assemble with the $F_6G_6(rR)_3R_2$ in aqueous solution to form well-dispersed NPs. These formed NPs exhibited low cytotoxicity and could readily penetrate the cell membranes. Moreover, the long-lived TADFs enabled their use for the time-resolved luminescence imaging in the oxygen-containing environments. These findings significantly expand the application of the cell-penetrating peptides for delivering molecules and NPs via non-covalent interactions, which are more flexible and easier than covalent modifications.

10.2.2 PAI

PAI is a powerful 3D imaging technique that utilizes the photoacoustic effect on biological tissues [47]. This effect occurs when an object absorbs light energy and converts it into thermal energy, which then causes thermoelastic expansion and generates an ultrasound signal that propagates outward. By combining the strengths of optical and ultrasound imaging, PAI enables the visualization of light absorption properties within the imaging area [48]. PAI is a non-ionizing biomedical imaging modality that offers higher resolution and imaging depth with greater sensitivity compared to FLI [49]. The integration of both imaging modalities has the potential to enhance the detection of cancer [50].

Targeted imaging and drug delivery in lung cancer face a significant challenge due to the lack of molecular targets and targeting probes. However, a recent study by Jung et al. has identified a novel targeting probe that effectively localized tumors in a lung cancer model using the in vivo phage display [51]. The CRQTKN peptide was found to home to tumor nodules at higher levels in the lungs of mutant mice compared to other candidate peptides that selected from five rounds of phage display. Photoacoustic tomography was utilized to detect lung tumors by labeling the CRQTKN peptides with NIR fluorescent dyes. The PAI result of isolated organs further demonstrated the tumor-homing ability of the CRQTKN peptide, with minimal accumulation observed in control organs, such as the liver. Additionally, the study showed that Adriamycin liposomes modified with the CRQTKN peptides on the surface delivered Adriamycin more efficiently and reduced the number or size of tumor lesions in mice compared to non-targeted liposomes and Adriamycin. The researchers identified tumor necrosis factor receptor superfamily member 19L (TNFRSF19L) as a receptor for the CRQTKN peptide, which was upregulated in lung tumors of mutant mice using an affinity identification approach.

In a recent study, Zhang et al. investigated the use of a dye-labeled cystine junction peptide that selectively recognizes the integrin $\alpha_v\beta_6$, a cell surface biomarker that is overexpressed in many different cancers [52]. The peptide was found to have high affinity for both PAI and FLI, making it a promising dual-mode probe for the diagnosis and intraoperative imaging of the integrin $\alpha_v\beta_6$-positive tumors. This innovative approach showed significant clinical applications and represented a major step forward in the development of targeted imaging strategies for cancer.

10.2.3 MRI

MRI is a clinical diagnostic technique that offers several advantages over other medical imaging methods [53]. It provides strong soft tissue contrast, does not use ionizing radiation, has high spatial resolution, and is not limited by the signal penetration depth [54]. As a result, it has become a widely used tool in the tumor treatment [55]. Studies have shown that MRI is effective in detecting and monitoring tumors, and can aid in the treatment planning and evaluation [55]. Its noninvasive nature and ability to provide detailed information about tumor morphology and function make it an important tool in the clinical setting [55].

During an MRI scan, a strong external magnetic field and a radiofrequency magnetic field are applied to the patient. The contrast agents, such as gadolinium (III), manganese (II), and superparamagnetic iron oxides (SPIONs), are used to alter the relaxation rate of nearby H^+ ions, thereby enhancing the contrast between different tissues. However, the commonly used MRI contrast agents have poor biocompatibility and lack the specificity, which limits their effectiveness. Fortunately, the use of bioactive peptides and peptide assemblies can improve the biocompatibility and enhance the imaging contrast at specific lesion sites. MRI has become an important tool in both cellular and molecular imaging due to its noninvasive and multidimensional laminar imaging capabilities. However, the development of novel contrast agents is necessary to enhance the MRI signal of tissues.

SPIONs are used as the signal-enhancing contrast agents because they display extraordinary magnetic properties at the nm size and have very low toxicity compared to other commercial contrast agents. Recent research by Sulek et al. has demonstrated a novel approach to functionalize the surface of the SPIONs using peptide amphiphiles (PAs) (Fig. 10.5a and b) [56]. This non-covalent wrapping of the SPIONs with PAs provided higher water solubility and enhanced the biocompatibility. The superparamagnetic properties of the peptide-SPION complexes and their ability to act as the contrast agents were demonstrated in vitro. Additionally, the cell culture experiments showed that the peptide-SPION complexes were biocompatible and localized around cells due to the peptide encapsulation. The use of the peptide-SPION complexes as MRI contrast agents was also confirmed. This promising research suggested that bioactive peptides may hold the key to improve the specificity and biocompatibility of MRI contrast agents in the future.

MRI is a widely used medical tool for and the disease diagnosis, and the MRI contrast agents are crucial for improving the detection sensitivity. Ultrasmall superparamagnetic iron oxides (USPIOs) have excellent magnetic response properties and are potential MRI T_2 enhancing contrast agents. However, how to confer them with specific tumor targeting and long-term cycling performance has been a hot topic in this field. Yin et al. have addressed this challenge by developing an elaborate chemical co-precipitation, surface modification, and peptide grafting process to prepare active tumor-targeting USPIOs@ F127-WSG [57]. Pluronic F127 and the peptide WSGPGVWGASVK (peptide-WSG) were selected as template agent and ovarian tumor-targeting ligand, respectively. The results showed that individual

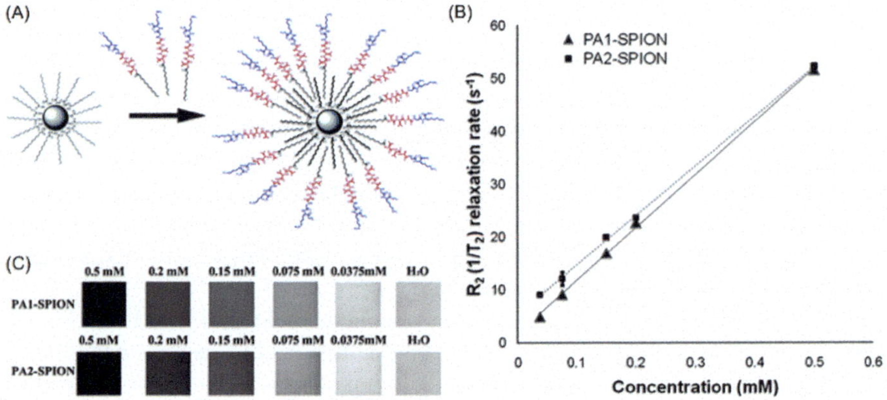

Fig. 10.5 **a** PA molecules for non-covalent functionalization of SPIONs for MRI. **b** Relaxation rates ($1/T_2$, s^{-1}) of PA1-SPION and PA2-SPION. **c** MRI images with peptide-SPION complexes. Reproduced with the permission from Ref. [56]. Copyright 2011, Materials Chemistry

USPIOs@F127-WSG particles were Fe_3O_4 NPs regulated by the micellar confinement of F127 with homogeneous spherical morphology and size (~ 9 nm) and the peptide-WSG was transplanted due to its tumor-targeting properties. The formed USPIOs@F127-WSG particles exhibited superparamagnetic behavior with high T_2 relaxation and the ability to target SKOV-3 cells in vitro, due to the peptide-WSG with SKOV-3-specific receptors. The results of in vivo experiments verified the targeting by the specific aggregation of the USPIOs@F127-WSG in the tumor region, leading to the enhancement of T_2-weighted MRI contrast.

Zhang et al. have developed a novel MRI contrast agent, Gd-DOTA-CC9, by coupling a cyclic peptide CC9 that targets the cell membrane of mesenchymal stem cells (MSCs) with Gd-DOTA (Fig. 10.6a) [58]. The agent was used to label human MSCs (hMSCs) via electroporation, which induced the cellular assembly of Gd-DOTA-CC9 nanoclusters in the cytoplasm, resulting in significantly improved cell labeling effectiveness and intracellular retention time. In vitro MRI of labeled hMSCs showed a significant signal reduction under T_2-weighted MRI, enabling long-term tracking of the migration of labeled cell transplants in vivo (Fig. 10.6b). The labeling strategy demonstrated a significant signal reduction in EP-labeled hMSCs under T_2-weighted MRI for more than 2 weeks in vitro and dark contrast after intracranial transplantation for more than 12 days in vivo (Fig. 10.6c), suggesting its potential for long-term follow-up of labeled cell transplantation. Furthermore, the EP labeling and cellular uptake of Gd-DOTA-CC9 were found to be safe in terms of cell viability, membrane damage, ROS induction, apoptosis and necrosis, and lipogenic and osteogenic differentiation potential.

10.2.4 CT Imaging

CT is a valuable biomedical imaging technique that utilizes X-rays to visualize the internal structures of the body [59–62], which is based on the attenuation of X-rays as the rays pass through the body [59–62]. The contrast agents can be used to enhance the attenuation of CT and improve the visualization of specific areas of interest. Typically, these agents contain iodinated compounds or bismuth or Au-based materials [59–62]. However, these materials lack the ability to target specific cells and often require high doses to be effective. Additionally, commonly used CT contrast agents can cause allergic reactions or serious side effects, such as the contrast nephropathy [59–62]. To address these limitations, disease-targeting peptides have been added to the CT contrast agents to reduce the required dose and avoid these adverse effects [59–62].

Zhang and colleagues have reported the use of AuNPs functionalized with a trans-activator transcription (TAT) peptide for tracking mesenchymal stem cells [63]. By utilizing the positively charged surface of the TAT peptide, AuNPs were able to be internalized into the stem cells, resulting in increased attenuation and allowing for their visualization and localization using CT scans. Successfully tracking these stem cells in mice over a period of 10 days using both CT imaging and bioluminescence

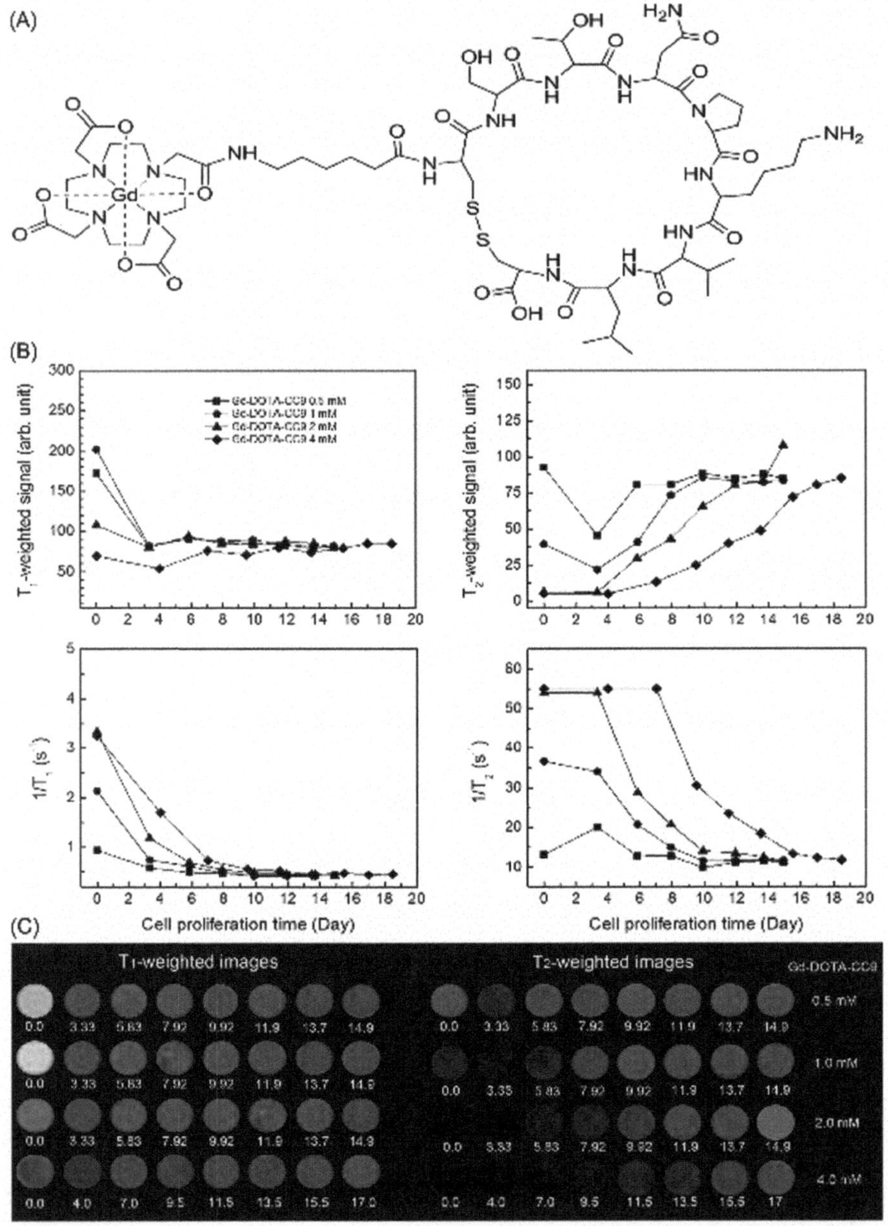

Fig. 10.6 **a** Schematic structure of Gd-DOTA-CC9. **b** T1-weighted and T2-weighted MRI signal (up), T1-relaxation and T2-relaxation rates (down) of human mesenchymal stem cell electroporation-labeled with Gd-DOTA-CC9 at different concentrations as functions of cell proliferation time. **c** In vitro T1-weighted (left) and T2-weighted (right) MR images of human mesenchymal stem cell pellet electroporation-labeled with Gd-DOTA-CC9 as functions of cell proliferation time at 11.7 T. Reproduced with the permission from Ref. [58]. Copyright 2018, Peptide Science

imaging has been achieved. Moreover, the spatial distribution of the transplanted stem cells in the lung was determined after 7 days using 3D CT imaging, demonstrating the potential of this method for studying the mechanisms of the stem cell therapy.

Tsvirkun et al. discussed the emerging field of CT molecular imaging, which relies on the detection of AuNPs and iodine-containing compounds to tag biomolecules [64]. They developed a new class of nanoscale cathepsin-targeted activity-based probes (ABPs) for functional CT imaging of cancer. These probes enabled the detection of elevated cathepsin activity within cancerous tissue, creating a direct link between biological processes and imaging signals. They further generated and evaluated a library of ABPs that tagged with different sized AuNPs and applied the most potent and stable AuNP-ABPs for noninvasive cancer imaging in mice. They found that the covalent linkage was key for the detection using low sensitive imaging modalities and that AuNP-ABPs was a promising tool for enzymatic-based CT imaging. In another case, Gaikwad and colleagues discussed the limitations of CT for molecular imaging due to its low contrast sensitivity and the development of a new class of iodinated nanoscale activity-based probes (IN-ABPs) that can accumulate at the target site by covalently binding cysteine cathepsins [65]. The obtained results showed that the 48-iodine bearing IN-ABP was optimal for detecting CT contrast and had potential as a contrast agent for molecular imaging of tumors using CT.

CT imaging combined with other imaging technique can improve the bioimaging efficiency of tumors. For instance, Yu et al. described the development of a multifunctional theragnostic agent using ultrasmall semimetal NPs of bismuth (Bi-LyP-1 NPs) conjugated with a tumor-homing peptide (LyP-1) [66]. These NPs had the ability to absorb both ionizing radiation and NIR-II laser radiation, making them suitable for dual-modal CT/PAI and synergistic NIR-II photothermal/radiotherapy of tumors. The materials were also rapidly cleared from the body through renal and fecal clearance and had low toxicity.

10.2.5 PET Imaging

PET offers excellent nM and even pM imaging sensitivity due to its high depth of penetration of the detection signal (511 keV γ protons) [67]. However, the short half-lives of conventional radionuclides in PET, such as ^{18}F, ^{15}O, ^{13}N, and ^{11}C, limited the longer detection times required for fast radiosynthesis. The PET/CT is currently the most successful multimodal imaging technique in both clinical practice and preclinical drug discovery [67, 68]. Specific molecular imaging is needed for both PET and PET/CT [69, 70]. Although antibodies with higher affinity and specificity have been successfully used in PET and PET/CT, targeted peptides have the advantages of being low cost, relatively easy to synthesize, mass producible, and most importantly, more biocompatible and practical for the radionuclide functionalization [71]. Therefore, targeted peptides are promising alternatives for specific molecular imaging in PET and PET/CT [71].

Britton and co-workers have developed a novel strategy for the ^{18}F labeling of unmodified peptides [72]. Their approach involves a direct, prosthesis-free ^{18}F labeling method for leucine-containing peptides (LCP), which utilizes a light-activated hydrogen atom extractant and a fluorine atom donor. The ^{18}F fluorination of various leucine-containing dipeptides and tetrapeptides was compared. This simplified procedure provided a direct method for the delivery of ^{18}F-labeled peptide PET platforms under mild aqueous conditions, enabling the rapid construction and discovery of peptide-based molecular imaging tools. Compared to conventional multi-step labeled ^{18}F fluorination, this approach offered a more straightforward and efficient means of producing ^{18}F-labeled peptides for PET imaging. In another case, Zhang and co-workers evaluated the effectiveness of a dual-targeting tracer, ^{68}Ga-BBN-RGD, for the PET/CT imaging of breast cancer and metastasis [73]. Twenty-two female patients were recruited and underwent PET/CT imaging after intravenous injection of the ^{68}Ga-BBN-RGD. The results showed positive accumulation of the ^{68}Ga-BBN-RGD in both primary cancer and metastases, with varying T/B ratios depending on the location.

Fluorinated polymers have become increasingly popular in the field of the delivery of gene and cytosolic protein. However, evaluating their biodistribution, clearance, and safety requires in vivo tracking, which is a significant challenge. To address this problem, Wang et al. developed a series of fluorinated poly-*L*-(lysine) (F-PLL) with high gene transfection efficiency and excellent biodegradation [74]. Then, they radiolabeled the F-PLL with radionuclide ^{18}F by the halogen replacement, allowing for in vivo tracking via PET imaging without altering the chemical structure of the polymer. Micro-PET imaging revealed the rapid clearance of the F-PLL and the DNA complexes from the liver and intestine after intravenous administration, indicating that the F-PLL did not accumulate in the liver and spleen when used as a gene carrier. This study presented a novel approach for tracking fluorinated polymers in vivo using PET imaging.

10.3 Conclusions

In summary, we presented recent advance in the design and synthesis of peptide-based nanomaterials for various bioimaging applications. Peptides offered a versatile platform for chemical modification, enabling the linking of different signaling molecules to achieve a range of biological functions. The structural diversity of peptides allowed them to self-assemble into well-organized nanostructures with tunable physicochemical properties. Moreover, peptides exhibited high biocompatibility and bioactivity due to their unique physicochemical properties. Modified peptide nanomaterials demonstrated reduced cytotoxicity and improved targeting, making them a promising tool for the tumor diagnosis and treatment. As such, the development of these peptide-based nanomaterials holds great potential for future biomedical applications.

Peptides can be modified to enhance their imaging capabilities by introducing various nanoscale components, such as the inorganic crystals, metal NPs, QDs, and polymers. The hybridization and functionalization of peptide nanomaterials with these components promoted their applications in tissue engineering, drug delivery, antibacterial activity, biosensing, and bioimaging greatly.

Acknowledgements Prof. Xin Zhou acknowledges the Natural Science Foundation of China (Nos. 21762045, 21911540466), Shandong Provincial Natural Science Foundation (No. ZR2019YQ12), China Postdoctoral Science Foundation (No. 219M652306), and Taishan Scholar Project (No. tsqn201812049) for supporting this work.

References

1. Q.H. Cui, Y.S. Zhao, J.N. Yao, Tailoring the structures and compositions of one-dimensional organic nanomaterials towards chemical sensing applications. Chem. Sci. **5**, 52–57 (2014)
2. F. De Leo, A. Magistrato, D. Bonifazi, Interfacing proteins with graphitic nanomaterials: from spontaneous attraction to tailored assemblies. Chem. Soc. Rev. **44**, 6916–6953 (2015)
3. M. Ovais, M. Guo, C. Chen, Tailoring nanomaterials for targeting tumor-associated macrophages. Eur. J. Immunol. **49**, 534–534 (2019)
4. Y.K. Zheng, L.M. Lai, W.W. Liu, H. Jiang, X.M. Wang, Recent advances in biomedical applications of fluorescent gold nanoclusters. Adv. Colloid. Interfac. Sci. **242**, 1–16 (2017)
5. J.J. Yang, F.L. Wang, H.Q. Yuan, L.S. Zhang, Y.Y. Jiang, X. Zhang, C. Liu, L. Chai, H. Li, M. Stenzel, Recent advances in ultra-small fluorescent Au nanoclusters toward oncological research. Nanoscale **11**, 17967–17980 (2019)
6. J.Y. He, C.C. Li, L. Ding, Y.N. Huang, X.L. Yin, J.F. Zhang, J. Zhang, C.J. Yao, M.R.N. Liang, R.P. Pirraco, J. Chen, Q. Lu, R. Baldridge, Y. Zhang, M.H. Wu, R.L. Reis, Y.L. Wang, Tumor targeting strategies of smart fluorescent nanoparticles and their applications in cancer diagnosis and treatment. Adv. Mater. **31**, 1902409 (2019)
7. J. Liu, C. Chen, S.L. Ji, Q. Liu, D. Ding, D. Zhao, B. Liu, Long wavelength excitable near-infrared fluorescent nanoparticles with aggregation-induced emission characteristics for image-guided tumor resection. Chem. Sci. **8**, 2782–2789 (2017)
8. K.H. Li, X.N. Zhao, G. Wei, Z.Q. Su, Recent advances in the cancer bioimaging with graphene quantum dots. Curr. Med. Chem. **25**, 2876–2893 (2018)
9. K.H. Li, W. Liu, Y. Ni, D.P. Li, D.M. Lin, Z.Q. Su, G. Wei, Technical synthesis and biomedical applications of graphene quantum dots. J. Mater. Chem. B **5**, 4811–4826 (2017)
10. M. Lv, Y. Liu, J.H. Geng, X.H. Kou, Z.H. Xin, D.Y. Yang, Engineering nanomaterials-based biosensors for food safety detection. Biosens. Bioelectron. **106**, 122–128 (2018)
11. Y. Wang, K. Xia, L.C. Wang, M.X. Wu, X.J. Sang, K.M. Wan, X.D. Zhang, X.M. Liu, G. Wei, Peptide-engineered fluorescent nanomaterials: structure design, function tailoring, and biomedical applications. Small **17**, 2005578 (2021)
12. S. Kim, J.H. Kim, J.S. Lee, C.B. Park, Beta-sheet-forming, self-assembled peptide nanomaterials towards optical, energy, and healthcare applications. Small **11**, 3623–3640 (2015)
13. C.J.C. Edwards-Gayle, I.W. Hamley, Self-assembly of bioactive peptides, peptide conjugates, and peptide mimetic materials. Org. Biomol. Chem. **15**, 5867–5876 (2017)
14. J.A. Hutchinson, S. Burholt, I.W. Hamley, Peptide hormones and lipopeptides: from self-assembly to therapeutic applications. J. Pept. Sci. **23**, 82–94 (2017)
15. X.Q. Yu, Z.P. Wang, Z.Q. Su, G. Wei, Design, fabrication, and biomedical applications of bioinspired peptide-inorganic nanomaterial hybrids. J. Mater. Chem. B **5**, 1130–1142 (2017)

16. W.S. Zhang, D.M. Lin, H.X. Wang, J.F. Li, G.U. Nienhaus, Z.Q. Su, G. Wei, L. Shang, Supramolecular self-assembly bioinspired synthesis of luminescent gold nanocluster-embedded peptide nanofibers for temperature sensing and cellular imaging. Bioconjugate Chem. **28**, 2224–2229 (2017)
17. R. Weissleder, Molecular imaging in cancer. Science **312**, 1168–1171 (2006)
18. R. Weissleder, M.J. Pittet, Imaging in the era of molecular oncology. Nature **452**, 580–589 (2008)
19. S.M. Park, A. Aalipour, O. Vermesh, J.H. Yu, S.S. Gambhir, Towards clinically translatable in vivo nanodiagnostics. Nat. Rev. Mater. **2**, 17014 (2017)
20. Y.Q. Yan, H. Wang, Y.L. Zhao, Radiolabeled peptide probe for tumor imaging. Chin. Chem. Lett. **33**, 3361–3370 (2022)
21. A. Koudrina, M.C. DeRosa, Advances in medical imaging: aptamer- and peptide-targeted MRI and CT contrast agents. ACS Omega **5**, 22691–22701 (2020)
22. P.C. Zhang, Y.G. Cui, C.F. Anderson, C.L. Zhang, Y.P. Li, R.F. Wang, H.G. Cui, Peptide-based nanoprobes for molecular imaging and disease diagnostics. Chem. Soc. Rev. **47**, 3490–3529 (2018)
23. Z.Y. Chen, Y.X. Wang, Y. Lin, J.S. Zhang, F. Yang, Q.L. Zhou, Y.Y. Liao, Advance of molecular imaging technology and targeted imaging agent in imaging and therapy. Biomed. Res. Int. **2014**, 1–13 (2014)
24. H.S. Peng, D.T. Chiu, Soft fluorescent nanomaterials for biological and biomedical imaging. Chem. Soc. Rev. **44**, 4699–4722 (2015)
25. G.B. Qi, Y.J. Gao, L. Wang, H. Wang, Self-assembled peptide-based nanomaterials for biomedical imaging and therapy. Adv. Mater. **30**, 1703444 (2018)
26. Y.H. Hu, W.J. Guo, H. Wei, Protein- and peptide-directed approaches to fluorescent metal nanoclusters. Isr. J. Chem. **55**, 682–697 (2015)
27. C. Zhang, W. Wu, R.Q. Li, W.X. Qiu, Z.N. Zhuang, S.X. Cheng, X.Z. Zhang, Peptide-based multifunctional nanomaterials for tumor imaging and therapy. Adv. Funct. Mater. **28**, 1804492 (2018)
28. C.H. Ren, Z.Y. Wang, Q. Wang, C.H. Yang, J.O.F. Liu, Self-assembled peptide-based nanoprobes for disease theranostics and disease-related molecular imaging. Small Methods, 4 (2020)
29. C.H. Cai, L.Q. Wang, J.P. Lin, Self-assembly of polypeptide-based copolymers into diverse aggregates. Chem. Commun. **47**, 11189–11203 (2011)
30. L.W. Dobrucki, A.J. Sinusas, PET and SPECT in cardiovascular molecular imaging. Nat. Rev. Cardiol. **7**, 38–47 (2010)
31. A. Signore, S.J. Mather, G. Piaggio, G. Malviya, R.A. Dierckx, Molecular imaging of inflammation/infection: nuclear medicine and optical imaging agents and methods. Chem. Rev. **110**, 3112–3145 (2010)
32. X.D. Meng, X.J. Pang, K. Zhang, C.C. Gong, J.Y. Yang, H.F. Dong, X.J. Zhang, Recent advances in near-infrared-II fluorescence imaging for deep-tissue molecular analysis and cancer diagnosis. Small **18**, 2202035 (2022)
33. Q. Lu, C.J. Wu, Z.Q. Liu, G.L. Niu, X.Q. Yu, Fluorescent AIE-active materials for two-photon bioimaging applications. Front. Chem. **8**, 617463 (2020)
34. S. Lee, J. Xie, X.Y. Chen, Peptides and peptide hormones for molecular imaging and disease diagnosis. Chem. Rev. **110**, 3087–3111 (2010)
35. Y. Hou, J. Zhou, Z.Y. Gao, X.Y. Sun, C.Y. Liu, D.H. Shangguan, W.S. Yang, M.Y. Gao, Protease-activated ratio metric fluorescent probe for pH mapping of malignant tumors. ACS Nano **9**, 3199–3205 (2015)
36. F.Q. Chen, D. Gerion, Fluorescent CdSe/ZnS nanocrystal-peptide conjugates for long-term, nontoxic imaging and nuclear targeting in living cells. Nano Lett. **4**, 1827–1832 (2004)
37. S.K. Li, Y.M. Liu, R.R. Xing, X.H. Yan, Covalently assembled dipeptide nanoparticles with adjustable fluorescence emission for multicolor bioimaging. ChemBioChem **20**, 555–560 (2019)

38. Z.C. Zhu, D. Tian, P.L. Gao, K. Wang, Y.C. Li, X.W. Shu, J.T. Zhu, Q. Zhao, Cell-penetrating peptides transport noncovalently linked thermally activated delayed fluorescence nanoparticles for time-resolved luminescence imaging. J. Am. Chem. Soc. **140**, 17484–17491 (2018)
39. Y.Y. Gao, Y.L. Liu, R. Yan, J.F. Zhou, H. Dong, X. Hua, P. Wang, Bifunctional peptide-conjugated gold nanoparticles for precise and efficient nucleus-targeting bioimaging in live cells. Anal. Chem. **92**, 13595–13603 (2020)
40. H.J. He, J.Q. Wang, H.M. Wang, N. Zhou, D. Yang, D.R. Green, B. Xu, Enzymatic cleavage of branched peptides for targeting mitochondria. J. Am. Chem. Soc. **140**, 1215–1218 (2018)
41. X.X. Li, L. Liu, Y. Fu, H.D. Chen, M.M.A. Abualrejal, H. Zhang, Z.X. Wang, H.M. Zhang, Peptide-enhanced tumor accumulation of upconversion nanoparticles for sensitive upconversion luminescence/magnetic resonance dual-mode bioimaging of colorectal tumors. Acta Biomater. **104**, 167–175 (2020)
42. Y.R. Wang, Y. Lei, J.Y. Wang, H. Yang, L.M. Sun, Tetrapeptide self-assembled multicolor fluorescent nanoparticles for bioimaging applications. Chin. Chem. Lett. **34**, 107915 (2023)
43. L.G. Yu, Q. Wang, R.C.H. Wong, S.R. Zhao, D.K.P. Ng, P.C. Lo, Synthesis and biological evaluation of phthalocyanine-peptide conjugate for EGFR-targeted photodynamic therapy and bioimaging. Dyes Pigments **163**, 197–203 (2019)
44. H. Li, H. Huang, J.J. Feng, X.L. Luo, K.M. Fang, Z.G. Wang, A.J. Wang, A polypeptide-mediated synthesis of green fluorescent gold nanoclusters for Fe^{3+} sensing and bioimaging. J. Colloid. Interf. Sci. **506**, 386–392 (2017)
45. M. Ni, S.M. Zhuo, C. Iliescu, P.T.C. So, J.S. Mehta, H. Yu, C.A.E. Hauser, Self-assembling amyloid-like peptides as exogenous second harmonic probes for bioimaging applications. J. Biophotonics **12**, 1–8 (2019)
46. Y. Zhou, Q.Q. Li, Y. Wu, X.Y. Li, Y. Zhou, Z. Wang, H. Liang, F.Q. Ding, S. Hong, N.F. Steinmetz, H. Cai, Molecularly stimuli-responsive self-assembled peptide nanoparticles for targeted imaging and therapy. ACS Nano **17**, 8004–8025 (2023)
47. Y.W. Zheng, M.Y. Liu, L.X. Jiang, Progress of photoacoustic imaging combined with targeted photoacoustic contrast agents in tumor molecular imaging. Front. Chem. **10**, 1077937 (2022)
48. V. Ntziachristos, D. Razansky, Molecular imaging by means of multispectral optoacoustic tomography (MSOT). Chem. Rev. **110**, 2783–2794 (2010)
49. Q. Liu, T. Jin, Q. Chen, L. Xi, Research progress of miniaturized photoacoustic imaging technology in biomedical field. Chin. J. Lasers 47 (2020)
50. E.Y. Park, H. Lee, S. Han, C. Kim, J. Kim, Photoacoustic imaging systems based on clinical ultrasound platform. Exp. Biol. Med. **247**, 551–560 (2022)
51. H. Jung, S. Park, G.R. Gunassekaran, M. Jeon, Y.E. Cho, M.C. Baek, J.Y. Park, G. Shim, Y.K. Oh, I.S. Kim, C. Kim, B. Lee, A peptide probe enables photoacoustic-guided imaging and drug delivery to lung tumors in K-ras(LA2) mutant mice. Cancer Res. **79**, 4271–4282 (2019)
52. C. Zhang, R. Kimura, L. Abou-Elkacem, J. Levi, L.Y. Xu, S.S. Gambhir, A cystine knot peptide targeting integrin alpha(v)beta(6) for photoacoustic and fluorescence imaging of tumors in living subjects. J. Nucl. Med. **57**, 1629–1634 (2016)
53. M. Mahmoudi, H. Hosseinkhani, M. Hosseinkhani, S. Boutry, A. Simchi, W.S. Journeay, K. Subramani, S. Laurent, Magnetic resonance imaging tracking of stem cells in vivo using iron oxide nanoparticles as a tool for the advancement of clinical regenerative medicine. Chem. Rev. **111**, 253–280 (2011)
54. P. Caravan, J.J. Ellison, T.J. McMurry, R.B. Lauffer, Gadolinium(III) chelates as MRI contrast agents: structure, dynamics, and applications. Chem. Rev. **99**, 2293–2352 (1999)
55. N. Lee, T. Hyeon, Designed synthesis of uniformly sized iron oxide nanoparticles for efficient magnetic resonance imaging contrast agents. Chem. Soc. Rev. **41**, 2575–2589 (2012)
56. S. Sulek, B. Mammadov, D.I. Mahcicek, H. Sozeri, E. Atalar, A.B. Tekinay, M.O. Guler, Peptide functionalized superparamagnetic iron oxide nanoparticles as MRI contrast agents. J. Mater. Chem. **21**, 15157–15162 (2011)
57. J. Yin, D.J. Yao, G.F. Yin, Z.B. Huang, X.M. Pu, Peptide-decorated ultrasmall superparamagnetic nanoparticles as active targeting MRI contrast agents for ovarian tumors. Acs Appl. Mater. Interface **11**, 41038–41050 (2019)

58. P.L. Zhang, Y.H. Zhang, B.B. Li, H.L. Zhang, H.X. Lin, Z.W. Deng, B. Tan, Cell-assembled nanoclusters of MSC-targeting Gd-DOTA-peptide as a T-2 contrast agent for MRI cell tracking. J. Pept. Sci. **24**, e3077 (2018)
59. F. Danhier, A. Le Breton, V. Preat, RGD-based strategies to target alpha(v) beta(3) integrin in cancer therapy and diagnosis. Mol. Pharmaceut. **9**, 2961–2973 (2012)
60. W.Y. He, K.L. Ai, L.H. Lu, Nanoparticulate X-ray CT contrast agents. Sci. China Chem. **58**, 753–760 (2015)
61. N. Lee, S.H. Choi, T. Hyeon, Nano-sized CT contrast agents. Adv. Mater. **25**, 2641–2660 (2013)
62. Y.L. Liu, K.L. Ai, L.H. Lu, Nanoparticulate X-ray computed tomography contrast agents: from design validation to in vivo applications. Acc. Chem. Res. **45**, 1817–1827 (2012)
63. H.Y. Bao, Y.Y. Xia, C.G. Yu, X.Y. Ning, X.Y. Liu, H. Fu, Z.J. Chen, J. Huang, Z.J. Zhang, CT/bioluminescence dual-modal imaging tracking of mesenchymal stem cells in pulmonary fibrosis. Small **15**, 1904314 (2019)
64. D. Tsvirkun, Y. Ben-Nun, E. Merquiol, I. Zlotver, K. Meir, T. Weiss-Sadan, I. Matok, R. Popovtzer, G. Blum, CT imaging of enzymatic activity in cancer using covalent probes reveal a size-dependent pattern. J. Am. Chem. Soc. **140**, 12010–12020 (2018)
65. H.K. Gaikwad, D. Tsvirkun, Y. Ben-Nun, E. Merquiol, R. Popovtzer, G. Blum, Molecular imaging of cancer using X-ray computed tomography with protease targeted iodinated activity-based probes. Nano Lett. **18**, 1582–1591 (2018)
66. X.J. Yu, A. Li, C.Z. Zhao, K. Yang, X.Y. Chen, W.W. Li, Ultrasmall semimetal nanoparticles of bismuth for dual-modal computed tomography/photoacoustic imaging and synergistic thermoradiotherapy. ACS Nano **11**, 3990–4001 (2017)
67. S.M. Ametamey, Molecular imaging with PET. Chem. Rev. **108**, 4036–4036 (2008)
68. K. Pant, O. Sedlacek, R.A. Nadar, M. Hruby, H. Stephan, Radiolabelled polymeric materials for imaging and treatment of cancer: quo Vadis? Adv. Healthc. Mater. **6**, 1601115 (2017)
69. D.E. Lee, H. Koo, I.C. Sun, J.H. Ryu, K. Kim, I.C. Kwon, Multifunctional nanoparticles for multimodal imaging and theragnosis. Chem. Soc. Rev. **41**, 2656–2672 (2012)
70. V. Kapoor, B.M. McCook, F.S. Torok, An introduction to PET-CT imaging. Radiographics **24**, 523–543 (2004)
71. W.Z. Wang, Z.Y. Hu, Targeting peptide-based probes for molecular imaging and diagnosis. Adv. Mater. **31**, 1804827 (2019)
72. Z.L. Yuan, M.B. Nodwell, H. Yang, N. Malik, H. Merkens, F. Benard, R.E. Martin, P. Schaffer, R. Britton, Site-selective, late-stage C-H F-18-fluorination on unprotected peptides for positron emission tomography imaging. Angew. Chem. Int. Ed. **57**, 12733–12736 (2018)
73. J.J. Zhang, F. Mao, G. Niu, L. Peng, L.X. Lang, F. Li, H.Y. Ying, H.W. Wu, B.J. Pan, Z.H. Zhu, X.Y. Chen, Ga-68-BBN-RGD PET/CT for GRPR and integrin alpha(v)beta(3) imaging in patients with breast cancer. Theranostics **8**, 1121–1130 (2018)
74. X.Y. Wang, G.Y. Rong, J.J. Yan, D.H. Pan, L.Z. Wang, Y.P. Xu, M. Yang, Y.Y. Cheng, In vivo tracking of fluorinated polypeptide gene carriers by positron emission tomography imaging. Acs Appl. Mater. Interface **12**, 45763–45771 (2020)

Chapter 11
Peptide Nanomaterials for Energy and Environmental Science Applications

Yan Wang, Keming Wan, Caiyu Liu, and Wenliang Zhang

Abstract Peptide-based nanomaterials with diverse structures and functionalities are considered as typical biomaterials for various applications. Due to their excellent biocompatibility and biological activity, peptide-based nanomaterials have been extensively used in different fields, especially in biomedical community, such as biosensing and bioimaging, cancer therapy, drug delivery, and tissue engineering. Apart from various biomedical applications, self-assembled peptide-based architectures can also be utilized as superior electrode materials for lithium-ion batteries and supercapacitors. In addition, their notable physical and chemical properties endow peptide-based nanomaterials as favorable adsorbents, catalysts, and detectors for both environmental remediation and monitoring. In this chapter, we focus on the design of peptide-based nanomaterials toward energy and environmental science applications.

Keywords Peptide nanomaterials · Environmental science · Energy storage · Hybrid materials

11.1 Peptide-Based Energy Storage Materials

Energy storage materials are the focus of peptide-directed synthesis approaches, especially those techniques access to biological interfaces that based on nanostructured inorganic metal oxide [1]. These metal oxide materials as energy storage materials need to have a high discharge capacity while minimizing capacity decay over time [2]. A priority of peptide-based synthesis is its capability of encoding the information to acquire desirable geometrical, physical, and chemical properties [3]. Until now, a variety of peptides and their derivatives have been confirmed to possess interesting hierarchical architectures [4, 5]. By precisely regulating the nanostructures, peptides could offer high surface area to be decorated with inorganic materials that endow novel functions. Self-assembled peptide nanofibers (PNFs) have been extensively applied to incorporate with electrochemically active materials to fabricate functional

---

Y. Wang (✉) · K. Wan · C. Liu · W. Zhang
College of Chemistry and Chemical Engineering, Qingdao University, Qingdao 266071, China
e-mail: yanwang@qdu.edu.cn

G. Wei (ed.), *Peptide Nano-Chemistry and Nanotechnology*,
https://doi.org/10.1007/978-981-96-2387-7_11

electrodes for lithium-ion batteries (LIBs) and supercapacitors with promoted performance, primarily due to their high specific surface area and nanostructure-induced effects [6–8].

Peptide-induced biomimetic synthesis have gained considerable interests for the design of functional nanomaterials due to their structural diversity, functional flexibility, and self-assembling capability [9, 10]. Recent studies indicate that the peptide self-assembly could be applied as effective bioinspired method to fabricate electrode materials by satisfying better battery performance requirements. For example, Ryu et al. [11] reported the fabrication of novel peptide/Co_3O_4 hybrid nanowires through the peptide self-assembly, in which the cobalt ions were reduced by $NaBH_4$, and diphenylalanine nanowires were decorated with Co_3O_4 nanocrystals at room temperature without any complicated procedures (Fig. 11.1a). The as-prepared peptide/ Co_3O_4 composite nanowires presented potential energy storage application as the cathodes for LIBs. By constructing the Swagelok-type cells with the peptide/Co_3O_4 nanowires as the negative electrode, their charge–discharge behavior was investigated. As shown in Fig. 11.1b, the hybrid nanowires were employed as a working electrode, in combination with a counter electrode, an electrolyte, as well as a separator. The charge–discharge analysis of the proposed hybrid nanowires was presented in Fig. 11.1c. It was found that pristine peptide nanowires presented electrochemical response with Li ions, showing a specific capacity of 80 mAh/g at the first discharge cycle, and finalized to 5 mAh/g at the equilibrium state. Differently, the peptide/ Co_3O_4 nanowires possessed a much higher specific capacity of around 280 mAh/ g at the first discharge cycle, while finalized to 80 mAh/g at the saturated state. This enhancement was attributed to the participation of Co_3O_4 NPs in the hybrid nanowires, which significantly promoted the specific capacity of the electrode. Moreover, the crystallinity of the pristine peptide nanowires with and without the discharge was evaluated to investigate their origin of the electrochemical activity.

The employment of peptide-based nanotubes and nanofibers as electrodes for the LIBs has been demonstrated by the thermal conversion of carbon-coated $FePO_4$ nanotubes from the $FePO_4$-mineralized PNFs. Briefly, $FePO_4$-mineralized PNFs were thermally treated at 350 °C to fabricate the $FePO_4$ nanotubes with the inner walls coated with a thin layer of conductive carbon by the carbonization of the peptide core [12]. The resultant $FePO_4$ nanotubes exhibited high reversible charge/discharge capacities and good capacity retention during the cycling as LIB cathodes. Bulk $FePO_4$ was extensively explored as promising cathode material in LIBS; nevertheless, key issues, such as low electric conductivity and slow insertion/extraction rate of Li ions, have not been well addressed [13]. In a further study, Ryu et al. [12] found that the utilization of nanostructured $FePO_4$ that decorated with conductive materials can solve the abovementioned problems. Compared with previously reported $FePO_4$ nanostructures, the as-synthesized $FePO_4$ nanotubes performed much better specific capacity and capacity retention as excellent cathode material for rechargeable LIBs, partially ascribed to the anisotropic structure of the $FePO_4$ nanotubes as well as small dimensions and carbon coating.

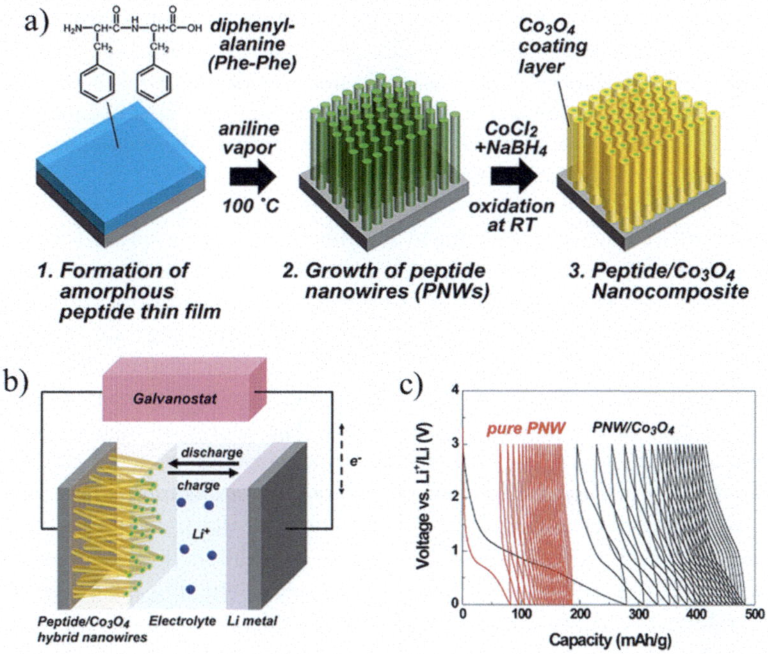

Fig. 11.1 **a** Synthesis process of peptide/Co_3O_4 hybrid nanowires. **b** Configuration of test cell, where prepared nanowires are used as an anode. **c** Li storage capability of pure peptide nanowires and peptide/Co_3O_4 hybrids [11]

There has been a bunch of recent research using material-binding peptides to fabricate other metal-ion batteries other than LIBs. Compared with LIBs, sodium-ion batteries (SIBs) using the material-binding peptides are still in their preliminary stage of exploration. Although the weight of SIBs limits their range of applications, sodium ions can provide a cheaper and safer energy storage solution as compared to LIBs [14]. Due to the fact that the electrode design requirements of SIBs are similar to those of LIBs, the peptide-mediated synthesis can be beneficially explored. To explore the cathode materials for SIBs, Moradi et al. [15] demonstrated an energy-efficient bio-templated route to synthesize nanostructured $FePO_4$ as the electrode material, in which self-assembled M13 viruses and single-walled carbon nanotubes (CNTs) were used as the templates to grow amorphous $FePO_4$ nanoparticles, so as to improve the electronic conductivity of the active material. Experimental tests confirmed the high discharge capacity of 166 mAh/g at C/10 rate, along with higher C-rate tests verifying the as-prepared material have the best capacity and power performance compared to reported $FePO_4$ electrodes.

Peptide-based nanomaterials have been verified to possess favorable electrochemical properties, which also make them potential electrode materials for supercapacitors [16]. The incorporation of peptide nanotubes (PNTs) in conventional carbon electrodes could not only enhance their capacitance even several orders of magnitude, but also significantly expend the charge–discharge cycles [1]. For example, Beker et al. [17] found that PNT-coated carbon electrode improved its double layer capacitance by enlarging the usable surface area of the electrode, revealing that the critical factor for the capacitance of the supercapacitor was the wetted surface area of the nanoscale hydrophilic channels of PNTs.

The design of self-assembling helical peptides with high controllability and tunability is still a challenging task due to the complexity and various self-assembling pathways of peptide molecules. For this aim, Hu et al. [18] demonstrated that helical peptides self-assembled into well-ordered nanostructures by employing the in-tether chiral center-induced helical peptides as a platform. They found that the self-assembling behavior was governed by the in-tether substitutional groups, where the chirality determined the formation of helical structures and the aromaticity provided the driving force for the peptide self-assembly. The energy storage properties of the peptide assemblies as active materials for supercapacitors were thereafter investigated by loading the assemblies onto a glassy carbon electrode. The capacitive charge–discharge curves of the peptide-loaded electrode illustrated the galvanostatic current densities ranging from 50 to 200 $\mu A/cm^2$. In addition, the capacitance retention was tested to be approx. 80% after 5000 cycles, implying the superior cycle stability and specific capacitance of the peptide-based nanomaterials.

11.2 Peptide-Based Energy Conversion Materials

Peptide-based materials are a new class of nanomaterials with high specific surface area, good stability, and modifiability, which make them with great interest for the energy conversion applications. In solar cells, fuel cells, electrochemical energy storage devices, and thermoelectric conversion devices, peptide-based nanomaterials have shown excellent performance and potential [16].

Conventional inorganic photoluminescent materials face severe limitations for the applications in bioimaging and bio-machine interfaces due to the inherent limitations of low biocompatibility and difficulty in the modulation. Therefore, the development of stable photoluminescent organic, tunable, flexible, and eco-friendly optoelectronic materials is ongoing. It has been shown that peptide-based nanomaterials typically consist of peptide chains with specific amino acid sequences and have a large surface area and abundant electron transfer capacity [19]. In the photochemical reactions, peptide-based nanomaterials can act as electronic donors or acceptors, converting light energy into chemical or electrical energy. When a peptide-based nanomaterial absorbs a photon, the excited-state electrons are excited into the conduction band and then transferred to the reducing or oxidizing agent via an electron transport chain. The current that generated by this electron transfer process can be used to drive the chemical reactions or to supply the electricity. As shown in Fig. 11.2a and b, Tao et al. [20] demonstrated how the W-based aromatic cyclo-dipeptide self-assemblies were modified to form quantum confined (QC) microstructures with the photoluminescent characteristics. By altering the self-assembly process, the emissions were adjusted specifically over the visible and near-infrared (NIR) spectrum. In particular, the emitted light was tuned from the visible region to the NIR region (420–820 nm) by tuning the peptide self-assembly process. No significant cytotoxic effects were observed for these nanostructures, and their application for the in vivo imaging and as the light emitting diode phosphors was demonstrated. These results suggested that the morphological and optical properties of the self-assembled aromatic cyclic dipeptides were tunable, making them potential candidates for supramolecular quantum confinement materials that offer biocompatible alternatives for a wide range of biomedical and optoelectronic applications.

Peptides are biologically active molecules, but recent research has shown that some peptides also have piezoelectric properties, which means that they can produce the charge separation and the potential differences when the mechanical forces are applied [21]. The mechanism by which peptides act as the piezoelectric materials to convert energy is through the introduction of polarizing groups into the peptide molecule. The peptide molecule undergoes structural changes when subjected to mechanical pressure in the presence of the polarizing group, and this structural change

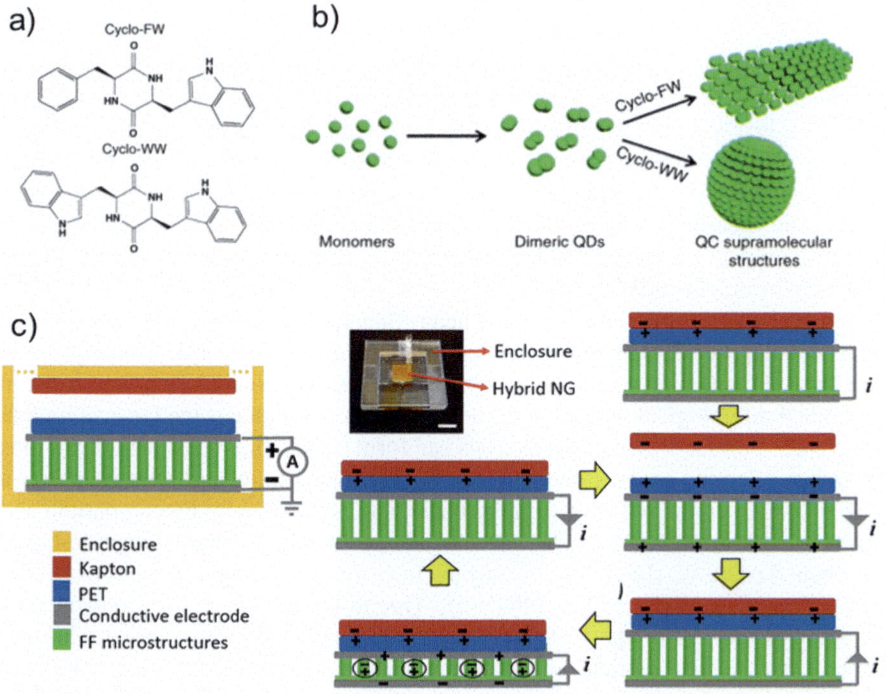

Fig. 11.2 **a** Molecular structures of cyclo-phenylalanine-tryptophan (cyclo-FW) and cyclo-WW [20]. **b** Schematic representing process of cyclo-dipeptides self-assembly [20]. **c** Composition and energy conversion process of PENG based on FF [22]

leads to the generation of the charge separation and potential difference, thus enabling the conversion of the mechanical force to electrical energy.

Vu et al. [22] obtained crystalline seed films with vertically aligned regions by subjecting amorphous diphenylalanine (FF) hydrogel films to vigorously circulating moist air. They controlled the diffusion of water during the formation of the seed film and then homogenized the polarization after applying an electric field. By using vertical FF peptide micro-rods, the generators with high electrical voltage and tunable polarization were prepared. Figure 11.2c shows the energy utilization process of the proposed nanogenerator. In brief, the motion of the top layer of the triboelectric nanogenerator (TENG) and the small deformation of the piezoelectric material caused by the pressure captures the energy. Therefore, the single-electrode friction TENG provided additional voltage and current output to the FF-based piezoelectric nanogenerator (PENG). The simple design of this hybrid nanogenerator (NG), which did not require additional conducting electrodes, simplified the fabrication process and reduced the size of the hybrid NG, providing an idea for the preparation of future peptide-based piezoelectric nanogenerators.

11.3 Peptide-Based Materials for the Pollutant Adsorption

Peptide-based nanomaterials can be used efficient adsorption materials for the environmental pollution control, for instance the removal of organic pollutants, gaseous pollutants, heavy metals, radioactive substances, and bacteria in water, air, and soil. Peptide-based nanomaterials have large specific surface areas and specific adsorption active sites, which can interact with the chemical bonds of target pollutants to achieve the specific adsorption [23]. In addition, peptide-based nanomaterials also have good biocompatibility and biodegradability, and will not have adverse effects on the environment and organisms [24].

In recent years, the pollution by toxic heavy metal ions in industrial wastewater has become a major environmental problem in many parts of the world. Although some heavy metal contaminants are metabolized and broken down by living organisms in a chemical or biological manner, their heavy metal contaminants tend to accumulate in the body. Peptide-based nanomaterials can form stable complexes with metal ions through their amino acid residues, have relatively high affinity and selectivity over a wide pH range compared to other adsorbents, and are also more biocompatible and degradable [25]. Shogo et al. [26] successfully prepared a functional peptide-based adsorbent AADAAC-$(FPGVG)_4$ by introducing a metal-binding sequence AADAAC at the N-terminal of short ELP $(FPGVG)_4$. Due to the temperature-dependent self-assembly properties of the ELP, the peptide molecule was transformed from the solution state to solid particles in the presence of metal ions like Cd^{2+} and Zn^{2+}, and was able to adsorb metal ions. Subsequently, the acidic solution-treated AADAAC-$(FPGVG)_4$ retained its metal-binding capacity and could be recycled for several times.

Water pollution caused by organic pollutants has become a key issue in perpetuating the imbalance of nature and life on earth [27]. Compared to other methods for the removal of organic pollutant, such as the chemical oxidation and activated carbon adsorption, peptides are natural biological macromolecules that are more environmentally friendly and less polluting to the environment when adsorbing the dyes in water. Moreover, peptide-based adsorption of dyes can be sequence-designed to achieve a high degree of selectivity for the target molecules. The development of biocompatible self-assembled peptide proto-fibers with high adsorption capacity for organic pollutants in water was discussed by Brar and co-workers [28], who prepared different morphological motifs of self-assembled dipeptide Fmoc-FW-OMe by solvent modulation and characterized them by multiple characterizations. The fibrous structure of the peptide assemblies has a specific adsorption capacity for selective adsorption of cationic dyes from aqueous solutions. The kinetic and adsorption isotherm studies were carried out to understand the adsorption mechanism of the dyes. The pH studies showed better adsorption under alkaline conditions. Moreover, the self-assembled peptide-based proto-fibers exhibited high reusability over five cycles, which showed a negligible effect on the dye adsorption capacity (Fig. 11.3a).

Peptide-based nanomaterials can adsorb carbon dioxide (CO_2) because some amino acid groups can chemically react with the gas. For instance, Zhang et al. [29] reported the synthesis and characterizations of a two-dimensional (2D) layered skeleton made of zinc ions and the dipeptide glycyl threonine [Zn(Gly-Thr)$_2$]CH_3OH. As shown in Fig. 11.3b, the peptide coordinates with the Zn^{2+} central in a chelated manner and produced a twisted octahedral geometry. The interaction between the threonine side chain and the terminal carboxylic acid group limited the twist freedom of the peptide and imposed an alcohol and methyl orientation. This chelating interaction also led to a smaller Zn–Zn distance within the layer compared to the non-chelated framework, thus reducing the intrinsic porosity. In the experiments, the material exhibited highly selective adsorption of CO_2 and showed one-dimensional (1D) porosity, while maintaining its crystallinity after the solvent removal. This flexibility and structural stability were mainly ascribed to the coordinated pattern of peptides and the precise control of supramolecular interactions. Their work illustrated that both flexible and rigid types of units could be used for the design of peptide-based metal–organic frameworks, which in turn had wider applications in the gas adsorption [29].

11.4 Peptide-Based Nanomaterials for Catalytic Degradation of Environmental Pollutants

In recent years, the application of peptide nano-composites and macroscopic materials as the templates or directly assembled for catalytic applications has attracted widespread attention [30, 31]. The design of functionalized nanoparticle and

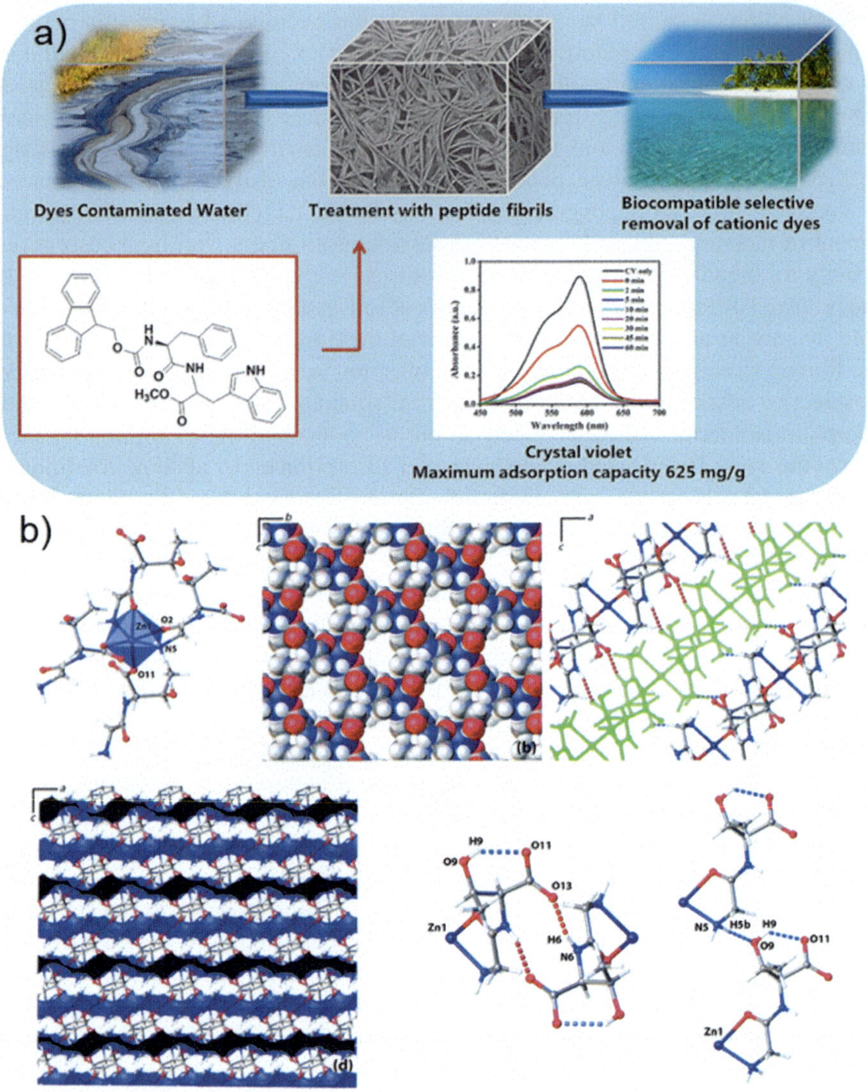

Fig. 11.3 **a** Adsorption of dye with PNFs [28]. **b** Chelation between peptide and Zn^{2+} for removal of metal ions [29]

advanced materials with specific properties has been a subject of great interest. Among them, biomimetic synthesis of noble metal nanoparticles supported by peptide templates has shown that peptides could effectively improve the performance of designed materials for catalytic applications [32].

Noble metal catalysts have received attention of many researchers due to their wide application in the electrocatalysis and photocatalysis [33, 34]. Noble metal catalysts have excellent characteristic functions, which are usually affected by the size, shape, surface, structure, and composition of materials. The peptide template has the ability to control the size of nanoparticles, adjust their shape and morphology, and establish the structural dimension of catalysts. Therefore, peptide template is well suitable for the synthesis of noble metal catalysts, which can significantly improve the catalytic activity and stability of hybrid materials.

Nitrobenzene compounds are chemical pollutants. Although their solubility is low, but they can be still dissolved in water if not properly treated, causing serious water pollution [35]. The reduction of 4-nitrophenol (4-NP) to the less toxic 4-aminophenol was catalyzed by Au, Pd, and several Ag-based catalysts [36–38]. Bhandari et al. [36] prepared a nanocomposite material by applying biomimetic technology for the catalysis and other technical applications. This material was prepared by using a self-assembled peptide template, which produced non-spherical Pd nanostructure. According to the change of the Pd/peptide ratio, the morphologic changes of the Pd/peptide hybrids were induced, and their catalytic performance was verified through different morphologic changes. That is, the 4-NP reduction was achieved, thereby reducing its toxicity. In another study, Xu et al. [39] studied the peptide assembly-mediated synthesis of metallic nanoparticles. AgNPs and AuNPs were infused with sodium borohydride in the peptide solution. Through the analysis of the catalytic performance of AuNPs, AgNPs, and AgNPs/nanofibers, it was found that three materials showed good catalytic activity for reducing the toxicity of 4-NP to 4-aminophenol. Organic dyes, such as the methylene blue (MB), methyl orange (MO), and Rhodamine B (RhB), are important pollutants causing water pollution and are widely used in papermaking, textile, medicine, and other fields [40]. The catalytic activity of AuNPs, AgNPs, and AgNPs/nanofibers for the dye degradation was then systematically evaluated by associating with MO, MB, and RhB. AuNPs showed a high reaction rate constant for MB of up to 1899 h^{-1} and could degrade more than 90% of MB within 10 s, which showed a very high degradation efficiency. AgNPs revealed a higher rate constant for MO, while the AgNPs/nanofiber compounds exhibited a higher rate constant for 4-NP, MO, and RhB. The reported AuNPs, AgNPs, and AgNPs/nanofiber hybrids showed excellent selective catalytic activity for different organic dyes and 4-NP.

Hydrogels are a kind of hydrophilic three-dimensional (3D) network structure gels, which can swell rapidly in water and keep a large volume of water without dissolving [41]. Recently, Ahmad et al. [30] prepared a gelatin peptide hydrogel biocomposite composed of copper oxide nanomaterials (GL-CuO). Figure 11.4a and b shows the preparation process of the GL-CuO nanocomposite hydrogels that prepared by CuO nanomaterials and gelatin composites, respectively. The catalytic reduction of two azo dyes (MO and Congo red (CR)) with these catalysts was tested. The mechanism of catalytic reduction of two azo dyes is illustrated in Fig. 11.4c. Borohydride ions ($BH_4{}^{-1}$) produced by the dissolution of sodium borohydride were adsorbed onto the surface of the catalyst, while the azo dye molecules were adsorbed onto the surface of the catalyst. When the electrons were transferred to the catalyst, $BH_4{}^{-1}$ species which act as the source of hydrogen released hydrogen to attack the azo bond in the adsorbed dye molecules at the same time, thus completing the catalytic reduction. Compared with other catalysts, it was found that the GL-CuO catalyst was more active than other catalysts and was able to catalyze the reduction of both dyes quickly.

11.5 Peptide-Based Materials for Environmental Pollutant Detection

Electrochemical sensor has the advantages of good selectivity, significant sensitivity, simple operation, and low cost [42]. Using electrochemical methods to detect pollutants in water has practical significance for environmental monitoring and human health [43]. Peptide-based nanomaterials are one of the most important biomaterials, with a variety of structures and functions. Peptide-based chemical sensors possess good low biotoxicity and biocompatibility, and are widely used in environmental detection.

For example, Zhu et al. [44] designed graphene oxide (GO) as raw material to fabricate a novel fluorescent sensor to detect bisphenol A (BPA) with fluorescein amide (FAM)-labeled anti-bisphenol A aptamers (FAM-ssDNA). As shown in Fig. 11.5a, BPA interacts with the FAM-ssDNA and changes its conformation to prevent the GO adsorption, resulting in a corresponding change in the fluorescence value in the sensing system. This method was simple and novel and could detect BPA quickly. In addition, the sensor was highly selective and specific toward the BPA analogues. A lower limit of detection of 0.05 ng/mL was obtained in the range of 0.1–10 ng/mL. Moreover, high sensitivity was used to analyze and determine BPA in actual water samples.

Liu et al. [45] recently reported mesoporous molecularly imprinted nanoparticles (DhHP-6-MIP) for the rapid and sensitive detection of phenolic substances. The synthesis process of the DhHP-6-MIP is shown in Fig. 11.5b, in which the peptide mimic of DhHP-6 was immobilized on the surface of mesoporous silica during the synthesis of molecularly imprinted polymer (MIP). Studies have shown that compared with free peptides, imprinted peptide nanoparticles were less affected by the environment, showing higher stability, better reusability, and improved practicality. Phenolic compounds are highly toxic to the human body and do not degrade easily. At present, this method has realized the detection and identification of a variety of phenolic compounds and was a low-cost, high-precision phenolic detection method.

In the aqueous environment, in addition to organic pollutants, inorganic pollutants, such as toxic metals and radionuclides, are not degradable and harmful to the environment. Because they are highly toxic to humans and other organisms [46], the detection and removal of these heavy metals play an increasingly important role in sustaining life and improving the quality of life [47]. Electrochemical sensors for detecting metals are relatively simple and do not require complex instrumentation [48]. Peptide-based chemical sensors have been increasingly reported to identify

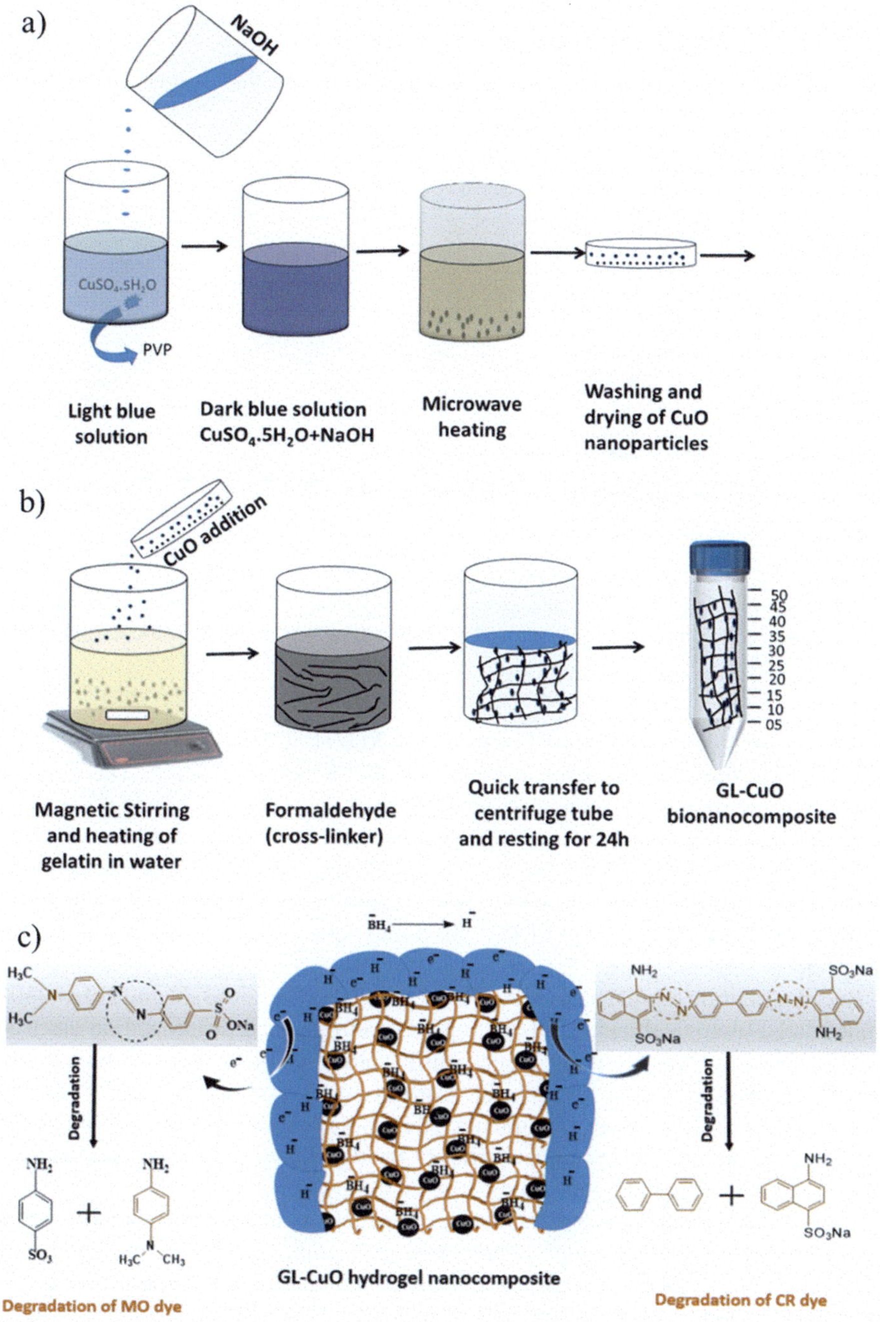

Fig. 11.4 **a** Preparation of peptide-based CuO nanomaterials. **b** Preparation of GL-CuO nanocomposite hydrogels. **c** Process strategy of reducing MO and CR with GL-CuO catalyst using sodium borohydride [30]

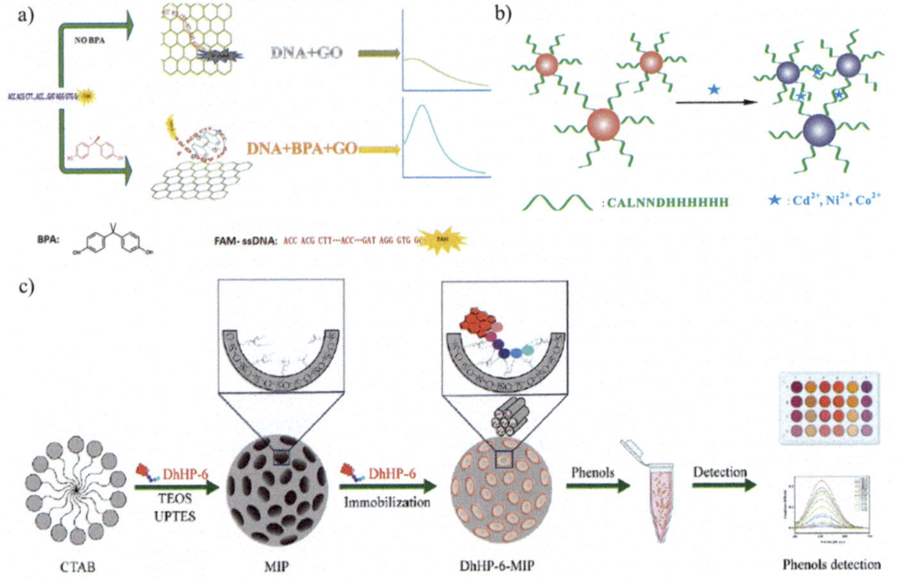

Fig. 11.5 **a** Peptide-based fluorescent sensor for detecting BPA [44]. **b** DhHP-6-MIP-based fluorescent sensor for detecting metal ions [45]. **c** Peptide-based sensors for colorimetric sensing Cd^{2+}, Ni^{2+}, or Co^{2+} [51]

metal ions in aqueous solutions. For instance, Xue et al. [49] synthesized a novel peptide-based fluorescent chemical sensor for the detection of Hg^{2+}. The fluorescence response time phenomenon showed that the sensor could detect Hg^{2+} quickly and stably, with a detection limit of 7.59 nM. The detection of Hg^{2+} by the sensor had high selectivity and excellent sensitivity in coexisting ions. Moreover, the peptide-based sensor exhibited the advantages of low cytotoxicity and good cell membrane permeability, and the detection of Hg^{2+} in living cells also had a good application prospect.

Because peptide biomaterials are able to conjugate metal ions, Viguier et al. [50] immobilized biological nanofibrils onto gold electrodes and developed an electrochemical biosensor for detecting copper ions in solution. The as-prepared biosensors showed a linear range up to 50 μM with a sensitivity of 0.68 $\mu A cm^{-2} \mu M^{-1}$. Moreover, the created sensors were manufactured that could be reused without significantly degrading the signal. These characteristics of the developed sensor showed that it is reliable in environmental applications.

Various metal ions often co-exist; hence, it is necessary to detect two or more metal ions in the meantime. For example, Zhang et al. [51] employed peptide-modified gold nanoparticles (P-AuNPs) as sensing elements, in which different responses of Cd^{2+}, Ni^{2+}, and Co^{2+} ions in aqueous solutions were detected and displayed simultaneously. As shown in Fig. 11.5c, functional peptide ligands (CALNNDHHHHHH) self-assembled on gold nanoparticles (AuNPs) to generate the P-AuNPs probes as the color indicators. Since the binding between metal ions and peptide ligands induced the aggregation of AuNPs, their color changed from red to blue or violet. The content of metal ions was also detected by the color change of the P-AuNPs probe. This new probe based on the P-AuNPs was very simple to prepare and operate, highly sensitive, and economically practical. The detection limits for the three ions were 0.05, 0.3, and 0.3 μM for Cd^{2+}, Ni^{2+}, and Co^{2+}, respectively. Therefore, the fabricated peptide-based colorimetric sensor exhibited potential application value in monitoring actual wastewater.

11.6 Conclusions

In this chapter, we summarized the applications of peptide-based nanomaterials in energy and environmental science. The advantageous features of self-assembled peptide nanomaterials with diverse architectures in combination with other functional materials offered good mechanical, electrochemical, thermal properties, and biocompatibility, making them unique and versatile materials for unconventional applications in the energy storage and conversion, adsorption and catalysis, as well as the sensing and detection of environmental pollutants. Future research on the energy and environmental applications of peptide-based nanomaterials is expected to focus on the engineering aspects of these assemblies required to achieve practical platforms, for example, the sustainability and circulatory stability of peptide-based nanomaterials under rigorous environments. Overall, albeit further exploration and utilization of peptide-based nanomaterials are still challenging, they present promising and achievable application potential in energy and environmental science.

References

1. L. Adler-Abramovich, D. Aronov, E. Gazit, G. Rosenman, Patterned arrays of ordered peptide nanostructures. J. Nanosci. Nanotech. **9**, 1701–1708 (2009)
2. J. Zhu, D. Yang, Z. Yin, Q. Yan, H. Zhang, Graphene and graphene-based materials for energy storage applications. Small **10**, 3480–3498 (2014)
3. T. Guan, J. Li, C. Chen, Y. Liu, Self-assembling peptide-based hydrogels for wound tissue repair. Adv. Sci. **9**, 2104165 (2022)
4. C. Yuan, W. Ji, R. Xing, J. Li, E. Gazit, X. Yan, Hierarchically oriented organization in supramolecular peptide crystals. Nat. Rev. Chem. **3**, 567–588 (2019)
5. W. Ji, Y. Tang, P. Makam, Y. Yao, R. Jiao, K. Cai, G. Wei, E. Gazit, Expanding the structural diversity and functional scope of diphenylalanine-based peptide architectures by hierarchical coassembly. J. Am. Chem. Soc. **143**, 17633–17645 (2021)
6. J.-H. Lee, J.H. Lee, Y.J. Lee, K.T. Nam, Protein/peptide based nanomaterials for energy application, *Cur.t Opin. biotech.* **24**, 599–605 (2013)
7. X. Chen, K. Gerasopoulos, J. Guo, A. Brown, C. Wang, R. Ghodssi, J.N. Culver, Virus-enabled silicon anode for lithium-ion batteries. ACS Nano **4**, 5366–5372 (2010)
8. P. Beker, G. Rosenman, Bioinspired nanostructural peptide materials for supercapacitor electrodes. J. Mater. Res. **25**, 1661–1666 (2010)

9. X. Gao, H. Matsui, Peptide-based nanotubes and their applications in bionanotechnology. Adv. Mater. **17**, 2037–2050 (2005)
10. I.W. Hamley, Peptide fibrillization. Angew. Chem. Int. Ed. **46**, 8128–8147 (2007)
11. J. Ryu, S.-W. Kim, K. Kang, C.B. Park, Synthesis of diphenylalanine/cobalt oxide hybrid nanowires and their application to energy storage. ACS Nano **4**, 159–164 (2010)
12. J. Ryu, S.W. Kim, K. Kang, C.B. Park, Mineralization of self-assembled peptide nanofibers for rechargeable lithium ion batteries. Adv. Mater. **22**, 5537–5541 (2010)
13. S.-Y. Chung, J.T. Bloking, Y.-M. Chiang, Electronically conductive phospho-olivines as lithium storage electrodes. Nat. Mater. **1**, 123–128 (2002)
14. V. Palomares, P. Serras, I. Villaluenga, K.B. Hueso, J. Carretero-González, T. Rojo, Na-ion batteries, recent advances and present challenges to become low cost energy storage systems. Energ. Environ. Sci. **5**, 5884–5901 (2012)
15. M. Moradi, Z. Li, J. Qi, W. Xing, K. Xiang, Y.-M. Chiang, A.M. Belcher, Improving the capacity of sodium ion battery using a virus-templated nanostructured composite cathode. Nano Lett. **15**, 2917–2921 (2015)
16. T. Li, X.-M. Lu, M.-R. Zhang, K. Hu, Z. Li, Peptide-based nanomaterials: self-assembly, properties and applications. Bioact. Mater. **11**, 268–282 (2022)
17. P. Beker, I. Koren, N. Amdursky, E. Gazit, G. Rosenman, Bioinspired peptide nanotubes as supercapacitor electrodes. J. Mater. Sci. **45**, 6374–6378 (2010)
18. K. Hu, Y. Jiang, W. Xiong, H. Li, P.-Y. Zhang, F. Yin, Q. Zhang, H. Geng, F. Jiang, Z. Li, Tuning peptide self-assembly by an in-tether chiral center, *Sci. Adv.* **4**, eaar5907 (2018)
19. Y. Chen, K. Tao, W. Ji, V.B. Kumar, S. Rencus-Lazar, E. Gazit, Histidine as a key modulator of molecular self-assembly: peptide-based supramolecular materials inspired by biological systems. Mater. Today **60**, 106–127 (2022)
20. K. Tao, Z. Fan, L.M. Sun, P. Makam, Z. Tian, M. Ruegsegger, S. Shaham-Niv, D. Hansford, R. Aizen, Z. Pan, S. Galster, J.J. Ma, F. Yuan, M.S. Si, S.N. Qu, M.J. Zhang, E. Gazit, J.B. Li, Quantum confined peptide assemblies with tunable visible to near-infrared spectral range. Nat. Commun. **9**, 3217 (2018)
21. S. Bera, S. Guerin, H. Yuan, J. O'Donnell, N.P. Reynolds, O. Maraba, W. Ji, L.J. Shimon, P.-A. Cazade, S.A.J. N.C. Tofail, Molecular engineering of piezoelectricity in collagen-mimicking peptide assemblies. *Nat. Commun.* **12**, 2634 (2021)
22. N. Vu, S. Kelly, R. Yang, Piezoelectric peptide-based nanogenerator enhanced by single-electrode triboelectric nanogenerator. APL Mater. **5**, 074108 (2017)
23. H. Wang, Z. Feng, B. Xu, Assemblies of peptides in a complex environment and their applications. Angew. Chem. Int. Ed. **58**, 10423–10432 (2019)
24. Y. Wang, W. Zhang, C. Gong, B. Liu, Y. Li, L. Wang, Z. Su, G.J.S.M. Wei, Recent advances in the fabrication, functionalization, and bioapplications of peptide hydrogels. Soft Matter **16**, 10029–10045 (2020)
25. R.V. Ulijn, A.M.J.C.S.R. Smith, Designing peptide based nanomaterials. Chem. Soc. Rev. **37**, 664–675 (2008)
26. S. Sumiyoshi, K. Suyama, D. Tatsubo, N. Tanaka, K. Tomohara, S. Taniguchi, I. Maeda, T. Nose, Metal ion scavenging activity of elastin-like peptide analogues containing a cadmium ion binding sequence. Sci. Rep. **2022**, 12 (1861)
27. V. Krstić, in *Water Pollution and Remediation: Heavy Metals* (Springer, 2020), pp. 313–440
28. S.K. Brar, N. Wangoo, R.K. Sharma, Enhanced and selective adsorption of cationic dyes using novel biocompatible self-assembled peptide fibrils. J. Environ. Manage **255**, 109804 (2020)
29. C. Martí-Gastaldo, J.E. Warren, K.C. Stylianou, N.L. Flack, M. Rosseinsky, Enhanced stability in rigid peptide-based porous materials. Angew. Chem. Int. Ed. **51**, 11044–11048 (2012)
30. S. Ahmad, S.B. Khan, A.M. Asiri, H.M. Marwani, T. Kamal, Polypeptide and copper oxide nanocomposite hydrogel for toxicity elimination of wastewater. J. Sol-Gel Sci. Technol. **96**, 382–394 (2020)
31. R.B. Pandey, H. Heinz, J. Feng, B.L. Farmer, J.M. Slocik, L.F. Drummy, R.R. Naik, Adsorption of peptides (A3, Flg, Pd^2, Pd^4) on gold and palladium surfaces by a coarse-grained Monte Carlo simulation. Phys. Chem. Chem. Phys. **11**, 1989–2001 (2009)

32. T. Jiang, C.F. Xu, Y. Liu, Z. Liu, J.S. Wall, X.B. Zuo, T.Q. Lian, K. Salaita, C.Y. Ni, D. Pochan, V.P. Conticello, Structurally defined nanoscale sheets from self-assembly of collagen-mimetic peptides. J. Am. Chem. Soc. **136**, 4300–4308 (2014)
33. H.T. Chung, C.M. Johnston, K. Artyushkova, M. Ferrandon, D.J. Myers, P. Zelenay, Cyanamide-derived non-precious metal catalyst for oxygen reduction. Electrochem. Commun. **12**, 1792–1795 (2010)
34. W.B. Li, J.X. Wang, H. Gong, Catalytic combustion of VOCs on non-noble metal catalysts. Catal. Today **148**, 81–87 (2009)
35. M. Zhang, L. Ding, J. Zheng, L.B. Liu, H. Alsulami, M.A. Kutbi, J.L. Xu, Surface modification of carbon fibers with hydrophilic Fe_3O_4 nanoparticles for nickel-based multifunctional composites. *Appl. Surf. Sci.* **509** (2020)
36. R. Bhandari, M.R. Knecht, Effects of the material structure on the catalytic activity of peptide-templated Pd nanomaterials. ACS Catal. **1**, 89–98 (2011)
37. R. Bhandari, M.R. Knecht, Synthesis, characterization, and catalytic application of networked Au nanostructures fabricated using peptide templates. Catal. Sci. Technol. **2**, 1360–1366 (2012)
38. B.D. Briggs, Y. Li, M.T. Swihart, M.R. Knecht, Reductant and sequence effects on the morphology and catalytic activity of peptide-capped Au nanoparticles. ACS Appl. Mater. Interfaces **7**, 8843–8851 (2015)
39. W. Xu, Y. Hong, Y. Hu, J. Hao, A. Song, Ultrafine Au and Ag nanoparticles synthesized from self-assembled peptide fibers and their excellent catalytic activity. ChemPhysChem **17**, 2157–2163 (2016)
40. M.O. Olagunju, X. Poole, P. Blackwelder, M.P. Thomas, B.S. Guiton, D. Shukla, J.L. Cohn, B. Surnar, S. Dhar, E.M. Zahran, L.G. Bachas, M.R. Knecht, Size-controlled $SrTiO_3$ nanoparticles photodecorated with Pd cocatalysts for photocatalytic organic dye degradation. ACS Appl. Nano Mater. **3**, 4904–4912 (2020)
41. J.A. Burdick, W.L. Murphy, Moving from static to dynamic complexity in hydrogel design. Nat. Commun. **3**, 1269 (2012)
42. S. Wu, K. Li, X. Dai, Z. Zhang, F. Ding, S. Li, An ultrasensitive electrochemical platform based on imprinted chitosan/gold nanoparticles/graphene nanocomposite for sensing cadmium (II) ions. Microchem. J. **155**, 104710 (2020)
43. C. Gong, S. Sun, Y. Zhang, L. Sun, Z. Su, A. Wu, G. Wei, Hierarchical nanomaterials via biomolecular self-assembly and bioinspiration for energy and environmental applications. Nanoscale **11**, 4147–4182 (2019)
44. Y. Zhu, Y. Cai, L. Xu, L. Zheng, L. Wang, B. Qi, C. Xu, Building An aptamer/graphene oxide FRET biosensor for one-step detection of bisphenol A. ACS Appl. Mater. Interfaces **7**, 7492–7496 (2015)
45. X. Liu, X. Zhang, Z. Fu, X. Cheng, L. Wang, L. Hu, Mesoporous molecularly imprinted nanoparticles with peptide mimics for the detection of phenolic compounds. Anal. Chim. Acta **1250**, 340970 (2023)
46. N. Khandelwal, G.K. Darbha, A decade of exploring MXenes as aquatic cleaners: covering a broad range of contaminants, current challenges and future trends. Chemosphere **279**, 130587 (2021)
47. A. Dhillon, N. Singh, M. Nair, D. Kumar, Analytical methods to determine and sense heavy metal pollutants using MXene and MXene-based composites: mechanistic prophecy into sensing properties. Chemosphere **303**, 135166 (2022)
48. J. Castillo, S. Gaspar, S. Leth, M. Niculescu, A. Mortari, I. Bontidean, V. Soukharev, S.A. Dorneanu, A.D. Ryabov, E. Csoregi, Biosensors for life quality—design, development and applications. Sens. Actua. B-Chem. **102**, 179–194 (2004)
49. S. Xue, P. Wang, K. Chen, A novel fluorescent chemosensor for detection of mercury(II) ions based on dansyl-peptide and its application in real water samples and living LNcap cells. Spectrochim. Acta A **226**, 117616 (2020)
50. B. Viguier, K. Zor, E. Kasotakis, A. Mitraki, C.H. Clausen, W.E. Svendsen, J. Castillo-Leon, Development of an electrochemical meta-ion biosensor using self-assembled peptide nanofibrils. ACS Appl. Mater. Interfaces **3**, 1594–1600 (2011)

51. M. Zhang, Y.-Q. Liu, B.-C. Ye, Colorimetric assay for parallel detection of Cd^{2+}, Ni^{2+} and Co^{2+} using peptide-modified gold nanoparticles. Analyst **137**, 601–607 (2012)

Chapter 12
Preparation and Applications of Antimicrobial Peptides

Han Wu, Bing Yu, and Hailin Cong

Abstract In the past few decades, antibiotics have significantly improved patient health and enhanced the effectiveness of various treatments, including cancer therapy and organ transplants. However, the overuse of antibiotics has led to the emergence of drug-resistant bacteria, resulting in a decline in the effectiveness of antibiotics, creating an urgent need for alternative antibacterial agents. Antimicrobial peptides (AMPs) have emerged as a promising solution, with nearly 5000 natural AMPs identified from animals and plants. These peptides typically consist of 5–15 amino acids, exhibiting broad-spectrum antimicrobial activity and low toxicity, making them suitable candidates for therapeutic applications. This chapter primarily introduces the research and application progress of antimicrobial peptides from the aspects of their origin, classification and mechanism of action.

Keywords Antibacterial peptide · Antimicrobial peptides · Cancer therapy · Drug delivery · Mechanisms

12.1 Introduction

Antibiotics are medicinal compounds derived primarily from fungi that have biocidal or biostatic activity against pathogenic microorganisms by inhibiting vital life processes in their cells. They have greatly improved patient health and enhanced the effectiveness of other treatments including cancer therapy, organ transplants, open-heart surgery, etc. Antibiotics implement their biocidal activity through various

H. Wu · B. Yu (✉) · H. Cong (✉)
College of Chemistry and Chemical Engineering, College of Materials Science and Engineering, Institute of Biomedical Materials and Engineering, Qingdao University, Qingdao 266071, China
e-mail: yubing198@qdu.edu.cn

H. Cong
e-mail: conghailin@sdut.edu.cn

H. Wu · H. Cong
School of Materials Science and Engineering, Shandong University of Technology, Zibo 255000, China

G. Wei (ed.), *Peptide Nano-Chemistry and Nanotechnology*,
https://doi.org/10.1007/978-981-96-2387-7_12

pathways; some of these methods include inhibition of cell wall, protein or nucleic acid synthesis, depolarisation of the cell membrane, synthesis of transmembrane pores and obstruction of metabolic pathways through inhibition of metabolites [1]. Prominent antibiotics include Penicillin, the first discovered antibiotic and Clotrimazole, an antifungal drug used to treat Candida albicans. Penicillin kills bacteria by specifically inhibiting the transpeptidase that catalyzes the final step in cell wall biosynthesis, the cross-linking of peptidoglycan Clotrimazole, on the other hand, inhibits the biosynthesis of provitamin D2 in a concentration-dependent manner by inhibiting the lanosterol 14-α-demethylase [2]. In just over one century since being discovered by Alexander Fleming in 1928, antibiotics have drastically changed allopathic medicine and have extended the average human lifespan worldwide by at least 23 years [3].

However, their popularity has seen a decline in recent decades, largely due to their reliability having been cast into doubt by the rise of antibiotic resistance in pathogens. Resistant infections occur when pathogens evolve immunity to the effects of these drugs on their metabolism; they are challenging, and sometimes impossible for healthcare providers to treat effectively [4]. The causative factor of antibiotic resistance in microorganisms is the misuse or overuse of antibiotics, such as being prescribed penicillin for a mere case of the common cold. This has caused the rise of Multidrug-Resistant (MDR) bacteria, whose infections are of much higher severity due to a lack of effective medications. The rising prominence of drug-resistant pathogen populations as well as a decline in the development of new antibiotics has caused medical researchers to consider alternative methods of medication to counter pathogen infection, and there is several alternatives to antibiotics which has been shown in Fig. 12.1 [5].

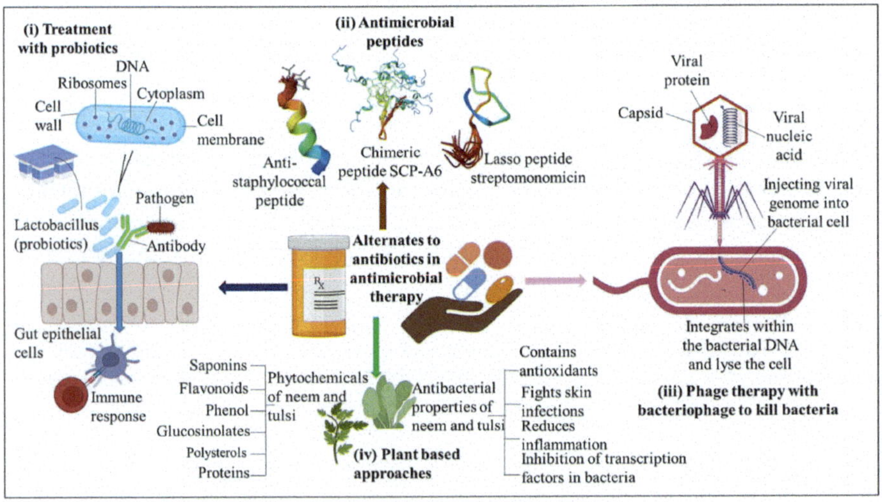

Fig. 12.1 Four main alternatives to antibiotics as mainstream medicine: (i) Probiotics; (ii) Antimicrobial peptides; (iii) Phage therapy; (iv) Traditional medicine [5]

In the recent years, antimicrobial peptides (AMPs) have attracted widespread attention from researchers both domestically and abroad as a novel class of antimicrobial agents. AMPs, also known as peptide antibiotics, are small-molecule short peptides with broad-spectrum antimicrobial activity, serving as essential molecular barriers in the host's natural immune defense against exogenous pathogens. Unlike traditional antibiotics, AMPs employ distinct bactericidal mechanisms that reduce the likelihood of pathogen resistance and cross-resistance [6]. They also exhibit multiple activities such as antifungal, antiviral, modulation of the host immune system, promotion of wound healing and anticancer effects, providing new avenues for addressing the challenge of pathogen resistance to conventional antibiotics. AMPs' specificity against pathogens and low toxicity or minimal toxicity to hosts have made them a research hotspot among scholars worldwide. Moreover, AMPs possess excellent stability under various conditions, including pH, heat and low-temperature storage, promising significant development and application prospects in medicine, food processing, animal feed and other fields [7].

As AMPs have excellent biocompatibility and a broad antimicrobial spectrum, they can replace antibiotics for clinical applications. Therefore, they have always been a hot spot in scientific research. Although antibiotics have their advantages, due to the continuous enhancement of the bacteria's drug resistance and the continuous abuse of antibiotics, drug-resistant bacteria continue to appear in nature. Therefore, finding an alternative new antibacterial drug is imminent. AMP has proven to be such an alternative [8, 9]. So far, scientists have discovered nearly 5000 natural AMPs from both animals and plants in nature. Among these AMPs, short peptides contain 5–15 amino acids and long peptides also contain hundreds of amino acids. Since about 50% of their amino acids contain the arginine and lysine residues, they are usually positively charged and resistant to different microorganisms such as gram bacteria [10, 11].

In addition to being the most promising alternative to antibiotics in the pharmaceutical field, AMPs have been also widely used in the food and feed industries. AMPs currently approved as food additives are Nisin [12] and ε-polylysine [13]. The addition of a small number of AMPs to the animal feed can achieve the effect of enhanced animal immunity and improved feed utilization. It can also reduce the use of antibiotics and reduce a series of problems that caused by the eating livestock and poultry products contaminated by antibiotics [14]. The great advantages of AMPs make them a research hotspot in the fields of food, biology, medicine and others [15, 16]. Since the Swedish scientist Boman discovered the first AMP cecropin from the celestial silkworm in the 1980s, more and more AMPs have been discovered [17]. Screening AMPs from materials with a wide range of sources and complex ingredients are the prerequisite and key to its research and wide application. Traditional screening methods are mostly guided by the in vitro antibacterial activity and follow steps such as the extraction, separation, purification and structural identification. The screening of antibacterial active components is often time-consuming, laborious and ineffective [18, 19]. In the recent years, rapid screening methods such as the cell membrane chromatography (CMC) have shown the advantages of high efficiency and strong targeting, which greatly improved the screening efficiency of AMPs [20].

Liu et al. successfully screened antimicrobial peptides with antimicrobial effects from a library containing 32 peptide chains using membrane chromatography. This method significantly reduced the screening time while ensuring accuracy [21]. To better understand antimicrobial peptides, this paper will outline their origins, classifications, mechanisms of action and current applications in antimicrobial peptide therapy.

12.2 Discovery of AMPs

In the early 1885, body fluids like saliva, plasma, sweat, granule extracts, blood and white blood cell secretions were known to have surprising antimicrobial properties [22]. Until 1939, Dubos discovered an AMP. He extracted AMP, named gramicidin, from Bacillus brevis and demonstrated protective effects of the extract on mice, suffering from pneumococci infection [23]. Later on, Hotchkiss and Dubos fractioned gramicidin extract and found that it was effective for the topical treatment of ulcers and wounds by successful trialing on guinea pig skin and become first commercialized AMP to be used as an antibiotic [24]. In 1939, an antimicrobial protein named lactoferrin was separated from bovine milk by Sorensen. After that in 1941, tyrocidine (AMP) with broad-spectrum activity against microorganisms was discovered but it showed cytotoxic effects in human blood cells. Another AMP, purothionin, was discovered in same year from *Triticum aestivum* (common wheat plant) which exhibited antimicrobial activity against some pathogenic bacteria and fungi. Afterward, first AMP of animal source was reported from rabbit leukocytes named defensins [23].

In 1962, bombinin (AMP) was discovered from orange speckled frog (*Bombina variegate*) [24]. Subsequently in 1981, a scientist named Hans Boman discovered an AMP from hemolymph of *Hyalophora cecropia* (silk moth) called cecropins that showed strong antimicrobial activity against bacteria and fungi [22]. It is also known to be the first reported α-helical AMPs. In 1987, Michael Zasloff isolated magainins from the skin of African clawed frog (*Xenopus laevis*) [25]. The discovery of magainins explores further knowledge in the study of AMPs that are not only present in lower invertebrates even so present in higher invertebrates. Scientists discovered more than 1400 AMPs from other invertebrates, birds, fishes, plants, bacteria, amphibians and mammals [26]. As reported in online AMPs database, a total of 2980 AMPs have been identified from different species such as bacteria, plants, animals and fungi. The natural AMPs not only act against microorganisms but are also able to perform other functions like wound healing, apoptosis and immunomodulation. However, over or under expression of these AMPs in human could lead to certain human diseases so, a balanced expression of AMPs is necessary [26].

12.3 Types of AMPs

AMPs are evolutionary conserved in the genome and produced by all life forms, from prokaryotes to humans [27]. In higher organisms, AMPs constitute important components of the innate immunity, protecting the host against infections. In contrast, bacteria produce AMPs in order to kill other bacteria competing for the same ecological niche [28]. Many AMPs exhibit an extraordinarily broad range of antimicrobial activity covering both Gram-positive and Gram-negative bacteria as well as fungi, viruses and unicellular protozoa. Besides having a direct antimicrobial activity, several AMPs display ability to modulate the innate immune responses of the host and thereby indirectly promote pathogen clearance [29]. The widespread distribution and abundance of AMPs in all multicellular organisms underscores their critical role in innate immunity [30]. Their importance is further demonstrated by the increased infection susceptibility of mice genetically modified to lack the gene encoding for the mouse analog of the human AMP LL-37 and of humans with diseases associated with reduced AMP production such as atopic dermatitis [31].

AMPs in nature are produced either by ribosomal translation of mRNA or by non-ribosomal peptide synthesis. While non-ribosomally synthesized peptides are mainly produced by bacteria, the ribosomally synthesized AMPs are genetically encoded and produced by all species of life, bacteria included. Compared to peptides of non-ribosomal origin that have been known for several decades and whereof many are used as antibiotics, the ribosomally synthesized AMPs have more recently been recognized for their critical role in innate immunity and for their therapeutic potential [32].

In mammals, AMPs are found primarily within granules of neutrophils and in secretions from epithelial cells covering skin and mucosal surfaces [33]. In many cases, AMPs are encoded in clusters in the genome and co-expressed, resulting in multiple AMPs accumulating at a single site [34]. Notably, many AMPs are produced as inactive precursors requiring proteolytic cleavage to become active [35]. Their regulation is therefore not only dependent on their own expression but also on the abundance of appropriate proteases [34]. In multicellular organisms, some AMPs are constitutively expressed, stored at high concentrations as inactive precursors in granules and released locally at infection and inflammation sites, whereas the expression of others is induced in response to pathogen-associated molecular patterns (PAMPs) or cytokines [36].

Several databases exist for natural AMPs, today covering more than 2000 peptides [37]. Most AMPs are relatively short, commonly consisting of 10–50 amino acids, display an overall positive charge ranging from + 2 to + 11, and contain a substantial proportion (typically 50%) of hydrophobic residues [38]. Structure of AMPs is foremost important for their antimicrobial activity. A slight mutation in their structure can greatly modify their activity. Most of the AMPs are cationic (have positive charge) while few are anionic (have negative charge) in nature. Cationic peptides consist of hydrophobic sequences at one side while hydrophilic sequences on the opposite side [39]. Highly cationic peptides or anionic peptides complexed with zinc

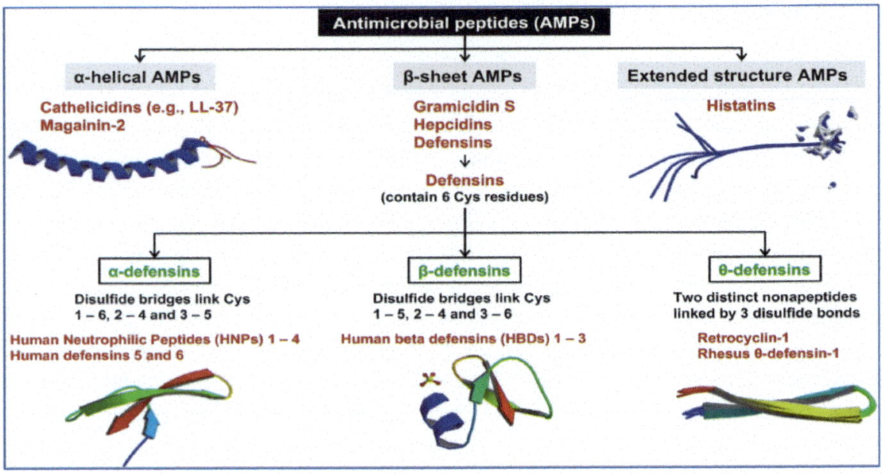

Fig. 12.2 Types of AMPs based on their structure [6]

are more active as compared to neutral peptides or peptides having low charge [40]. As AMPs are diverse in nature but based on their secondary structure (Fig. 12.2), they are classified as; α-helix, β-sheet, hairpin or loop, Extended peptides [41], in addition, there are structures such as Beta turn, 3_10 helix, and random coil, but the main structures are α-helix, β-sheet, β-turn, and random coil.

12.3.1 Alpha Helical (α-helical) Peptides

The largest and most widely studied AMPs are α-helical peptides [42]. About 250 peptides belong to α-helical peptide category isolated from vertebrates, plants and invertebrates [39]. Moreover, amphipathic α-helical AMPs are more active as compared to other secondary structures which are less defined [40].

In the α-helix conformation, the distance between two adjacently attached amino acids is about 0.15 nm distance forming a cationic amphipathic helix, and from top view, the angle among them from the center is about 100° forming a hinge or kink in the middle [43]. Cationic peptides are rich in histidine, lysine, proline, tryptophan and arginine. Magainin is one of the most commonly studied antimicrobial cationic alpha helical peptide. Anionic peptides are rich in glutamic acid and aspartic acid. Examples of α-helical peptides include cyclic indolicin, alamethicin, gramicidin, cecropin and protegrin. In an aqueous solution, these peptides do not have any structural configuration but upon interaction with microbial membrane AMPs fold themselves into α-helical structure. Most of the α-helical peptides that kill microorganisms by pore formation are bacteriostatic in nature [39].

12.3.2 *β-sheets Peptides*

In the β-sheets conformation, at least 2 β-strands attach to each other with the help of disulfide bonds forming a dimer. AMPs adopt cyclic conformation with conserved pattern of 6 cysteine residues constituting 3 disulfide bonds. The arrangement of the cysteine residues in β-helical peptide is so crucial that replacement with other amino acids such as aspartic acid and alanine leads to inactivation of these peptides, while in some studies, cysteine replacement with analogs such as tyrosine, phenylalanine, methionine and valine maintains its antimicrobial activity [44]. Moreover, cyclic conformation is requisite for their antimicrobial activity. In an aqueous solution, AMPs continue to maintain β-sheet configuration and upon interaction with microbial membrane, stabilize themselves. AMPs display antibacterial but, in some cases, also have antifungal activity, e.g., arenicin, gramicidine S, human β-defensin-2, tyrocidine, tachyplesins, polymyxin B, lactoferricin and protegrin [42, 45]. Wu et al. designed an antimicrobial peptide chain LKARI. CD spectroscopy revealed that this peptide chain exhibits a β-folding conformation. The peptide chain demonstrates excellent biocompatibility, stability in various environments, and excellent antimicrobial resistance against 12 generations of bacteria. It has shown remarkable therapeutic effects in both in vivo sepsis and in vitro wound healing treatments [46].

12.3.3 *Loop/Hair Pin Peptides*

These AMPs are composed of three strands in which two strands consists of β-hairpin structure or it may be that these consist of both α-helix and β-sheets. The peptides have rich amount of proline and arginine, and due to overexpression of proline residues and despite forming amphipathic structure, they undergo formation of hairpin or loop structure. In the loop or hairpin peptides conformation, loop structure stabilized by 1–4 disulfide bonds. Stabilization with disulfide bonds facilitates proteolytic resistance. Examples include thanatin, protegrins, androctonin, gomesin, arenicin-3, tigerinin, bactenecin [47, 48].

12.3.4 *Extended Peptides*

Extended peptides are linear peptides; therefore, these cannot pleat into typical secondary structure and contain high ratio of amino acids like histidine, tryptophan, arginine and proline. The peptides penetrate in the cell membrane of the organism and exert their antimicrobial activities through interaction with different intracellular proteins such as the heat shock proteins such as GroEL to inhibit chaperon-assisted protein folding and DnaK to suppress DnaK ATPase activity [49]. In an aqueous solution, these peptides are highly flexible. There are approximately 90 peptides belong to this category [39]. Examples include metchnikowin, abaecin, lebocin, drosocin and pyrrhocoricin [43].

12.4 Mode of Action of AMPs

According to mechanism of action, AMPs are categorized into 2 main categories: membrane disruptive AMPs and non-membrane disruptive AMPs. Membrane disruptive AMPs interact with the cell membrane of the target organism leading to deliquescence of the membrane or by producing lesions on membrane or permeabilize phospholipid bilayer which leads to cell death. Non-membrane disruptive AMPs penetrate the cell membrane and cause intracellular disruption of universal cellular processes where the targets are cellular organelles, proteins or nucleic acid [50, 51].

However, there are several variables on which mode of action of AMPs depends, such as peptide: lipid ratio of cell membrane, peptide structure, lipid membrane properties, affinity for negatively charged phospholipids, amphipathic conformation of AMPs and amphiphilic nature[39]. It also depends upon cell membrane topology, structure and aggregation. Each AMP can interact through several modes of actions [52]. Figure 12.3 describes the general mechanisms adopted by AMPs against microbial infections [53].

12.4.1 Membrane Disruptive Peptides

Membrane disruptive AMPs target bacterial cell membrane and cause lysis of the cell. Permeability of AMPs varies in Gram-positive or Gram-negative bacteria due to differences in their cell wall structure. In case of Gram-positive bacteria, there is 40–60 nm thick peptidoglycan layer containing negatively charged teichoic acid which surrounds the cytoplasmic membrane and very small nano-sized pores are present in the matrix of peptidoglycan layer through which AMPs diffuses into the membrane. In case of Gram-negative bacterial cell, not only peptidoglycan but another outer membrane lipopolysaccharide is also present which represent a barrier for the AMPs to target the inner cytoplasmic membrane. The inner membrane of Gram-negative bacteria mainly consists of phospholipids, while in the outer membrane, lipopolysaccharide is present and it is rich in negatively charged phosphate groups which interact with divalent cations, such as Mg^{2+} and Ca^{2+}, leading to the formation of electrostatic network. These electrostatic regions serve as barriers to molecules with hydrophobic nature including antibiotics, resulting in low permeabilities. The killing of Gram-negative bacteria involves penetration and disruption of both outer and inner membranes. Therefore, without penetrating the outer membrane, the molecules cannot exert its antimicrobial properties. So, AMPs can easily penetrate and disrupt the cell membrane of Gram-positive bacterial cells as compared to Gram-negative

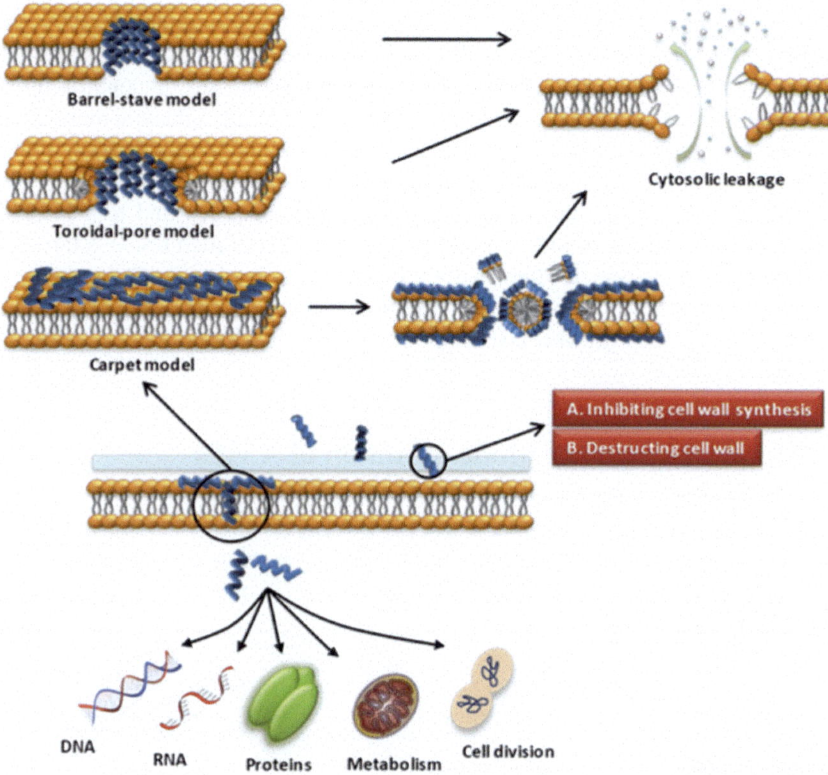

Fig. 12.3 Schematic presentation of the antibacterial mechanism of AMPs. Some AMPs act on cell membranes through different modes of action, increasing membrane permeability, leading to leakage of cell contents and cell death. Some AMPs act on the cell wall and exert antibacterial effects by affecting the synthesis of cell wall components and destroying the cell wall structure. Some AMPs enter the cell through direct penetration or endocytosis and exert antimicrobial effects by targeting the nucleus, organelles, present in fungi or intracellular proteins [53]

bacteria [52]. Gao et al. synthesized an antimicrobial peptide chain using a solid-phase synthesis method. SEM and TEM images clearly show that the antimicrobial peptide acts on bacterial cell membranes, causing membrane rupture [54]. Guo et al. also designed an antimicrobial peptide chain, SALSP. It was observed that this peptide caused noticeable holes on the surface of *Escherichia coli* and Staphylococcus aureus, leading to cell membrane rupture and bacterial death [55].

12.4.2 Non-membrane Disruptive AMPs

Most of the AMPs are non-lytic, generally, proline-rich AMPs invade microbial cells by disrupting membrane to a lesser extent and target intracellular organelles or processes. Buforin II, e.g., diffuses through cell membrane and targets RNA and DNA. Polyphemusin invades *E. coli* without damaging cell membrane. Some AMPs synchronically targets cell membrane and intracellular processes, depending upon enforced concentration of AMPs, such as indolicidin disrupt cytoplasmic membrane and inhibit synthesis of DNA [22, 56]. The exact mode of action of non-lytic AMPs is unclear yet but it is proposed that AMPs attach on the surface of cell membrane on account of their amphipathic structure, form transient pores and translocated the membrane. In case of proline-rich AMPs, translocation of membrane is through SbmA or MdtM transporters. Proline-rich AMPs are less active against Gram-positive bacteria but more active against Gram-negative bacteria because Gram-positive bacteria lack SbmA and MdtM transporters [56, 57].

12.5 Models Explaining Mode of Action of AMPs

Several models have been proposed which depicts the mechanism of action of AMPs, i.e., torroidal pore or wormhole model, barrel-stave model or helical bundle model, carpet model, molecular electroporation model, sinking raft model and aggregate model [58–60] (Fig. 12.4) [5].

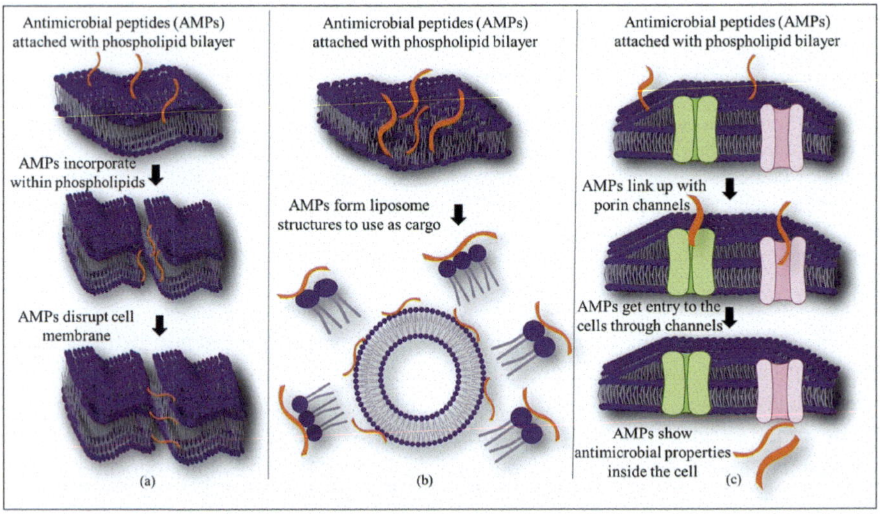

Fig. 12.4 The three main mechanisms through which antimicrobial peptides (AMPs) destroy cell membranes: **a** Wormhole model-antimicrobial peptides (AMPs) penetrate and perforate the bilayer by altering their conformation in large aggregations; **b** Carpet model-antimicrobial peptides (AMPs) with hydrophobic amino acid sequences bind to hydrophilic heads of the outer region of the membrane and form micelles, thereby eroding the membrane; **c** Barrel-Stave model-antimicrobial peptides (AMPs) cluster together and create a barrel-shaped channel protein that causes leakage of the cytosol into the external environment [5]

12.5.1 Torroidal Pore Model

Torroidal pore or wormhole model is the most defined membrane-peptide interaction to explain the mode of action of magainin-2 isolated from *Xenopus laevis*. Other peptides which are known to act by this mechanism include protegrins, mellitin and PGLa. In this model, AMPs align themselves parallel to cell membrane, attached in functionally inactive form and inserted into the membrane. When peptide-lipid threshold exceeded (estimated to be 1:30 for magainin), peptides are transformed into functional state and induce to bend the lipid monolayer continuously through pore in order to drawn away water core by both lipid head groups and inserted peptides results the formation of irreversible rift in hydrophobic regions of cell membrane. These changes build dynamic supramolecular peptide-lipid complex and therefore this model is also known as supramolecular complex model. The transformation of peptide from inactive to active form depends upon phospholipid composition of lipid bilayer and concentration of AMPs [40, 61, 62].

12.5.2 Barrel-Stave Model

Amphipathic α-helical AMPs are attached to the target cell membrane forming a bundle and inserted into the membrane. Upon binding with membrane, the peptides may be present in monomer or linear form, first undergo conformational changes in structure then align themselves in a manner that hydrophobic regions of the peptide face outward toward the acyl chains of lipid bilayer of the membrane and hydrophilic regions remain interior and form lining of the pore. Barrel-stave model is named due to the formation of "barrel like" rings in membrane channels by the activity of AMPs and in this barrel each transmembrane spoke is referred to the term "stave". The transmembrane spokes are formed either by one peptide or by peptide complexes and cause membrane depolarization and ultimately cell death [63, 64]. This model is also known as helical bundle model and it explains the mode of action of alamethicin. However, alamethicin produces smaller and less variable pore size as compared to magainin induced torroidal pores [40].

12.5.3 Carpet Model

This model was suggested for the first time to explain antimicrobial activity of dermaseptin S. Afterward, it was also proposed for antimicrobial activity of other peptides [65]. In this model, cationic AMPs arrange and accumulate on the surface of the cell membrane. Due to electrostatic interaction between negatively charged phospholipids of membrane and positively charged peptides, these peptides cover the external layer of cell membrane without penetrating, forming the dense carpet that is why it is named as carpet model. When threshold limit is attained, peptides rotate and realigned themselves toward hydrophobic region of the cell membrane, cause disintegration of membrane and ultimately lysis of cell membrane. In this model, pores are formed due to interaction between hydrophobic center of lipid bilayer and hydrophobic region of peptides, but channels are not formed as in case of other peptide models. Moreover, negatively charged lipids helps to decrease repulsive electrostatic forces formed between positively charged peptides. Most of the cationic peptides follow this type of mechanism for their antimicrobial activity, e.g., cecropin, trichogin GA IV, LL-37 found in humans, caerin and dermaseptin. The peptides following this model have high affinity for acidic lipids as compared to zwitterionic lipids [61, 65, 66].

12.5.4 Molecular Electroporation Model

This model was suggested for cationic AMPs which attached to cell membrane by producing electric potential difference over it. When potential difference extends to 0.2 V, then pores are produced in the membrane. NK-lysin induces pores by this mechanism [67].

12.5.5 Sinking Raft Model

This model suggested that attachment of amphipathic AMPs makes mass imbalance and enhance membrane curvature which stimulates the translocation of peptides across lipid bilayer. Upon self-association, peptides sink into cell membrane through transient pores that is why this is called sinking raft model [58].

12.5.6 Channel Forming/Aggregate Model

According to aggregate model, cationic peptides bind to the cell membrane surface. When enough peptide concentration is attained, peptides reorient and form micelle-like complexes with the lipids, ultimately spanning bilayer in lipid-peptide complex. This phenomenon of lipid-peptide aggregation causes the leakage of ions through membrane as well as translocation of peptides in the cytoplasm, consequently leading to intracellular damage and finally cell lysis. Indolicidin and polyphemusin are the examples of suggested AMPs for aggregate model [60].

12.6 Applications of AMPs

While more than five thousand AMPs have been discovered since 1939, the number of peptides used in medications is still very low [68]. Most are still undergoing clinical trials to evaluate their potential in medicine and those which are used in medicine have only done so in topical ointments and similar formulations. The cause of this is due to various limitations of AMPs in terms of the drug delivery routes they can be introduced into the patient.

12.6.1 Antibacterial Applications

AMPs can not only be used to target ESKAPE bacteria but can be incorporated into medications to treat other bacterial infections as well, particularly respiratory diseases. Hepcidin and LL-37 have been observed to be effective against TB-causing *Mycobacterium tuberculosis* in the lungs. They have been observed to do so by suppressing enzymes and chaperone proteins vital for replication and transcription; LL-37 is also responsible for protein degradation activity inside bacterial cytosol [69]. Similarly, two modified extracts of Tilapia piscidin 4 (TP4) were shown to be effective in controlling *Acinetobacter baumanii* infections in pneumonic mice by upregulating immune cell activity [70]. Furthermore, AMPs can be used to target bacterial STDs, with many AMPs such as cathelicidins, dermaseptins and Pep-1 being effective against chlamydial and gonorrhoeal strains by inhibiting their ability to form biofilms and inclusion bodies as well as inhibiting mechanisms involved in bacterial growth and cell division [71]. Bacterial diseases such as cholera (*Vibrio cholerae*), tetanus (*Clostridium tetani*), and dysentery (*Salmonella enterica* and *Shigella dysenteriae*) are common and recurring issues in many underdeveloped and some developing regions in the world where sanitation and primary healthcare facilities are scarce. The use of AMPs as treatment for these diseases could also improve the natural immunity of the affected population to these pathogens due to their immunomodulatory properties. Thus, their use would be beneficial to the affected population in the long term, which cannot be said for antibiotics.

Amphibian AMP is cationic α-helical peptides encoded by the genes expressed in amphibian skin cells. These host defense peptides exhibit broad-spectrum antimicrobial, antiviral and immunomodulatory activities. One such amphibian-derived AMP, Ocellatin-3N, demonstrates both antibacterial and antifungal activity. It inhibits both Gram-positive and Gram-negative bacteria as well as Candida parapsilosis, an infectious yeast species [72]. However, amphibian-derived host defense peptides present difficulties in clinical development as they can be susceptible to degradation, limited bioavailability, and show hemolytic (cytotoxic) activity [73]. Cao [74] and Zhang [75], among others, have worked on developing antimicrobial peptide chains with excellent antibacterial effects. These peptide chains, in addition to their antibacterial properties, also exhibit outstanding biocompatibility and stable resistance to drug resistance.

Hydrogels are recognized as viable wound dressings due to their unique antibacterial properties and biocompatibility. Natural polysaccharides are ideal candidates for preparing antibacterial hydrogels because of their distinct biological characteristics, biodegradability and non-toxicity. Polysaccharide-based antibacterial hydrogels serve as protective barriers that create a sterile microenvironment over the wound, keep the wound moist, accelerate healing and prevent bacterial accumulation and invasion [76]. Consequently, hydrogels loaded with antimicrobial peptides are a current research focus. For instance, Gao et al. designed antimicrobial peptides and incorporated them into oxidized dextran (ODEX) to create antibacterial hydrogels, which have shown excellent antibacterial properties and wound healing promotion [54] (Fig. 12.5).

12.6.2 Antifungal Applications

Cathelicidins have also been attributed with antifungal applications, being able to kill species of the Aspergillus and *Candida* genus, which comprise the most prevalent fungal pathogens. They do so by interfering with processes involving cell growth and reproduction. Eijk et al. suspected fungal mitochondria to be a target site for their fungicidal AMPs [77]. Peptides like ToAP2 and NDBP-5.7 are able to effectively kill Candida albicans colonies, which are resistant to most drugs due to their ability to form biofilms. These proteins not only help in degrading the biofilm layer but also degrade the cell membrane as well. ToAP2 further affects the inner membranes of organelles to affect the cell morphology and also inhibits growth and reproduction by inhibiting the ability of the fungus to filament [78]. Other AMPs with fungicidal properties include cystatins, lyngbyabellins and polyoxins. Cystatins damage cell envelopes, lyngbyabellins disrupt actin metabolism via polymerization and polyoxins inhibit cell wall formation by suppressing chitin synthase activity [79].

12.6.3 Antiviral Applications

The viricidal application of AMPs includes a wide spectrum of viral species which is still yet to be fully fathomed. The antiviral activity of amphibian AMP against a range of DNA and RNA viruses makes them potential antiviral agents. For example, Temporin G is effective against the influenza virus, effectively repressing replication in the early stages of viral growth; the same holds for defensins against dengue virus and herpesvirus by preventing their entry into cells [80]. Temporin G, B and L have significant inhibitory effects against viruses including herpes simplex virus type-1 (HSV-1), John Cunningham polyomavirus (JPCyV), paramyxoviruses, influenza viruses and SARS-CoV-2. These peptides act on the early stages of the viral life

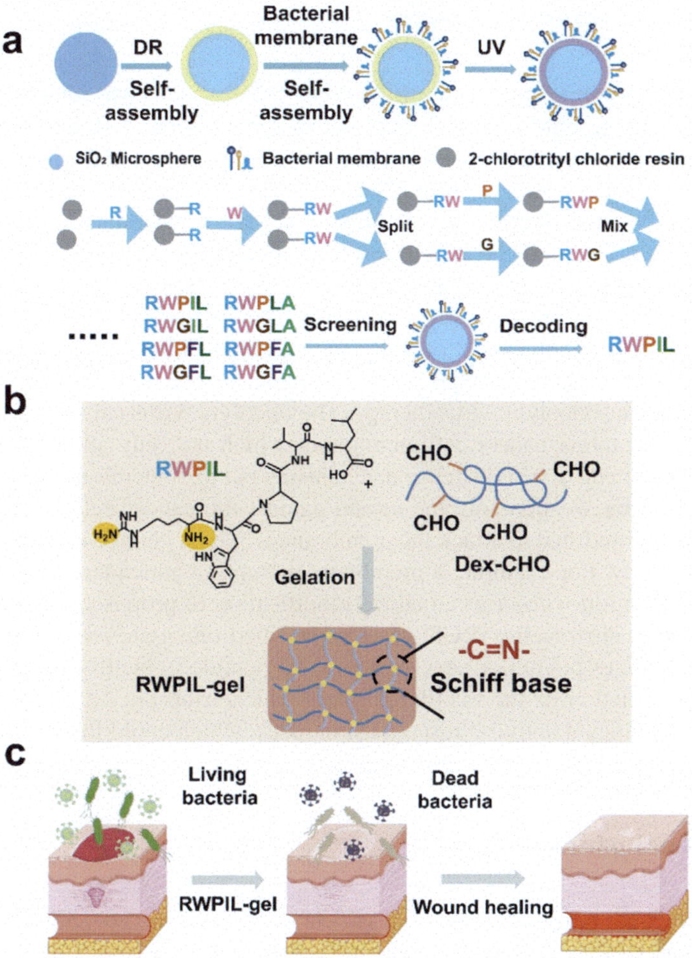

Fig. 12.5 Schematic diagram of screening and application of antimicrobial peptides; **a** Schematic diagram of the preparation of bacterial membrane chromatography and peptide library synthesis screening; **b** Schematic diagram of the preparation of antibacterial hydrogels; and **c** Description of the hydrogel treatment for skin infections [54]

cycle, including DNA replication, viral envelope disruption and interaction with viral glycoproteins and capsid proteins [81–83]. Recent discoveries have attested that the AMP Tachyplesin is effective against *Hepatitis B* [84]. Its mechanism involves the inhibition of HBV proteins are virus for infection. Other AMPs such as Lactoferrin used by adenoviruses to bind and infect respiratory epithelial cells [85]. Thus, the suppression of lactoferrin expression in patients affected by HAdVs could control the spread of the disease. LL-37 and defensins, stimulated by Vitamin D intake, have also been found to inhibit SARS-CoV infections as well [86].

12.6.4 Applications in Cancer

Apart from controlling pathogenic infections, AMPs also have tumor-repressing properties by being able to induce necrosis in cancer cells. TP4 has been shown to do so by inhibiting mitochondrial activity in glioblastomas and aggressive brain cancers [87]. Cancer cells are generally more vulnerable to AMP action due to the differing composition of their cell membranes from normal cells. These variations include different levels of cholesterol resulting in higher membrane fluidity, higher abundance of microvilli as well as a greater negative charge in the cell membrane due to higher levels of phosphatidylserine [88]. This increased anionicity allows for selective targeting by anticancer peptides such as Epinecidin-1 and Paradaxin [89]. The cell membrane-penetrating power of AMPs can also be used in other anti-tumor treatments such as oncolytic virus therapy. The oncolytic virus capsid can be coated with AMPs that target cancer cell membranes which not only allows the virus to evade the cancer cell's virus-defense mechanisms but also increases the potency of biocidal action by weaponizing the capsid against the cancer cell. Therefore, the virus' capsid is modified to attack the membrane while its genetic material destroys the cell internally. For example, a membrane-disrupting anticancer peptide can be used in combination with or as an added modification to proposed oncolytic viral-vector therapies such as the iRGD peptide-modified oncolytic adenovirus therapy [90], which utilizes the tumor-penetrating iRGD peptide to facilitate drug delivery. While the modified virus shows efficacy in preclinical trials [91], it faces challenges such as the significant antiviral response it induces, which could lower its efficacy.

12.7 Conclusions

In this chapter, we introduced the development of antimicrobial peptides, their types and secondary structures as well as their antimicrobial mechanisms and current applications. Compared to antibiotics, antimicrobial peptides have significant advantages in stability and resistance to drug resistance. However, their high cost and low stability in enzymatic environments make them difficult to replace antibiotics, limiting their widespread use in medical applications. Nevertheless, due to the recent promising advancements in biotechnology, these limitations may quickly disappear in the coming years, allowing antimicrobial peptides to become mainstream therapeutic agents rather than just alternatives for specific cases [92].

Acknowledgements This study was supported by the National Natural Science Foundation of China (21874078, 22074072, 22274083), the Innovation and Development Joint Fund of Natural Science Foundation of Shandong Province (ZR2022LZY022, ZR2023LZY005), the National Key R&D Program of China—"Intergovernmental International Cooperation on Science and Technology Innovation" Key Special Project (2024YFE0104100), the Science and Technology Planning Project of South District of Qingdao City (2022-4-005-YY).

References

1. M.A. Kohanski, D.J. Dwyer, B. Hayete, C.A. Lawrence, J.J. Collins, A common mechanism of cellular death induced by bactericidal antibiotics. Cell **130**(5), 797–810 (2007)
2. P.D. Crowley, H.C. Gallagher, Clotrimazole as a pharmaceutical: past, present and future. J. Appl. Microbiol. **117**(3), 611–617 (2014)
3. M.I. Hutchings, A.W. Truman, B. Wilkinson, Antibiotics: past, present and future. Curr. Opin. Microbiol. **51**, 72–80 (2019)
4. Y. Nami, B. Haghshenas, N. Abdullah, A. Barzegari, D. Radiah, R. Rosli, A. Yari Khosroushahi, Probiotics or antibiotics: future challenges in medicine. **64**(2), 137–146 (2015)
5. A. Mulukutla, R. Shreshtha, V. Kumar Deb, P. Chatterjee, U. Jain, N. Chauhan, Recent advances in antimicrobial peptide-based therapy. Bioorg. Chem. **145**, 107151 (2024)
6. S. Nayab, M.A. Aslam, S. ur Rahman, Z.D. Sindhu, S. Sajid, N. Zafar, M. Razaq, R. Kanwar, Amanullah, A review of antimicrobial peptides: its function, mode of action and therapeutic potential. Int. J. Pept. Res. Ther. **28**(1), 46 (2022)
7. S. Datta, A. Roy, Antimicrobial peptides as potential therapeutic agents: a review. Int. J. Pept. Res. Ther. **27**(1), 555–577 (2021)
8. C. Wang, T. Hong, P. Cui, J. Wang, J. Xia, Antimicrobial peptides towards clinical application: delivery and formulation. Adv. Drug Deliv. Rev. **175**, 113818 (2021)
9. M. Magana, M. Pushpanathan, A.L. Santos, L. Leanse, M. Fernandez, A. Ioannidis, M.A. Giulianotti, Y. Apidianakis, S. Bradfute, A.L. Ferguson, A. Cherkasov, M.N. Seleem, C. Pinilla, C. de la Fuente-Nunez, T. Lazaridis, T. Dai, R.A. Houghten, R.E.W. Hancock, G.P. Tegos, The value of antimicrobial peptides in the age of resistance. Lancet. Infect. Dis **20**(9), e216–e230 (2020)
10. C.E. Valdez-Miramontes, J. De Haro-Acosta, C.F. Aréchiga-Flores, L. Verdiguel-Fernández, B. Rivas-Santiago, Antimicrobial peptides in domestic animals and their applications in veterinary medicine. Peptides **142**, 170576 (2021)
11. L. Kang, T. Han, H. Cong, B. Yu, Y. Shen, Recent research progress of biologically active peptides. **48**(3), 575–596 (2022)
12. J.C.P. Santos, R.C.S. Sousa, C.G. Otoni, A.R.F. Moraes, V.G.L. Souza, E.A.A. Medeiros, P.J.P. Espitia, A.C.S. Pires, J.S.R. Coimbra, N.F.F. Soares, Nisin and other antimicrobial peptides: production, mechanisms of action, and application in active food packaging. Innov. Food Sci. Emerg. Technol. **48**, 179–194 (2018)
13. J. Hiraki, T. Ichikawa, S.-I. Ninomiya, H. Seki, K. Uohama, H. Seki, S. Kimura, Y. Yanagimoto, J.W. Barnett, Use of ADME studies to confirm the safety of ε-polylysine as a preservative in food. Regul. Toxicol. Pharmacol. **37**(2), 328–340 (2003)
14. R.F. Silveira, C.A. Roque-Borda, E.F. Vicente, Antimicrobial peptides as a feed additive alternative to animal production, food safety and public health implications: an overview. Anim. Nutr. **7**(3), 896–904 (2021)
15. A. Sultana, H. Luo, S. Ramakrishna, Antimicrobial peptides and their applications in biomedical sector. **10**(9), 1094 (2021)
16. Y. Wang, F. Wang, H. Zhang, B. Yu, H. Cong, Y. Shen, Antibacterial material surfaces/interfaces for biomedical applications. Appl. Mater. Today **25**, 101192 (2021)
17. H. Steiner, D. Hultmark, Å. Engström, H. Bennich, H.G. Boman, Sequence and specificity of two antibacterial proteins involved in insect immunity. Nature **292**(5820), 246–248 (1981)
18. E.F. Haney, S.C. Mansour, R.E.W. Hancock, Antimicrobial peptides: an introduction, in *Antimicrobial peptides: methods and protocols*. ed. by P.R. Hansen (Springer New York, New York, NY, 2017), pp.3–22
19. S.-S. Tang, Z.H. Prodhan, S.K. Biswas, C.-F. Le, S.D. Sekaran, Antimicrobial peptides from different plant sources: isolation, characterisation, and purification. Phytochemistry **154**, 94–105 (2018)
20. W. Tang, H. Zhang, L. Wang, H. Qian, New cationic antimicrobial peptide screened from boiled-dried anchovies by immobilized bacterial membrane liposome chromatography. J. Agric. Food Chem. **62**(7), 1564–1571 (2014)

21. S. Liu, H. Cong, B. Yu, Y. Shen, Screening of a short chain antimicrobial peptide-LKLHI and its application in hydrogels for wound healing. Int. J. Biol. Macromol. **238**, 124056 (2023)
22. L.T. Nguyen, E.F. Haney, H.J. Vogel, The expanding scope of antimicrobial peptide structures and their modes of action. Trends Biotechnol. **29**(9), 464–472 (2011)
23. A.K. Marr, W.J. Gooderham, R.E.W. Hancock, Antibacterial peptides for therapeutic use: obstacles and realistic outlook. Curr. Opin. Pharmacol. **6**(5), 468–472 (2006)
24. N.Y. Yount, A.S. Bayer, Y.Q. Xiong, M.R. Yeaman, Advances in antimicrobial peptide immunobiology. **84**(5), 435–458 (2006)
25. S. Azim, D. McDowell, A. Cartagena, R. Rodriguez, T.F. Laughlin, Z. Ahmad, Venom peptides cathelicidin and lycotoxin cause strong inhibition of Escherichia coli ATP synthase. Int. J. Biol. Macromol. **87**, 246–251 (2016)
26. A.A. Bahar, D. Ren, Antimicrobial peptides. **6**(12), 1543–1575 (2013)
27. R.E.W. Hancock, Cationic antimicrobial peptides: towards clinical applications. Expert Opin. Investig. Drugs **9**(8), 1723–1729 (2000)
28. M. Hassan, M. Kjos, I.F. Nes, D.B. Diep, F. Lotfipour, Natural antimicrobial peptides from bacteria: characteristics and potential applications to fight against antibiotic resistance. J. Appl. Microbiol. **113**(4), 723–736 (2012)
29. A.T.Y. Yeung, S.L. Gellatly, R.E.W. Hancock, Multifunctional cationic host defence peptides and their clinical applications. Cell. Mol. Life Sci. **68**(13), 2161–2176 (2011)
30. R.E.W. Hancock, A. Nijnik, D.J. Philpott, Modulating immunity as a therapy for bacterial infections. Nat. Rev. Microbiol. **10**(4), 243–254 (2012)
31. P.Y. Ong, T. Ohtake, C. Brandt, I. Strickland, M. Boguniewicz, T. Ganz, R.L. Gallo, D.Y.M. Leung, Endogenous antimicrobial peptides and skin infections in atopic. Dermatitis **347**(15), 1151–1160 (2002)
32. M. Mahlapuu, J. Håkansson, L. Ringstad, C. Björn, Antimicrobial peptides: an emerging category of therapeutic agents. **6** (2016)
33. H.G. Boman, Peptide antibiotics and their role in innate immunity. **13**(13), 61–92 (1995)
34. Y. Lai, R.L. Gallo, AMPed up immunity: how antimicrobial peptides have multiple roles in immune defense. Trends Immunol. **30**(3), 131–141 (2009)
35. R. Bals, Epithelial antimicrobial peptides in host defense against infection. Respir. Res. **1**(3), 5 (2000)
36. R.E.W. Hancock, G. Diamond, The role of cationic antimicrobial peptides in innate host defences. Trends Microbiol. **8**(9), 402–410 (2000)
37. G. Wang, Improved methods for classification, prediction, and design of antimicrobial peptides, in *Computational peptidology*. ed. by P. Zhou, J. Huang (Springer New York, New York, NY, 2015), pp.43–66
38. M. Pasupuleti, A. Schmidtchen, M. Malmsten, Antimicrobial peptides: key components of the innate immune system. Crit. Rev. Biotechnol. **32**(2), 143–171 (2012)
39. A.J. Battersby, J. Khara, V.J. Wright, O. Levy, B. Kampmann, Antimicrobial proteins and peptides in early life: ontogeny and translational opportunities. **7** (2016)
40. B. Bechinger, S.-U. Gorr, Antimicrobial peptides: mechanisms of action and resistance. **96**(3), 254–260 (2017)
41. B. Bechinger, K. Lohner, Detergent-like actions of linear amphipathic cationic antimicrobial peptides. Biochim. Biophys. Acta (BBA)—Biomembranes **1758**(9), 1529–1539 (2006)
42. Y.J. Gordon, E.G. Romanowski, A.M. McDermott, A review of antimicrobial peptides and their therapeutic potential as anti-infective drugs. Curr. Eye Res. **30**(7), 505–515 (2005)
43. K. Berthelot, F. Peruch, S. Lecomte, Highlights on Hevea brasiliensis (pro)hevein proteins. Biochimie **127**, 258–270 (2016)
44. K.A. Brogden, Antimicrobial peptides: pore formers or metabolic inhibitors in bacteria? Nat. Rev. Microbiol. **3**(3), 238–250 (2005)
45. N. Bruni, M.T. Capucchio, E. Biasibetti, E. Pessione, S. Cirrincione, L. Giraudo, A. Corona, F. Dosio, Antimicrobial activity of lactoferrin-related peptides and applications in human and veterinary medicine. **21**(6), 752 (2016)

46. H. Wu, Y. Wang, Z. Ren, X. Liu, M. Yu, Y. Cao, H. Cong, B. Yu, Y. Shen, Screening of short-chain antimicrobial peptide LKARI with broad-spectrum bactericidal properties and its application in promoting wound healing. ACS Appl. Mater. Interfaces. **16**(25), 32087–32103 (2024)
47. M.S. Butler, M.A. Blaskovich, M.A. Cooper, Antibiotics in the clinical pipeline in 2013. J. Antibiot. **66**(10), 571–591 (2013)
48. S.C. Shin, I.H. Ahn, D.H. Ahn, Y.M. Lee, J. Lee, J.H. Lee, H.-W. Kim, H. Park, Characterization of two antimicrobial peptides from Antarctic fishes (Notothenia coriiceps and Parachaenichthys charcoti). PLoS ONE **12**(1), e0170821 (2017)
49. J.M. García-Beltrán, M. Arizcun, E. Chaves-Pozo, Antimicrobial peptides from photosynthetic marine organisms with potential application in aquaculture. **21**(5), 290 (2023)
50. J. Li, J.-J. Koh, S. Liu, R. Lakshminarayanan, C.S. Verma, R.W. Beuerman, Membrane active antimicrobial peptides: translating mechanistic insights to design. **11** (2017)
51. J.-W. Liou, Y.-J. Hung, C.-H. Yang, Y.-C. Chen, The antimicrobial activity of gramicidin A is associated with hydroxyl radical formation. PLoS ONE **10**(1), e0117065 (2015)
52. B.A. Lipsky, K.J. Holroyd, M. Zasloff, Topical versus systemic antimicrobial therapy for treating mildly infected diabetic foot ulcers: a randomized controlled, double-blinded, multicenter trial of pexiganan cream. Clin. Infect. Dis. **47**(12), 1537–1545 (2008)
53. X. Li, S. Zuo, B. Wang, K. Zhang, Y. Wang, Antimicrobial mechanisms and clinical application prospects of antimicrobial peptides. Molecules **27**(9), 2675 (2022)
54. F. Gao, A. Ahmed, H. Cong, B. Yu, Y. Shen, Effective strategies for developing potent, broad-spectrum antibacterial and wound healing promotion from short-chain antimicrobial peptides. ACS Appl. Mater. Interfaces. **15**(27), 32136–32147 (2023)
55. Y. Guo, F. Gao, M. Rafiq, B. Yu, H. Cong, Y. Shen, Preparation of antimicrobial peptides and their combination with hydrogels for wound healing applications. Int. J. Biol. Macromol. **274**, 133494 (2024)
56. P.H. Mygind, R.L. Fischer, K.M. Schnorr, M.T. Hansen, C.P. Sönksen, S. Ludvigsen, D. Raventós, S. Buskov, B. Christensen, L. De Maria, O. Taboureau, D. Yaver, S.G. Elvig-Jørgensen, M.V. Sørensen, B.E. Christensen, S. Kjærulff, N. Frimodt-Moller, R.I. Lehrer, M. Zasloff, H.-H. Kristensen, Plectasin is a peptide antibiotic with therapeutic potential from a saprophytic fungus. Nature **437**(7061), 975–980 (2005)
57. E. Nilsson, C. Björn, V. Sjöstrand, K. Lindgren, M. Münnich, I. Mattsby-Baltzer, M.-L. Ivarsson, K. Olmarker, M. Mahlapuu, A novel polypeptide derived from human lactoferrin in sodium hyaluronate prevents postsurgical adhesion formation in the rat. **250**(6), 1021–1028 (2009)
58. S. Liu, L. Fan, J. Sun, X. Lao, H. Zheng, Computational resources and tools for antimicrobial peptides. 23(1), 4–12 (2017)
59. E.L.-M. Joel, A.A. Alejandra Ochoa-Zarzosa José, D.L.-L. Pedro, Antimicrobial peptides: diversity and perspectives for their biomedical application, in *Biomedical Engineering*, eds. K. Malgorzata Anna, O.-J. Sylwia (IntechOpen, Rijeka, 2011), p. Ch. 12
60. N. Malanovic, R. Leber, M. Schmuck, M. Kriechbaum, R.A. Cordfunke, J.W. Drijfhout, A. de Breij, P.H. Nibbering, D. Kolb, K. Lohner, Phospholipid-driven differences determine the action of the synthetic antimicrobial peptide OP-145 on Gram-positive bacterial and mammalian membrane model systems. Biochim. Biophys. Acta (BBA)—Biomembranes **1848**(10, Part A), 2437–2447 (2015)
61. A. León-Buitimea, C.R. Garza-Cárdenas, J.A. Garza-Cervantes, J.A. Lerma-Escalera, J.R. Morones-Ramírez, The demand for new antibiotics: antimicrobial peptides, nanoparticles, and combinatorial therapies as future strategies in antibacterial agent design. **11** (2020)
62. M.E. Marcocci, D. Amatore, S. Villa, B. Casciaro, P. Aimola, G. Franci, P. Grieco, M. Galdiero, A.T. Palamara, M.L. Mangoni, L. Nencioni, The amphibian antimicrobial peptide temporin B inhibits *in vitro* herpes simplex virus 1 infection. **62**(5) (2018). https://doi.org/10.1128/aac.02367-17
63. J.F. Marcos, M. Gandía, Antimicrobial peptides: to membranes and beyond. Expert Opin. Drug Discov. **4**(6), 659–671 (2009)

64. M. Mardirossian, R. Grzela, C. Giglione, T. Meinnel, R. Gennaro, P. Mergaert, M. Scocchi, The host antimicrobial peptide Bac71-35 binds to bacterial ribosomal proteins and inhibits protein synthesis. Chem. Biol. **21**(12), 1639–1647 (2014)
65. D.K. Mercer, C.S. Stewart, L. Miller, J. Robertson, V.M.S. Duncan, D.A. O'Neil, Improved methods for assessing therapeutic potential of antifungal agents against dermatophytes and their application in the development of NP213, a novel onychomycosis therapy candidate. **63**(5) (2019) https://doi.org/10.1128/aac.02117-18
66. M. Mladic, J. Slagboom, J. Kool, F.J. Vonk, G.P. van Wezel, M.K. Richardson, Detection and identification of antibacterial proteins in snake venoms using at-line nanofractionation coupled to LC-MS. Toxicon **155**, 66–74 (2018)
67. H. Moravej, Z. Moravej, M. Yazdanparast, M. Heiat, A. Mirhosseini, M. Moosazadeh Moghaddam, R. Mirnejad, Antimicrobial peptides: features, action, and their resistance mechanisms in bacteria. Microb. Drug Resist. **24**(6), 747–767 (2018)
68. S. Ramazi, N. Mohammadi, A. Allahverdi, E. Khalili, P. Abdolmaleki, A review on antimicrobial peptides databases and the computational tools. Database **2022** (2022)
69. K. Mehta, P. Sharma, S. Mujawar, A. Vyas, Role of antimicrobial peptides in treatment and prevention of mycobacterium tuberculosis: a review. Int. J. Pept. Res. Ther. **28**(5), 132 (2022)
70. C.-J. Jung, Y.-D. Liao, C.-C. Hsu, T.-Y. Huang, Y.-C. Chuang, J.-W. Chen, Y.-M. Kuo, J.-S. Chia, Identification of potential therapeutic antimicrobial peptides against Acinetobacter baumannii in a mouse model of pneumonia. Sci. Rep. **11**(1), 7318 (2021)
71. H. Madanchi, M. Shoushtari, H.H. Kashani, S. Sardari, Antimicrobial peptides of the vaginal innate immunity and their role in the fight against sexually transmitted diseases. New Microbes New Infections **34**, 100627 (2020)
72. J.M. Conlon, L. Hunter, S. Attoub, B. Casciaro, M. Mechkarska, Y.H.A. Abdel-Wahab, Antimicrobial, cytotoxic, and insulin-releasing activities of the amphibian host-defense peptide ocellatin-3N and its L-lysine-substituted analogs. **29**(4), e3463 (2023)
73. J. Shi, J. Wu, G. Feng, H. Yang, L. Mu, Natural amphibian-derived host defense peptides: peptide immunomodulators with potential therapeutic value. Protein Pept. Lett. **30**(7), 562–573 (2023)
74. R. Zhang, H. Yan, X. Wang, H. Cong, B. Yu, Y. Shen, Screening of a short chain antimicrobial peptide-FWKFK and its application in wound healing. Biomater. Sci. **11**(5), 1867–1875 (2023)
75. Y. Cao, L. Kang, Y. Wang, Z. Ren, H. Wu, X. Liu, H. Cong, B. Yu, Y. Shen, Screening and investigation of a short antimicrobial peptide: AVGAV. J. Mater. Chem. B **11**(45), 10941–10955 (2023)
76. R. Zhang, B. Yu, Y. Tian, L. Pang, T. Xu, H. Cong, Y. Shen, Diversified antibacterial modification and latest applications of polysaccharide-based hydrogels for wound healthcare. Appl. Mater. Today **26**, 101396 (2022)
77. M. van Eijk, S. Boerefijn, L. Cen, M. Rosa, M.J.H. Morren, C.K. van der Ent, B. Kraak, J. Dijksterhuis, I.D. Valdes, H.P. Haagsman, H. de Cock, Cathelicidin-inspired antimicrobial peptides as novel antifungal compounds. Med. Mycol. **58**(8), 1073–1084 (2020)
78. J. Talapko, T. Meštrović, M. Juzbašić, M. Tomas, S. Erić, L. Horvat Aleksijević, S. Bekić, D. Schwarz, S. Matić, M. Neuberg, I. Škrlec, Antimicrobial peptides—mechanisms of action, antimicrobial effects and clinical applications. **11**(10), 1417 (2022)
79. G.B.D. Cesare, S.A. Cristy, D.A. Garsin, M.C. Lorenz, Antimicrobial peptides: a new frontier in antifungal therapy. **11**(6) (2020). https://doi.org/10.1128/mbio.02123-20
80. A. Ahmed, G. Siman-Tov, G. Hall, N. Bhalla, A. Narayanan, Human antimicrobial peptides as therapeutics for viral infections. **11**(8), 704 (2019)
81. M.E. Marcocci, B.G. Jackowska, C. Prezioso, V. Protto, M. De Angelis, F.S. Di Leva, B. Casciaro, A. Carotenuto, M.L. Mangoni, A.T. Palamara, V. Pietropaolo, G. De Chiara, L. Nencioni, The inhibition of DNA viruses by the amphibian antimicrobial peptide temporin G: a virological study addressing HSV-1 and JPCyV. **23**(13), 7194 (2022)
82. C. Zannella, A. Chianese, L. Palomba, M.E. Marcocci, R. Bellavita, F. Merlino, P. Grieco, V. Folliero, A. De Filippis, M. Mangoni, L. Nencioni, G. Franci, M. Galdiero, Broad-spectrum antiviral activity of the amphibian antimicrobial peptide temporin L and its analogs. **23**(4), 2060 (2022)

83. M. De Angelis, B. Casciaro, A. Genovese, D. Brancaccio, M.E. Marcocci, E. Novellino, A. Carotenuto, A.T. Palamara, M.L. Mangoni, L. Nencioni, Temporin G, an amphibian antimicrobial peptide against influenza and parainfluenza respiratory viruses: insights into biological activity and mechanism of action. **35**(2), e21358 (2021)
84. P. Narula, S. Kiruthika, S. Chowdhari, P. Vivekanandan, A. Chugh, Inhibition of hepatitis B virus (HBV) by Tachyplesin, a marine antimicrobial cell-penetrating peptide. **15**(2), 672 (2023)
85. B.D. Persson, A. Lenman, L. Frängsmyr, M. Schmid, C. Ahlm, A. Plückthun, H. Jenssen, N. Arnberg, Lactoferrin-hexon interactions mediate CAR-independent adenovirus infection of human respiratory cells. **94**(14) (2020). https://doi.org/10.1128/jvi.00542-20
86. S. Laneri, M. Brancaccio, C. Mennitti, M.G. De Biasi, M.E. Pero, G. Pisanelli, O. Scudiero, R. Pero, Antimicrobial peptides and physical activity: a great hope against COVID 19. **9**(7), 1415 (2021)
87. B.-C. Su, C.-Y. Pan, J.-Y. Chen, Antimicrobial peptide TP4 induces ROS-mediated necrosis by triggering mitochondrial dysfunction in wild-type and mutant p53 glioblastoma cells. **11**(2), 171 (2019)
88. W. Chiangjong, S. Chutipongtanate, S. Hongeng, Anticancer peptide: physicochemical property, functional aspect and trend in clinical application (review). Int. J. Oncol. **57**(3), 678–696 (2020)
89. J. Tan, J. Tay, J. Hedrick, Y.Y. Yang, Synthetic macromolecules as therapeutics that overcome resistance in cancer and microbial infection. Biomaterials **252**, 120078 (2020)
90. C. Puig-Saus, L.A. Rojas, E. Laborda, A. Figueras, R. Alba, C. Fillat, R. Alemany, IRGD tumor-penetrating peptide-modified oncolytic adenovirus shows enhanced tumor transduction, intratumoral dissemination and antitumor efficacy. Gene Ther. **21**(8), 767–774 (2014)
91. A.A. Al-Zaher, R. Moreno, C.A. Fajardo, M. Arias-Badia, M. Farrera, J. de Sostoa, L.A. Rojas, R. Alemany, Evidence of anti-tumoral efficacy in an immune competent setting with an iRGD-modified hyaluronidase-armed oncolytic adenovirus. Molecular Therapy—Oncolytics **8**, 62–70 (2018)
92. L. Tajer, J.-C. Paillart, H. Dib, J.-M. Sabatier, Z. Fajloun, Z. Abi Khattar, Molecular mechanisms of bacterial resistance to antimicrobial peptides in the modern era: an updated review. **12**(7), 1259 (2024)

Chapter 13
Peptide Nanomaterials for Tissue Engineering and Regenerative Medicine

Huiqiang Hu, Yanni Sun, Zhengang Sun, and Yongming Xi

Abstract Peptide nanomaterials have shown extensive and promising applications in materials science, disease diagnosis and therapy, drug carrier, tissue engineering, regenerative medicine, biomedicine, and other fields. In this chapter, we summarize the latest progress in the applications of peptide nanomaterials in tissue engineering and regenerative medicine. To achieve this goal, the application of synthetic peptide nanomaterials in bone tissue, tendon, skin, eye, heart, and other tissue engineering and tissue regeneration is demonstrated, and the mechanism is briefly explained. It is believed that this work will help readers to understand the preparation method of peptide nanomaterials for tissue engineering and tissue regeneration, the effects, as well as the mechanism of tissue regeneration, which will provide inspiration for the subsequent design and synthesis of new peptide nanomaterials and arouse great interest in peptide nanotechnology and nano-chemistry, biomedical science, and clinical science.

Keywords Peptide nanofibers · Hybrids · Tissue engineering · Regenerative medicine

13.1 Introduction

With the increasing incidence of tumor, trauma, infection, diabetes, and so on, the incidence of tissue and organ trauma and defect increases, which seriously endangers the health of patients [1–3]. Large-scale tissue and organ defects caused by multiple pathogenic factors are an important challenge for clinicians. Following a traumatic injury, local and systemic responses develop, which help to quickly restore

H. Hu (✉)
Department of Orthopedic Surgery, Qingdao University Medical College, Qingdao 266071, China
e-mail: hhqalliswell@163.com

H. Hu · Y. Sun · Z. Sun · Y. Xi (✉)
Department of Orthopedic Surgery, The Affiliated Hospital of Qingdao University, Qingdao 266035, China
e-mail: xym700118@163.com

G. Wei (ed.), *Peptide Nano-Chemistry and Nanotechnology*,
https://doi.org/10.1007/978-981-96-2387-7_13

the integrity and function of tissues and organs. However, this process of repair and regeneration is not always successful, especially in the case of large tissue and organ defects, which are difficult to heal spontaneously [4–6]. This significantly increases the difficulty of the treatment, reduces the quality of life of patients, and puts them at higher risk of infection, disability, and death. In order to solve these problems, there is an urgent need for the synthesis of materials that can promote the healing and regeneration of tissues and organs, so as to accelerate the healing of defects and achieve the purpose of tissue repair, regeneration, and reconstruction of biological functions [7, 8]. Therefore, the research progress of tissue engineering and regenerative medicine is crucial.

In clinical practice, there are three commonly used tissue engineering repair pathways, including the autologous tissue transplantation, allogeneic tissue transplantation, and artificial replacement material transplantation [9, 10]. The autologous tissue transplantation has incomparable advantages in the tissue repair and is recognized as the gold standard of the tissue repair. As the name suggests, the donor of autologous tissue transplantation comes from other parts of the individual, which indicates that its source is limited, and it will inevitably bring new trauma in the process of obtaining donor tissue. Compared with the autologous tissue transplantation, the sources of the allograft tissue transplantation are more extensive, but the repair effect of the allograft tissue transplantation remains a certain gap with that of the autologous tissue transplantation, which inevitably has the risk of the immune rejection and allograft infection. Therefore, in the recent years, more and more researchers have turned their attention to artificial substitute transplantation. First, artificial materials are widely available and easily obtained. Secondly, the disadvantages of the autologous tissue transplantation and allogeneic tissue transplantation can also be improved. At present, many research results showed that artificial substitute materials have shown strong potential for clinical application in the tissue engineering and regeneration fields, among which peptide nanomaterials have attracted wide attention as an important class of materials for both tissue engineering repair and regeneration [1, 11, 12].

The peptide nanomaterials mentioned in this chapter are peptides that can be edited as base materials and some specific peptides in vivo as signaling molecules for the tissue regeneration. Because of different tissue engineering requirements for the peptide properties, the sequence, structure, and content of peptides are often tailored according to the needs of peptide nanomaterials. At present, a variety of functional peptides have been applied in the tissue engineering and regeneration [2, 3, 6, 13]. In these peptide-containing nanomaterials, the biological function of the peptide is amplified due to its superior bio-affinity. The in vitro and in vivo experiments indicated that peptide nanomaterials usually have the following functions, such as carrying and releasing inducing factors into the host body, inducing and recruiting the host's stimulating factors, and directly acting as stimulating factors. Peptide nanomaterials are non-toxic, biodegradable, low immunogenicity, and easy to be modified. These properties make peptide materials have very promising applications in the field of regenerative medicine and tissue engineering.

Clinical tissue engineering fields include the aspects that related to bone, blood, nerve, skin, oral, tendon, eye, and heart tissues. In this chapter, we focus on the applications of peptide nanomaterials in these tissue engineering and regeneration fields.

13.2 Peptide Nanomaterials for Tissue Engineering and Regenerative Medicine

13.2.1 Bone and Cartilage Tissue Engineering and Regeneration

Biomaterials are often required to have high mechanical strength and a longer period of stability that matches the bone regeneration in application in bone tissue engineering and regeneration, especially when applied to bone defects in load-bearing parts of the body [6, 9, 14]. However, peptide materials usually lack sufficient mechanical strength and long-term stability, so the combination of peptide nanomaterials with other materials is an effective method to address this problem. Peptide nanomaterials can be directly used as the stimulating factors to stimulate the reaction with tissue cells, and can also be modified to make the composite peptide nanomaterials have a variety of biological functions [12].

RADA16 has been widely used for the formation of self-assembling peptide hydrogels, [15] which have been used for rat and human bone marrow mesenchymal stem cells. It has been observed that when these two types of cells were co-cultured with RADA16, it could significantly promote the osteogenic differentiation and exhibit obvious mineralization effects. After filling the defect with RADA16 in vivo, the regenerated bone tissue had both better volume and strength compared to the control group. Combining the RADA16 peptide with functional sequences such as the RGD, DGR, and PRG significantly promoted the cell adhesion, migration, and osteogenic differentiation. These functional sequences promoted the cell adhesion by binding with their corresponding membrane receptors [16]. Studies have shown that self-assembling peptides, such as RADA16, as a packaging carrier, promoted the proliferation of bone marrow mesenchymal stem cells by slowly releasing the wrapped TGF-β1, thus achieving the goal of bone and cartilage regeneration [17, 18].

In addition to the direct use of peptide nanomaterials as materials to induce the bone formation and promote the bone defect healing, many peptides do not have a direct function in the bone regeneration but play a key role in the bone tissue engineering and bone regeneration through indirect effects. Peptide nanomaterials were cross-linked with polymers, hydroxyapatite (HAp), and other materials to form composite materials, in which peptides would play roles such as recruiting and adsorbing host induction factors, carrying and releasing induction factors, carrying target genes for efficient transfection of host cells.

The development of a gene-activated matrix (GAM) as a potential solution for treating large segmental bone defects was discussed [19]. The GAM is a composite scaffold consisting of an electrospun poly(ε-caprolactone) (PCL) fiber sheath and an alginate hydrogel core that carries plasmid DNA encoding bone morphogenetic protein 2 (pBMP2) and vascular endothelial growth factor (pVEGF), respectively, as shown in Fig. 13.1. The authors constructed a cell-penetrating peptide (CPP) sequence called KALA, which was used to modify the scaffold as a gene carrier and give the scaffold cross-linked by Kala the ability to penetrate cells. The transfection efficiency of the Kala scaffolds was proved to be higher than that of the PEI scaffolds, which was generally considered as the gold standard transfection agent with a high transfection rate. However, the transfection rate was higher and the transfection activity was longer in the Kala group. Moreover, it was observed that the survival rate of cells transfected with Kala was even higher than that of the control group, demonstrating its good biocompatibility. The GAM delivered genes locally and efficiently, offering an effective therapeutic approach for the critical-sized bone defect repair. The study demonstrated the potential of cell-free scaffold-based gene therapy for tissue engineering by promoting the osteogenic differentiation of pre-osteoblasts and accelerating the bone healing by activating the angiogenesis and osteogenesis in vivo.

Ding et al. used a synthetic growth factor (GF)-binding peptide HSNGLPL to covalently link with a GelMA hydrogel to construct a bioactive scaffold for in vivo cartilage regeneration in rats [20]. The GF-binding peptide HSNGLPL (also called affinity peptide) can specifically recruit and adsorb host transforming GFs without the dependence on exogenous GFs. The 3D-printed bi-layered scaffold was composed of a peptide-GelMA layer and a hydroxyapatite/GelMA layer for the osteochondral regeneration. The in vitro experiments verified the peptide has the ability of the cartilage induction through selective adsorption of transforming growth factor β1 (TGF-β1) on the scaffold. Subsequently, the in vivo experiments in rat models proved

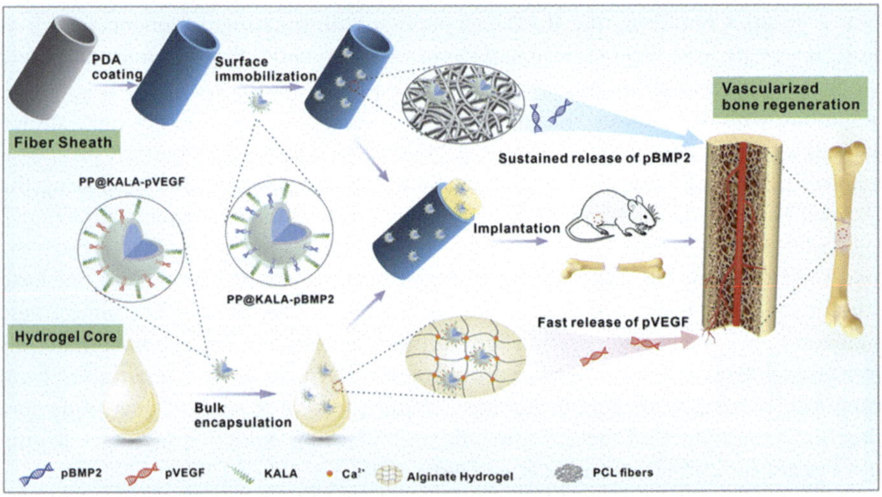

Fig. 13.1 Preparation of PP@KALA-pVEGF hybrids for bone regeneration application. Reprinted with the permission from Ref. [19]. Copyright 2022, Wiley–VCH

the efficacy in the cartilage and osteogenic induction without apparent inflammation. The effect of the scaffold on promoting tissue regeneration and motor recovery was verified by the gait analysis after repairing joint defects in rats. Therefore, the results of this study provided a new way to recruit the endogenous GFs and provide a new treatment for the cartilage injury and arthritis.

Proper mechanical stress is an indispensable and important condition in the bone tissue engineering. At present, peptide nanomaterials cannot provide mechanical support for the bone regeneration, and usually cannot play the role of load-bearing scaffolds, but only perform their biological functions. In the composite material systems, it is usually necessary to have a frame that can withstand a certain strength pressure. In the future, the research of peptide nanomaterials may need to find a reasonable strategy to enhance the intermolecular force and mechanical strength, improve the deficiency of compressive resistance, and maintain long-term stability and time–space matching degradability.

13.2.2 Tendon Tissue Engineering and Regeneration

The tendon injuries are common after sudden large tension caused by complex injuries, overuse, and fatigue. Compound injuries usually occur with the bone injuries. Compared to the repair of bone injuries, tissue engineering treatment for the tendon injuries involves not only conventional transplantation but also suturing. Since tendons are special connective tissues that attach to bones and connect muscles as a power transmission part, they usually need to store large elastic potential energy during the movement and withstand strong tension, which requires high tensile strength. Therefore, peptide nanomaterials used for the tendon repair need to have high tensile and anti-fatigue properties to adapt to special structure and function [5]. In the study of Kannus et al., it was believed that more than 1/3 of the patients in sports injuries suffer from the tendon injuries. It is known that the United States spends over $10 billion annually on the tendon and ligament injuries. Autografts face significant donor shortages and potential complications on the donor side when it comes to reconstructing tendon biology. Therefore, the development of effective tendon tissue engineering materials is extremely important in the tendon repair process [21].

Researchers have used a 29-mer peptide that derived from pigment epithelial-derived factor (PEDF) and synthesized it with acetylation at the NH_2 termini and the amidation at the COOH termini for high stability [13]. They investigated the potential application of the peptide to treat acute tendon ruptures in animals and found that it promoted the tendon regeneration by inducing the clonogenicity of CD146 + TSPCs, implying the tendon stem/progenitor cells (TSPC) stemness retention during the TSPC expansion in culture. The expanded TSPCs differentiated readily into tenocyte-like cells after the removal of the 29-mer from the culture. In the rats with the tendon injury, 29-mer treatment caused extensive expansion of CD146 + TSPCs followed by the infiltration of the CD146+/BrdU-TSPCs into the injured tendon. The CD146+/BrdU-TSPCs increased the tensile strength and more organized collagen fiber regeneration compared to control tendons. Their work demonstrated that substantial expansion of resident TSPCs by the exogenous GFs is beneficial to the tendon healing and that the synthetic 29-mer peptide could be an innovative therapy for acute tendon rupture.

RADA is a self-assembling polypeptide that has the potential as an injectable cell delivery carrier for stem cells for facilitating the in vivo tendon repair. Specifically, RADA nanofiber hydrogel provides a suitable microenvironment for the survival and tenogenesis of human tendon stem/progenitor cells (TSPC). Zhang et al. demonstrated that the combination of TSPC and nanofiber hydrogel (RADA + TSPC group) boosted the in vivo patellar tendon repair with better ambulatory function recovery compared to the control groups, in which the tendon defects were untreated (Defect group) or treated with the RADA hydrogel alone (RADA group). Better macroscopic appearance and improved matrix organization in the repaired tendon with less cell amount and reduced adipocyte accumulation and blood vessel formation were observed in the RADA + TSPC group. Moreover, the TSPC-laden RADA hydrogel resulted in diminished heterotopic ossification (HO) at eight weeks post-operation. [22]

The RADA hydrogels are formed in the presence of a physiological salt solution and remain stable when loaded with cells in culture medium [23]. RADA was functionalized by directly adding the RGD motif, an integrin-binding domain found in ECM proteins, to form the designer RADA-RGD peptide (Ac-$(RADA)_4$-GG-RGDS-$CONH_2$). Both the RADA and RADA/RGD peptides underwent spontaneous assembly and gelation into hydrogels with nanofiber structure in the presence of a physiological salt solution. The challenges associated with stem cell-based therapies due to the changes in aging stem cells were highlighted, and how the peptide nanofiber structure of the 3D microenvironment affected the TSPCs' biological behavior was explored. By analyzing the cell survival, proliferation, apoptosis, morphology, F-actin organization, cell alignment, and gene expression profiles of TSPCs with both RADA and RADA/RGD hydrogels, the study concluded that the use of the hydrogel matrix helped rejuvenate aged tendon stem cells and suggested that focusing on the ECM microenvironment could help develop novel rejuvenation schemes for stem/progenitor cell functional improvement.

13.2.3 Muscle Tissue Engineering and Regeneration

Muscle is an important component of the musculoskeletal system, accounting for a large portion of body weight. However, muscles are highly susceptible to the injury and damage. Conservative treatment is usually the preferred method, but the therapeutic effect is often unsatisfactory. Therefore, tissue engineering and regenerative medicine are expected to become a new and effective method for treating the muscle injuries and defects.

Alheib and co-workers used the RKRLQVQLSIRT peptide for the chemical modification of divinyl sulfone-modified gellan gum [24]. They prepared injectable laminin-biofunctionalized gellan gum hydrogels loaded with myoblasts for skeletal muscle regeneration. They combined cells and/or GFs into appropriate biomaterials (scaffolds/matrices) to build a regenerative construct for the defect repair. Different hydrogel formations were developed and tested on mice to establish the most suitable hydrogel formulation with localized cross-linking to promote skeletal muscle regeneration. The results of the hydrogel bio-functionalization, the hydrogel injectability, and cognitive expansion have been reported, establishing that such injectable hydrogel had many benefits, including providing a native-like environment for the implanted/host cells to aid in tissue regeneration by contributing to higher myogenesis in the muscle defects.

Cimenci et al. utilized a laminin mimetic peptide nanofiber (PNF) scaffold system for their research on effectively regenerating damaged skeletal muscle [25]. Peptide amphiphile (PA) molecules were synthesized manually by standard solid-phase peptide synthesis method, and positively charged bioactive LM-PA and negatively charged E-PA were mixed to form nanofibers. They tested the efficacy of the technique in promoting the regeneration and differentiation of skeletal muscle fibers through the behavioral, histological/physiological, and molecular analyses. The PNF treatment accelerated skeletal muscle regeneration by enhancing satellite cell recruitment and muscle fiber enlargement, and the ECM-mimetic structure of PNFs provided mechanical support, creating a suitable environment for the satellite cell differentiation. In a similar study, Baek et al. found that self-transducible LRS-UNE-L peptide can also enhance the muscle regeneration [26].

13.2.4 Oral Tissue Engineering and Regeneration

Tooth loss, caries, chronic inflammation, and bone damage are often faced by dentists. Because the mouth is connected to the outside world, the growth of bacterial infections is also a major challenge [3].

Li and co-workers reported a new approach to regenerate dental pulp in a full-length human tooth root, which is a complex and challenging task in regenerative endodontics [27]. The dental pulp is highly vascularized tissue situated in a rigid environment that surrounded by the dentin walls and has a blood supply from a small apical foramen of a tooth root, which makes regenerating pulp tissue in a full-length tooth root difficult. They used the dental pulp stem cells (DPSCs) transfected with a multifunctional peptide-conjugated, pH-sensitive, non-viral gene vector to induce the over-expression of VEGF and promote new blood vessel formation for the pulp regeneration. The VEGF-transfected DPSC group was found to promote fast revascularization and regenerate vascularized pulp tissue throughout the canal of the tooth root. Their study demonstrated that the DPSCs were a promising cell source for the pulp regeneration and multifunctional gene carrier.

Compared with the application of peptides in the tooth defects and bone loss, peptides are also commonly used in tooth antibacterial, periodontitis, and tooth caries prevention. Due to the limitations of current chemical methods for the fluoride synthesis and the limitations of fluoride in re-mineralized enamel, the problems posed by dental caries have not been solved. Wang et al. proposed a biomimetic strategy inspired by amelogenin to create a coating [28]. The coating a phase-transited lysozyme (PTL) film mimicking an N-terminal amelogenin with central domain (N-Ame) combined with synthetic peptide (C-AMG) based on the functional domains of C-terminal telopeptide (C-Ame) was designed, which promoted the enamel remineralization beyond the hardness of native enamel. The PTL/C-AMG nanofilm was formed on the substrate through rapid interfacial adsorption of PTL oligomeric NPs with the implantation of the C-AMG. The coating was typically formed in 1 min by immersing a silicon wafer in a buffer containing lysozyme, TCEP, and C-AMG. The amyloid-like structure of the PTL/C-AMG served as an adhesive to support the strong bonding of newly formed HAp crystals with the underlying substrate. The formation of HAp was induced by the PTL/C-AMG film on a silicon wafer, and the formed hybrid materials had the potential for effective repair of abraded enamel or dental caries by improving the aesthetics, hardness, and enamel function.

Zhou et al. developed a peptide preparation for antibacterial and mineralized balance as an anti-caries agent [29]. They discussed five bioactive peptides including the P-113, P-113-DPS, P-113-DA, P-113-SK, and Sp-H5, synthesized using the standard 9-fluorenylmethoxycarbonyl (Fmoc) solid-phase method as the dentotropic moiety to construct an anti-biofouling and mineralizing dual-bioactive tooth surface. They evaluated the bioactive properties of these peptides such as the binding affinity to the tooth surface, anti-biofouling activity, ability to suppress demineralization, and the ability to promote the remineralization. The constructed bioactive peptide was found to inhibit the cariogenic bacteria S. mutans adhesion, kill planktonic S. mutans, and destroy the S. mutans biofilm on the tooth surface. It also protected teeth from the demineralization in acidic environments and induces the self-healing regeneration in the remineralization environment. The peptide exhibited good cytocompatibility and was found to be a promising functional agent to prevent dental caries and is effective in inducing in situ self-healing remineralization for the treatment of decayed teeth.

13.2.5 Neural Tissue Engineering and Regeneration

Nerve regeneration is difficult after damage due to its unique anatomy. Some neurotrophic factors play an important role in the development and recovery of nerve cells after the nerve injury, and peptide nanomaterials provide a biocompatible carrier for these factors, which can release neurotrophic factors in a spatiotemporal order to achieve the purpose of nerve recovery [2, 30]. In addition, peptide hydrogel is a potential material for transporting neural stem cells due to the structure of peptide hydrogel similar to ECM.

RADA-based self-assembling peptides have shown good potential for the bone tissue regeneration application in the above sections. In the nerve tissue engineering, RADA and its modified products also have great potential for the nerve repair. Sun and co-workers investigated the self-assembling peptide (SAP) RADA16-I and its potential for creating nanoscale biomaterials from the bottom-up self-assembly [30]. They discovered that this peptide had a low pH characteristic that could damage cells and host tissues upon direct exposure. However, they developed a strategy for preparing nanofiber hydrogels from two designer SAPs, RADA16-RGD, and RADA16-IKVAV, which showed opposite net charges at neutral pH levels, resulting in a nanofiber hydrogel with significant viscoelasticity. This research revealed potential applications in 3D cell culture, nerve regeneration, and nerve recovery following injury.

Wei and co-workers used a silk fibroin peptide (SF16) as a component to develop a novel peptide-based compound 3D scaffold for promoting peripheral nerve regeneration [31]. The in vitro cell culture analysis showed that the SF16 hydrogel was an effective matrix to maintain the viability, morphology, and proliferation of PC12 cells. The in vivo electrophysiology and walking track analysis demonstrated that the use of the SF16 hydrogel scaffold resulted in a significant improvement in the amplitude recovery compared to the negative control. The SF16 hydrogel scaffold may, therefore, promote excellent axonal regeneration and functional recovery after peripheral nerve injury, making it a promising candidate for producing 3D scaffolds to enhance the nerve regeneration.

Lu et al. used functionalized PNF hydrogel with dual-functionalized VEGF and brain-derived neurotrophic factor (BDNF)-mimetic peptide epitopes for the peripheral nerve reconstruction [32]. The VEGF-mimetic peptide motif used was KLTWQELYQLKYKGI, while the BDNF-mimetic peptide motif was RGIDKRHWNSQ. These motifs were introduced to a designer SAP nanofiber scaffold RAD/KLT/RGI, which was used to construct an artificial neurovascular microenvironment in the lesion area to enhance peripheral nerve regeneration. The VEGF-mimetic peptide motif was designed to bind to and activate the VEGF receptors, while the BDNF-mimetic peptide motif was designed based on the solvent-exposed Loops 3 and 4 of BDNF, which could promote the neurotrophy and neurite outgrowth by binding to the p75NTR and TrkB receptors. The study suggested that the hydrogel scaffold created a 3D neurovascular microenvironment for the endothelial cell and neural cell growth with the potential for promoting the nerve tissue engineering and regeneration, not only in the peripheral nervous system (PNS) but also in other tissues.

Liu et al. provided a method for repairing peripheral nerves using a catheter prepared from chitosan and coated with DNA-peptide nanomaterials on the surface of the catheter [33]. Among them, the coating consisted of single-strand DNA (ssDNA) and its complementary DNA-peptide mimics. Through the two-step cross-linking, the DNA and chitosan were first cross-linked, and then the peptide was cross-linked with the DNA/chitosan catheter through the principle of complementary base pairing to form a hybrid nanomaterial. It promoted the proliferation of the human umbilical vein endothelial cells (HUVECs) and axon outgrowth of the dorsal root ganglion. Effective nerve regeneration and angiogenesis in vivo were verified by measuring the recovery of muscle function in the nerve control area. Therefore, the authors suggested that the chitosan catheter modified by the short peptide sequence could be an effective method to achieve the nerve repair.

In addition to performing direct nerve repair, Nam et al. described the preparation and characterization process of a supramolecular β-peptide-based hydrogel that serves as a biocompatible, conductive, and biostable neural interface [34]. The β-peptide (KVKEVFFVKEVFFVK-EVY) was synthesized using the Fmoc chemistry. Cyanine3-fluorophores were conjugated to the N-terminal of the β-peptide using a CEM Discover microwave synthesizer. The final hydrogel was prepared by mixing the purified β-peptide with carbon nanotubes (CNTs) solution for conductive hydrogel or with β-peptide-Cy3 to yield Cy3-containing hydrogel. Supramolecular β-peptide-based hydrogels provided a biocompatible, conductive, and biostable neural interface that allows for signal amplification via a tight neural/hydrogel contact without the neuroinflammation. The β-peptide nanofibers formed a 3D electrical network in the hydrogel, enabling the signal transmission and preventing chronic neuroinflammation and neural signal loss. Additionally, the hydrogel mimicked the viscoelastic properties of neural tissue and had moldable characteristics for the use in both acute and chronic intercortical and epidural neural signal recording, eliciting neural signal amplification with minimum signal dissipation.

13.2.6 Skin Tissue Engineering and Regeneration

Peptide nanomaterials have strong application prospects in skin tissue engineering due to their reversible and dynamic properties. In addition, the skin healing is a replicating biological process. In addition to stopping the bleeding and repairing the damaged tissue structure, it is also necessary to restore the skin's barrier function. Skin tissue engineering materials require less mechanical strength than bone tissue and tendon tissue engineering, which can be better achieved in the purpose of application during the preparation process. Moreover, the self-healing properties of peptide nanomaterials are necessary because the skin environment is susceptible to pulling. The peptide self-assembly is an effective strategy for the skin tissue engineering. The synthetic SAP hydrogels have excellent biocompatibility and biodegradability, which can form nanofiber structure networks that similar to natural ECM.

Hsu et al. used SAPs composed of RADA16-I to develop a thin-film coating for the hemostatic bandages [35]. The films were assembled by layer-by-layer (LbL) assembly under acidic conditions with naturally derived polyanions like hyaluronic acid, chondroitin sulfate, and dextran sulfate. These films were capable of forming nanofiber clots after the exposure to extreme temperatures from − 80 to 60 °C and remained biodegradable and biocompatible. They demonstrated the effectiveness of the hemostatic bandages in rat kidney and liver models, achieving fast clotting and hemostasis. It was seen that the SAPs had the advantage of hemostasis and coagulation during the skin incision healing.

Skin is an important defense line for human body to resist external injury, and the anti-infection is also an important demand for the skin tissue engineering. Therefore, polypeptides with antibacterial effects have important application value in the skin tissue engineering and regeneration. Previous studies have shown that peptide materials containing poly-L-lysine exhibited certain antimicrobial properties [36, 37]. And with the increase of the peptide content, the antibacterial activity was stronger. The peptide material provided a high content of water and a breathable environment for the wound healing, which promoted rapid wound healing. In the face of skin defect wounds caused by tumors and chronic infections, good tumor-killing, and anti-infection effects by developing nanocomposite hydrogels loaded with antibacterial peptides have been achieved [38].

Comune et al. used LL37 peptide-conjugated gold nanoparticles (LL37-Au NPs) as therapeutic agents for the skin wound healing [39]. The NPs were prepared using a one-step synthetic scheme, where the peptide acted as a capping agent and affected the kinetics of AuNP formation. The study conducted the bioactivity and receptor channel assays, as well as the electrophysiological recordings and scratch assays to demonstrate that the LL37-Au NPs revealed enhanced wound-healing capabilities compared to the peptide alone. The study also indicated that most of the NPs were eliminated from the skin after 15 days post-application and the residual amount of NPs was found to be $< 0.05\%$ in major organs. The results showed that the LL37-Au NPs exhibited high skin regenerative potential and could be used for the development of AMP-based nanotherapeutics for the skin disorders.

Lou and co-workers developed an injectable peptide hydrogel to promote the tissue regeneration of chronic wounds [40]. The authors synthesized SAPs, KGH and KLD12, using standard Fmoc chemistry. The biocompatibility and biodegradability of the synthesized peptides were evaluated through the in vitro cytotoxicity assays, which confirmed the cytocompatibility of the SAP hydrogels. The KGH hydrogel was characterized to accelerate overall wound repair in diabetic mice. The in vitro and in vivo results revealed that the KGH hydrogel enhanced the synthesis and secretion of ECM proteins and GFs through the Rho/ROCK and TGF-β/MEK/MAPK pathways, promoting the wound tissue regeneration.

In addition to the skin wound healing, a lot of efforts have been carried out in balancing the wound healing and scar hyperplasia. For instance, Park and co-workers used a polypeptide, poly ethylene argininylaspartate diglyceride (PEAD), to prepare coacervates for the delivery of GFs, IL-10, and TGF-b3 [41]. They investigated the regenerative effect of IL-10/TGF-b3 on in vitro and in vivo scarless wound healing using a rat skin wound defect model. The results showed that dual IL-10 and TGF-b3 delivery improved the skin healing and reduced the scar formation. They hypothesized that dual IL-10 and TGF-b3 administration using the Coa platform would help in the scar reduction. In another study, Bai et al. developed a multifunctional single-component polymeric hydrogel using dopamine-modified poly(*l*-glutamate) at an ultralow concentration, which exhibited high-performance hemostasis and scarless wound-healing attributes [42]. The hydrogel was constructed by physical treatments and was covalent/H-bonding co-cross-linked and highly negative-charged networks, which enabled it to be thermo-, salt-, urea-resistant, self-healing, injectable, and adhesive yet detachable.

Xi and co-workers constructed a biomimetic elastomeric multifunctional PCL-PCE nanofibrous matrix using electrospinning [43]. The functionalized hybrid nanofibrous matrix showed biomimetic elastomeric behavior and excellent biocompatibility. The nanofibrous system effectively prevented the MDR bacteria-derived wound infection and significantly enhanced complete skin-thickness wound healing in a mouse model. They also conducted the in vitro antibacterial assays, cytotoxicity, and hemocompatibility evaluation, western blotting analysis, gene, and protein expression in wound healing, and photoluminescent properties evaluations to further validate the efficacy of the PCL-PCE nanofibrous matrix. The results obtained suggested that the PCL-PCE nanofibrous matrix was a competitive multifunctional dressing for the bacteria-infected wound healing and skin regeneration and could be useful in overcoming the MDR bacterial infections.

For skin wounds with special conditions, such as non-healing of wounds caused by diabetes. In the model of diabetic wound, some scholars have verified the application of peptide nanomaterials for the reconstruction of blood circulation and the treatment of multidrug-resistant bacterial infection. In a typical study, Xing and co-workers designed an artfully designed chiral gel (HA-LM2RMR) with enhanced revascularization properties, which featured stereoselective interaction and a specific mechanism for killing multidrug-resistant bacteria [44]. As shown in Fig. 13.2, the gel consisted of L-phenylalanine and cationic hexapeptide co-assembled helical nanofibers cross-linked with HA via hydrogen bonding. This chiral gel possessed abundant chiral and cationic sites, specifically killing multidrug-resistant bacteria rather than host cells since the cationic hexapeptides selectively interact with negatively charged microbial membranes. The chiral dressing was constructed for diabetic wound healing by removing AGEs from the diabetic wound bed and promoting the wound healing by improving the angiogenesis in infected full-thickness skin defects of diabetic mice. The chiral dressing presented superior wound-healing efficiency over traditional wound gels in angiogenesis and re-epithelialization. The HA-LM2-RMR fibers presented an attractive potential to activate the neovascularization of human umbilical vein endothelial cells by upregulating the expression of VEGF and OPA1.

Griffin et al. developed a hydrogel scaffold made of microporous annealed particles (MAP) that can accelerate the wound healing and induce the tissue regeneration [45]. They switched the chirality of the cross-linking peptides from L to D amino acids to slow the degradation rate of the scaffold. However, they unexpectedly found that D-peptide cross-linked MAP hydrogels sped up the degradation in vivo and resulted in the tissue regeneration, including increased tensile strength and hair neogenesis. They also elicited significant antigen-specific immunity, and an intact adaptive immune response was required for the hydrogel-induced skin regeneration. They demonstrated that an adaptive immune response can be generated from a biomaterial to induce regenerative healing, despite faster scaffold degradation.

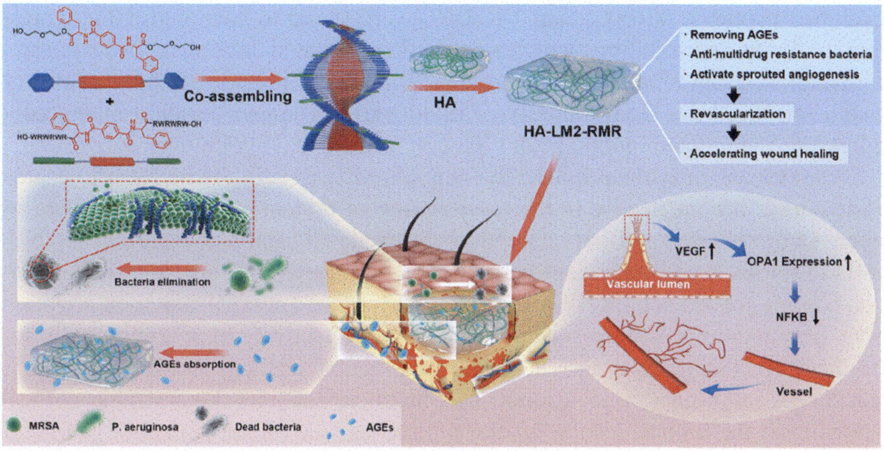

Fig. 13.2 Preparation of HA-LM2-RMR fibers for tissue engineering and wound healing. Reprinted with the permission from Ref. [44]. Copyright 2023, American Chemical Society

13.2.7 Ocular Tissue Engineering and Regeneration

Uzunalli et al. used PA molecules carrying bioactive peptide sequences, YIGSR, and RGD derived from laminin and fibronectin, to induce the corneal regeneration [46]. They incorporated YIGSR and RGD peptide motifs derived from laminin and fibronectin, respectively, into the bioactive peptide nanofiber scaffold to mimic the ECM and promote the cell attachment. This scaffold successfully induced the healing of injured corneas while mimicking proper corneal function cellular processes. The materials regulated cellular behaviors simultaneously, which helped in maintaining normal corneal function. The bioactive peptide nanofibers have a promising potential for pharmaceutical and clinical applications.

Jangamreddy and co-workers used collagen-like peptides (CLP) as the material and conjugated them to an inert, mechanically tougher multifunctional PEG in the preparation of hydrogel implants for regenerative corneal implants [47]. The hydrogel implants were tested under pre-clinical conditions in mini-pigs, where they emulated the function of a collagen-based implant for promoting the regeneration, using the cornea as a model system. The safety and efficacy of the implant were compared to the implants made from clinically tested full-length recombinant human collagen III (RHCIII)-MPC. The study indicated that the short peptide analog implant was a promising alternative to the RHCIII-MPC implants for regenerative corneal implants.

Fungal keratitis is often associated with the corneal injury and regeneration. Amit et al. proposed a new solution to address this problem, which has a good therapeutic effect on fungal keratitis caused by *Candida albicans* and *Fusarium* [48]. They designed corneal specific CPPs through a subtractive proteomic approach for the treatment of fungal keratitis. They enhanced the bioavailability and delivery of peptides using a gelatin hydrogel-based delivery system. Two out of the four developed peptides, VRF005 and VRF007, were found to have high CPP scores, amphiphilicity, and solubility, which were further validated through secondary structure simulations and molecular dynamic simulations. The hydrogel-based delivery system improved the bioavailability of the drug for a longer duration, potentially addressing a significant public health issue in tropical and agricultural countries.

In the disease of corneal injury, the new blood vessels in the non-vascular area of the cornea are unfriendly to the regeneration of the cornea and the recovery of vision. Chu et al. used gp91 ds-tat peptide-encapsulated gelatin nanoparticles (GNP-gp91) to inhibit the corneal angiogenesis [49]. As shown in Fig. 13.3, they developed peptide-loaded gelatin NPs as eye drops for treating corneal neovascularization (NV). The gelatin NPs were selected as they can effectively treat corneal diseases and their size facilitates drug retention on the ocular surface. The study aimed to treat NV by inhibiting the invasion of new blood vessels into the avascular corneal regions from the limbus, leading to chronic corneal edema, inflammation, and scar formation. They reported a 7.89% reduction in the corneal vessel area when treated with the GNP-gp91 in chemically burned corneal neovascularization models. The nanomedicine GNP-gp91 had the potential to reduce ROS content in cells and it reduced the protein expression of Nox2, VEGF, MMP2, and MMP9, thereby inducing a cascade inhibition of angiogenesis.

In addition, optic neuropathy is also an important cause of vision loss. Based on this, Danesh-Meyer et al. reported a peptide nanomaterial with protective effects on retinal ganglion cells [50]. The authors used a rat connexin43 mimetic peptide (sequence VDCFLSRPTEKT), which was designed to block the connexin43 gap junction channels following the CNS injury. They conducted the experiments in vivo and in vitro to investigate the potential neuroprotective effects of the connexin43 mimetic peptide in an animal model of retinal ischemia–reperfusion. It was found that retinal ischemia–reperfusion induced significant vascular leakage and disruption, and the connexin43 mimetic peptide significantly reduced the dye leak and increased

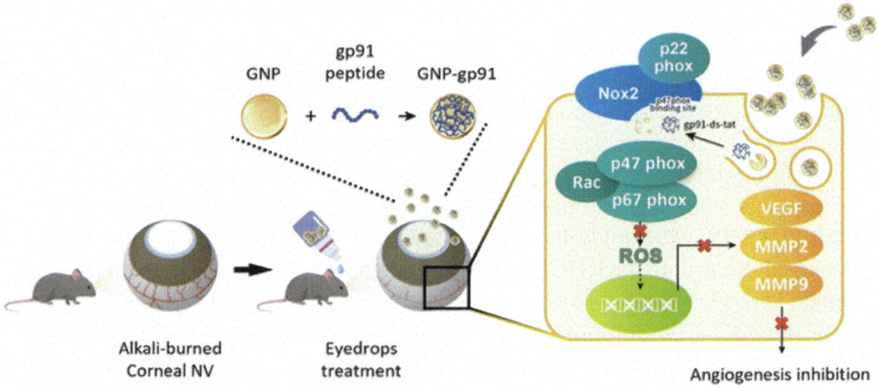

Fig. 13.3 Preparation of peptide-encapsulated gelatin nanoparticles (GNP-gp91) for treating NV. Reprinted with the permission from Ref. [49]. Copyright 2023

retinal ganglion cell survival at 7 and 21 days. The in vitro studies on endothelial cells showed that endothelial cell death following hypoxia can be mediated directly by the opening of the connexin43 hemichannels.

13.2.8 Cardiac Tissue Engineering and Regeneration

Myocardial infarction (MI) is an irreversible myocardial injury and is one of the most serious ischemic heart diseases worldwide. Surgical treatments, such as the stent implantation and coronary artery bypass surgery, are preferred for heart attack patients. However, surgery can only relieve the symptoms and cannot regenerate the already infarcted myocardium.

In response to this, Zhang and co-workers have attempted to use injectable hydrogels loaded with peptides to treat the myocardial infarction [51]. They used a 7-amino-acid peptide (7Ap) derived from histone deacetylase 7 (HDAC7) and loaded it into a collagen hydrogel for the MI therapy. The collagen I hydrogel was fabricated into an injectable form and utilized as a delivery system to support the tissue regeneration and restrain left ventricular (LV) remodeling. The 7Ap-collagen hydrogel promoted the neo-microvessel formation, enhanced the stem cell recruitment, and differentiation, decreased the cellular apoptosis, and promoted the cardiomyocyte cycle progression. It restricted the LV wall fibrosis, reduced infarct wall thinning, and improved cardiac performance in a rat model of MI. The injectable collagen hydrogel provided a favorable microenvironment for the cell survival and proliferation. It also constituted a functional platform to prolong the retention of bioactive elements, such as peptides. This study demonstrated the promising role of biomaterials, specifically collagen hydrogels, in MI therapy.

In another case, Rufaihah et al. used glycosaminoglycan (GAG)-mimetic PAs, which mimic heparan sulfate glycosaminoglycans found in the ECM, to induce the angiogenesis and repair damaged myocardial tissues in a rat cardiac infarct model [52]. The GAG-mimetic PAs were synthesized and self-assembled into nanofibers. The authors injected the angiogenic GAG-mimetic PNFs and showed that PNFs led to the preservation of more cardiac muscle and the formation of new blood vessels after myocardial infarction in rats, resulting in improved cardiac function. They also observed a significantly high level of endogenous GF expression and localization in the infarct area.

13.2.9 Vascular Tissue Engineering and Regeneration

In the repair of defective tissues and organs, angiogenesis is indispensable, and the initial angiogenesis is very important. Vascular tissue engineering is not only a problem faced in the repair of vascular damage, but also a common problem faced by all tissue engineering repair [8]. If any tissue engineering material has a strong angiogenesis ability, the repair effect will be greatly improved.

Jakub et al. have loaded polyisocyanopeptide (PIC) onto polymers to prepare peptidomimetic-polymer thermo-responsive hydrogels that support blood vessel formation, demonstrating broad potential applications for 3D cell, tissue, and organ culture [53]. As a thermo-responsive and biocompatible material, PIC could mimic the integrin-like cell binding sites by controlling temperature and adding GRGDS peptides. The created hydrogels decorated with peptides exhibited excellent compatibility by mimicking the binding sites of certain stimulating factors. By adjusting the concentration of GRGDS peptides, various human cell types, including the fibroblasts, endothelial cells, adipose-derived stem cells, and melanoma cells, could proliferate, migrate, and differentiate in the PIC hydrogels. Additionally, through optimizing the hybrid PIC polymer to adjust the properties of hydrogels, the formation of ECM-like structures that allow the in vitro spontaneous human HDMEC angiogenesis was supported. They also demonstrated successful transplantation of the organ structure generated by human-sourced cells in the PIC hydrogel, which survived and integrated into the surrounding tissue.

Considering that high doses of vascular stimulating factors may increase the risk of systemic side effects, as well as low drug delivery efficiency, Huang et al. proposed a biomimetic SAP hydrogel with proteoglycan-assisted GF delivery for the treatment of angiogenesis [54]. The SAP hydrogel was designed with two customized sequences, RADA16-GGQQLK (QLK) and RADA16-GGLRKKLGKA (LRK), that mimic the 3D fibrous skeleton of the natural ECM. The SAP hydrogel conjugated with the QLK functional motif could be cross-linked by endogenous transglutaminase, and the mechanical property of the hydrogel was enhanced without the need for external support. The LRK sequence exhibited a good binding affinity with the proteoglycan heparan sulfate and could act as a cofactor by sustaining the release of embedded GFs. The scaffold could support the survival of encapsulating cells and promote the capillary-like tubular structure formation by dual angiogenic GFs. The GF-loaded hydrogel promoted the sprout of surrounding vessels in a spoke-wheel pattern compared to GF-free counterparts. The SAP hydrogel-based therapy may be useful for minimally invasive surgery, ischemic tissue disorders, and chronic wound healing.

Sarker and co-workers introduced the use of biodegradable acellular scaffolds, conjugated with the SAP scaffolds to induce the angiogenesis, as a promising strategy to deliver angiogenic signals in functional tissue regeneration [8]. They have discussed the design of synthetic tissue-mimetic scaffolds with biological signals embedded in their structure and the bioactive factors required for the angiogenesis regulation. Furthermore, they have highlighted the potential of angiogenic SAP scaffolds in treating ischemic pathologies such as the peripheral vascular disease, stroke, and myocardial infarction.

13.3 Conclusion

Currently, a large number of peptide nanomaterials and peptide derivatives have been developed and studied, greatly enhancing our understanding of tissue engineering and regenerative medicine. Through a detailed analysis of the above cases, it can be concluded that the applications of peptide nanomaterials in the tissue engineering and regenerative medicine can greatly improve the tissue regeneration efficiency and success rate. However, most of the current validation is based on the cellular and animal levels, with limited clinical evidence. As our understanding of peptides increases, peptide nanomaterials will play a better role in clinical biomedical applications in the future.

References

1. B. Pramanik, M.M. Islam, H.K. Patra, Rational design of peptide-based implants for corneal bioengineering. Curr. Opin. Biotechnol. **81**, 102947 (2023)
2. K.M. Koss, L.D. Unsworth, Neural tissue engineering: bioresponsive nanoscaffolds using engineered self-assembling peptides. Acta Biomater. **44**, 2–15 (2016)
3. A.N. Moore, J.D. Hartgerink, Self-assembling multidomain peptide nanofibers for delivery of bioactive molecules and tissue regeneration. Acc. Chem. Res. **50**, 714–722 (2017)
4. D. Stern, H. Cui, Crafting polymeric and peptidic hydrogels for improved wound healing. Adv. Healthc. Mater. **8**, 1900104 (2019)
5. S. Ruiz-Alonso, M. Lafuente-Merchan, J. Ciriza, L. Saenz-del-Burgo, J.L. Pedraz, Tendon tissue engineering: cells, growth factors, scaffolds and production techniques. J. Control. Release **333**, 448–486 (2021)
6. R. Visser, G.A. Rico-Llanos, H. Pulkkinen, J. Becerra, Peptides for bone tissue engineering. J. Control. Release **244**, 122–135 (2016)
7. K. Aoki, N. Alles, N. Soysa, K. Ohya, Peptide-based delivery to bone. Adv. Drug Deliv. Rev. **64**, 1220–1238 (2012)
8. B. Sarkar, P.K. Nguyen, W. Gao, A. Dondapati, Z. Siddiqui, V.A. Kumar, Angiogenic self-assembling peptide scaffolds for functional tissue regeneration. Biomacromol **19**, 3597–3611 (2018)
9. J. Zhang, D. Tong, H. Song, R. Ruan, Y. Sun, Y. Lin et al., Osteoimmunity-regulating biomimetically hierarchical scaffold for augmented bone regeneration. Adv. Mater. **34**, 2202044 (2022)
10. S. Akhlaghpoor, A. Aziz Ahari, A. Arjmand Shabestari, M.R. Alinaghizadeh, Radiofrequency ablation of osteoid osteoma in atypical locations: a case series. Clin. Orthop. Relat. Res. **468**, 1963–1970 (2010)
11. M. Watanabe, M. Okamoto, S. Komichi, H. Huang, S. Matsumoto, K. Moriyama et al., Novel functional peptide for next-generation vital pulp therapy. J. Dent. Res. **102**, 322–330 (2023)
12. I. Pountos, M. Panteli, A. Lampropoulos, E. Jones, G.M. Calori, P.V. Giannoudis, The role of peptides in bone healing and regeneration: a systematic review. BMC Med. **14**, 103 (2016)
13. T.-C. Ho, Y.-P. Chiang, C.-K. Chuang, S.-L. Chen, J.-W. Hsieh, Y.-W. Lan et al., PEDF-derived peptide promotes skeletal muscle regeneration through its mitogenic effect on muscle progenitor cells. Am. J. Physiol. Cell Physiol. **309**, 159–168 (2015)
14. J. Tuckermann, R.H. Adams, The endothelium-bone axis in development, homeostasis and bone and joint disease. Nat. Rev. Rheumatol. **17**, 608–620 (2021)

15. J. Chen, Z.D. Shi, X. Ji, J. Morales, J. Zhang, N. Kaur et al., Enhanced osteogenesis of human mesenchymal stem cells by periodic heat shock in self-assembling peptide hydrogel. Tissue Eng. Part A **19**, 716–728 (2013)
16. P.C. Chang, Z.J. Lin, H.T. Luo, C.C. Tu, W.C. Tai, C.H. Chang et al., Degradable RGD-functionalized 3D-printed scaffold promotes osteogenesis. J. Dent. Res. **100**, 1109–1117 (2021)
17. P. Yu, L. Duan, Z. Yan, J. Li, D.Z. Cai, RADA-16-based self-assembled peptide nanofiber scaffolds loaded with TGF-β1 enhance the chondrogenic differentiation potential of BMSCs in vitro. Curr, Stem Cell Res, Ther. (2023). https://doi.org/10.2174/1574888X1866623031611 2847.
18. W. Ye, Z. Yang, F. Cao, H. Li, T. Zhao, H. Zhang et al., Articular cartilage reconstruction with TGF-β1-simulating self-assembling peptide hydrogel-based composite scaffold. Acta Biomater. **146**, 94–106 (2022)
19. S. He, J. Fang, C. Zhong, M. Wang, F. Ren, Spatiotemporal delivery of pBMP2 and pVEGF by a core-sheath structured fiber-hydrogel gene-activated matrix loaded with peptide-modified nanoparticles for critical-sized bone defect repair. Adv. Healthc. Mater. **11**, 2201096 (2022)
20. X. Ding, J. Gao, X. Yu, J. Shi, J. Chen, L. Yu et al., 3D-printed porous scaffolds of hydrogels modified with TGF-β1 binding peptides to promote in vivo cartilage regeneration and animal gait restoration. ACS Appl. Mater. Interfaces **14**, 15982–15995 (2022)
21. D. Docheva, S.A. Müller, M. Majewski, C.H. Evans, Biologics for tendon repair. Adv. Drug Deliv. Rev. **84**, 222–239 (2015)
22. H. Zhang, Y. Dai, H. Long, R. Cao, L. Shi, J. Zhao et al., Tendon stem/progenitor cell-laden nanofiber hydrogel enhanced functional repair of patellar tendon. Tissue Eng. Part A **29**, 150–160 (2023)
23. H. Yin, F. Strunz, Z. Yan, J. Lu, C. Brochhausen, S. Kiderlen et al., Three-dimensional self-assembling nanofiber matrix rejuvenates aged/degenerative human tendon stem/progenitor cells. Biomaterials **236**, 119802 (2020)
24. O. Alheib, L.P. da Silva, M.A. da Silva, K.A. Mesquita, R.P. Pirraco, R.L. Reis et al., Injectable laminin-biofunctionalized gellan gum hydrogels loaded with myoblasts for skeletal muscle regeneration. Acta Biomater. **143**, 282–294 (2022)
25. C. Eren Cimenci, G. Uzunalli, O. Uysal, F. Yergoz, E. Karaca Umay, M.O. Guler et al., Laminin mimetic peptide nanofibers regenerate acute muscle defect. Acta Biomater. **60**, 190–200 (2017)
26. M.-O. Baek, H.-J. Cho, D.S. Min, C.S. Choi, M.-S. Yoon, Self-transducible LRS-UNE-L peptide enhances muscle regeneration. J. Cachexia. Sarcopenia Muscle **13**, 1277–1288 (2022)
27. Q. Li, Z. Hu, Y. Liang, C. Xu, Y. Hong, X. Liu, Multifunctional peptide-conjugated nanocarriers for pulp regeneration in a full-length human tooth root. Acta Biomater. **127**, 252–265 (2021)
28. D. Wang, J. Deng, X. Deng, C. Fang, X. Zhang, P. Yang, Controlling enamel remineralization by amyloid-like amelogenin mimics. Adv. Mater. **32**, 2002080 (2020)
29. L. Zhou, Q.L. Li, H.M. Wong, A novel strategy for caries management: constructing an antibiofouling and mineralizing dual-bioactive tooth surface. ACS Appl. Mater. Interfaces **13**, 31140–31152 (2021)
30. Y. Sun, W. Li, X. Wu, N. Zhang, Y. Zhang, S. Ouyang et al., Functional self-assembling peptide nanofiber hydrogels designed for nerve degeneration. ACS Appl. Mater. Interfaces **8**, 2348–2359 (2016)
31. G.J. Wei, M. Yao, Y.S. Wang, C.W. Zhou, D.Y. Wan, P.Z. Lei et al., Promotion of peripheral nerve regeneration of a peptide compound hydrogel scaffold. Int. J. Nanomedicine **8**, 3217–3225 (2013)
32. J. Lu, X. Yan, X. Sun, X. Shen, H. Yin, C. Wang et al., Synergistic effects of dual-presenting VEGF- and BDNF-mimetic peptide epitopes from self-assembling peptide hydrogels on peripheral nerve regeneration. Nanoscale **11**, 19943–19958 (2019)
33. S. Liu, L. Zhou, C. Li, T. Min, C. Lu, S. Han et al., Chitin conduits modified with DNA-peptide coating promote the peripheral nerve regeneration. Biofabrication **14**, 015013 (2021)
34. J. Nam, H.-K. Lim, N.H. Kim, J.K. Park, E.S. Kang, Y.-T. Kim et al., Supramolecular peptide hydrogel-based soft neural interface augments brain signals through a three-dimensional electrical network. ACS Nano **14**, 664–675 (2020)

35. B.B. Hsu, W. Conway, C.M. Tschabrunn, M. Mehta, M.B. Perez-Cuevas, S. Zhang et al., Clotting mimicry from robust hemostatic bandages based on self-assembling peptides. ACS Nano **9**, 9394–9406 (2015)
36. L. Cai, S. Liu, J. Guo, Y.-G. Jia, Polypeptide-based self-healing hydrogels: design and biomedical applications. Acta Biomater. **113**, 84–100 (2020)
37. M.P. Bevilacqua, D.J. Huang, B.D. Wall, S.J. Lane, C.K. Edwards, J.A. Hanson et al., Amino acid block copolymers with broad antimicrobial activity and barrier properties. Macromol. Biosci. **17**, 1600492 (2017)
38. R. Wang, Q. Li, B. Chi, X. Wang, Z. Xu, Z. Xu et al., Enzyme-induced dual-network ε-poly-l-lysine-based hydrogels with robust self-healing and antibacterial performance. Chem. Commun. **53**, 4803–4806 (2017)
39. M. Comune, A. Rai, K.K. Chereddy, S. Pinto, S. Aday, A.F. Ferreira et al., Antimicrobial peptide-gold nanoscale therapeutic formulation with high skin regenerative potential. J. Control. Release **262**, 58–71 (2017)
40. P. Lou, S. Liu, Y. Wang, C. Pan, X. Xu, M. Zhao et al., Injectable self-assembling peptide nanofiber hydrogel as a bioactive 3D platform to promote chronic wound tissue regeneration. Acta Biomater. **135**, 100–112 (2021)
41. U. Park, M.S. Lee, J. Jeon, S. Lee, M.P. Hwang, Y. Wang et al., Coacervate-mediated exogenous growth factor delivery for scarless skin regeneration. Acta Biomater. **90**, 179–191 (2019)
42. Q. Bai, L. Teng, X. Zhang, C.-M. Dong, Multifunctional single-component polypeptide hydrogels: the gelation mechanism, superior biocompatibility, high performance hemostasis, and scarless wound healing. Adv. Healthc. Mater. **11**, 2101809 (2022)
43. Y. Xi, J. Ge, Y. Guo, B. Lei, P.X. Ma, Biomimetic elastomeric polypeptide-based nanofibrous matrix for overcoming multidrug-resistant bacteria and enhancing full-thickness wound healing/skin regeneration. ACS Nano **12**, 10772–10784 (2018)
44. C. Xing, H. Zhu, X. Dou, L. Gao, S. Baddi, Y. Zou et al., Infected diabetic wound regeneration using peptide-modified chiral dressing to target revascularization. ACS Nano **17**, 6275–6291 (2023)
45. D.R. Griffin, M.M. Archang, C.-H. Kuan, W.M. Weaver, J.S. Weinstein, A.C. Feng et al., Activating an adaptive immune response from a hydrogel scaffold imparts regenerative wound healing. Nat. Mater. **20**, 560–569 (2021)
46. G. Uzunalli, Z. Soran, T.S. Erkal, Y.S. Dagdas, E. Dinc, A.M. Hondur et al., Bioactive self-assembled peptide nanofibers for corneal stroma regeneration. Acta Biomater. **10**, 1156–1166 (2014)
47. J.R. Jangamreddy, M.K.C. Haagdorens, M. Mirazul Islam, P. Lewis, A. Samanta, P. Fagerholm et al., Short peptide analogs as alternatives to collagen in pro-regenerative corneal implants. Acta Biomater. **69**, 120–130 (2018)
48. C. Amit, S. Muralikumar, S. Janaki, M. Lakshmipathy, K.L. Therese, V. Umashankar et al., Designing and enhancing the antifungal activity of corneal specific cell penetrating peptide using gelatin hydrogel delivery system. Int. J. Nanomedicine **14**, 605–622 (2019)
49. Y.C. Chu, H.W. Fang, Y.Y. Wu, Y.J. Tang, E.H. Hsieh, Y.Z. She et al., Functional peptide-loaded gelatin nanoparticles as eyedrops for cornea neovascularization treatment. Int. J. Nanomed. **18**, 1413–1431 (2023)
50. H.V. Danesh-Meyer, N.M. Kerr, J. Zhang, E.K. Eady, S.J. O'Carroll, L.F. Nicholson et al., Connexin43 mimetic peptide reduces vascular leak and retinal ganglion cell death following retinal ischaemia. Brain **135**, 506–520 (2012)
51. Y. Zhang, D. Zhu, Y. Wei, Y. Wu, W. Cui, L. Liuqin et al., A collagen hydrogel loaded with HDAC7-derived peptide promotes the regeneration of infarcted myocardium with functional improvement in a rodent model. Acta Biomater. **86**, 223–234 (2019)
52. A.J. Rufaihah, I.C. Yasa, V.S. Ramanujam, S.C. Arularasu, T. Kofidis, M.O. Guler et al., Angiogenic peptide nanofibers repair cardiac tissue defect after myocardial infarction. Acta Biomater. **58**, 102–112 (2017)
53. J. Zimoch, J.S. Padial, A.S. Klar, Q. Vallmajo-Martin, M. Meuli, T. Biedermann et al., Polyisocyanopeptide hydrogels: a novel thermo-responsive hydrogel supporting pre-vascularization and the development of organotypic structures. Acta Biomater. **70**, 129–139 (2018)

54. L.-C. Huang, H.-C. Wang, L.-H. Chen, C.-Y. Ho, P.-H. Hsieh, M.-Y. Huang et al., Bioinspired self-assembling peptide hydrogel with proteoglycan-assisted growth factor delivery for therapeutic angiogenesis. Theranostics **9**, 7072–7087 (2019)

Chapter 14
Peptide-Based Nanomaterials for Sensors and Biosensors

Hani Nasser Abdelhamid

Abstract This book chapter offered an overview of the several types of peptide-based nanomaterials as (bio)sensors. It highlighted the recent advancements in related technologies during recent years. Peptide-based nanomaterials can be applied to several analytical techniques including electrochemical methods, electrochemiluminescence, and optical methods (UV-Vis-NIR spectroscopy, colorimetric, fluorescence, Raman spectroscopy, and surface plasmon resonance). They offered the quantification of target analytes in the sub-nanomolar range. They can be integrated into advanced nanotechnology. Peptides are effective nanomaterials for the development of sensitive and convenient biosensors. Peptide-based biosensors were applied for the detection of analytes, e.g., biomarkers, metal ions, pollutants, explosive materials, and antibodies. Precise chemical engineering of a peptide can serve as an alternative to an antibody offering significant advantages including enhanced tolerance to environmental conditions, and customizable sequences for detecting a diverse array of targets. Peptide-based nanomaterials can be used for large-scale sensor manufacturing. Peptide-based biosensors advanced clinical diagnosis.

Keywords Biosensors · Peptides · Sensors · Biosensors · Nanomaterials

14.1 Introduction

The market of analyte detection using devices is paramount for clinical diagnosis and medical health care [1]. IUPAC defined biosensors as ''A device that uses specific biochemical reactions mediated by isolated enzymes, immunosystems, tissues, organelles or whole cells to detect chemical compounds usually by electrical, thermal or optical signals". Biosensors are typically characterized as devices comprising a bio-recognition element (BRE) and a transducer. They are used for qualitative and quantitative/semi-quantitative analytical data.

H. N. Abdelhamid (✉)
Department of Chemistry, College of Science, Imam Mohammad Ibn Saud Islamic University (IMSIU), Riyadh 11623, Saudi Arabia
e-mail: hnabdelhamid@imamu.edu.sa

G. Wei (ed.), *Peptide Nano-Chemistry and Nanotechnology*,
https://doi.org/10.1007/978-981-96-2387-7_14

Peptides are supramolecular amino acid monomers with a shorter length than proteins [2–6]. They can be synthetic or naturally occurring. They offered dynamic structures resulting in nanoscale formations. They can undergo self-assembly via non-covalent interactions, including hydrogen bonds, electrostatic, π-stacking, hydrophobic, and van der Waals forces. There is increased interest in peptides due to their simple and efficient production process. Peptides may undergo self-assembly in the fabrication of organic templates and nanoscale components for advanced biosensors [7]. Short peptides can be employed as active components for the identification of their natural receptors. For example, pathogenic bacteria were detected using cell-penetrating peptides [8].

This book chapter discussed the principle of peptide-based nanomaterials and their applications for sensors and biosensors. Peptide-based nanomaterials were used for the detection of wide analytes including pollutants, explosives, biomarkers, heavy metal ions, and other important analytes. They can be integrated into several analytical and bio-analytical techniques such as electrochemical-based methods (e.g., cyclic voltammetry (CV), differential pulse voltammetry (DPV), and electrochemical impedance spectroscopy (EIS)), optical-based methods (ultraviolet (UV)-Visible (Vis)-Near infrared (NIR) spectroscopy, colorimetric, fluorescence, Raman spectroscopy, surface enhanced Raman spectroscopy (SERS), surface plasmon resonance spectroscopy (SPR), and other techniques.

14.2 Peptides-Based Nanomaterials

Peptides are short biopolymers of amino acid as monomers (Table 14.1). They are composed of a small number of amino acids. They exhibit superior chemical and structural stability compared to proteins (Table 14.1). Twenty natural amino acids constitute the essential building blocks in the synthesis of peptide (Table 14.1). Peptides adopt specific structures depending on the spatial arrangement of the groups inside a given peptide chain. They contain mainly functional groups of amino acids, i.e., carboxyl groups ($-COO^-$) and free/protonated amine groups ($-NH_2$/$-NH_3^+$). The negative and positive charges offered the zwitterionic properties of peptides (Table 14.1). The ionized functional groups can form a hydration layer on the surface via hydrogen bonding. This confers a hydrophilic property to the peptides, allowing them to diminish protein adsorption by alleviating their interactions. When the net charge of some peptide sequences nears neutrality, they can impede the adsorption of both cationic and anionic proteins.

The peptides structure plays an important role in the interaction with target analytes for analysis (Table 14.1). Certain peptides such as aptamers display strong interaction or high-affinity toward certain analytes, e.g., ions, proteins, and cells [9, 10]. The changes in the structure due to analyte interactions can be evaluated using in vivo or in vitro measurements. Aptamers peptides offer multiple functions, e.g., targeting specific compounds as recognition elements, inhibitors, and acting as biological therapeutic agents.

Table 14.1 Amino acids used for peptide formation including their abbreviations, chemical structure, and functional groups

Amino acid	3-letter abbreviation	1-letter abbreviation	Chemical structure	Functional group
Alanine	Ala	A	CH_3–$CH(NH_2)$–COOH	Amino (–NH_2), carboxylic acid (–COOH)
Arginine	Arg	R	NH_2–CH_2–CH_2–CH_2–CH_2–$CH(NH_2)$–COOH	Amino (–NH_2), guanidino (–NH–C(=NH)–NH_2)
Asparagine	Asn	N	NH_2–CO–CH_2–$CH(NH_2)$–COOH	Amide (–$CONH_2$), amine (–NH_2), carboxylic acid (–COOH)
Aspartic acid	Asp	D	HOOC–CH_2–$CH(NH_2)$–COOH	Carboxylic acid (–COOH), amine (–NH_2)
Cysteine	Cys	C	HS–CH_2–$CH(NH_2)$–COOH	Thiol (–SH), amino (–NH_2), carboxylic acid (–COOH)
Glutamine	Gln	Q	NH_2–CO–CH_2–CH_2–$CH(NH_2)$–COOH	Amide (–$CONH_2$), amine (–NH_2), carboxylic acid (–COOH)
Glutamic acid	Glu	E	HOOC–CH_2–CH_2–$CH(NH_2)$–COOH	Carboxylic acid (–COOH), amine (–NH_2)
Glycine	Gly	G	NH_2–CH_2–COOH	Amino (–NH_2), carboxylic acid (–COOH)
Histidine	His	H	CH_3–$CH(NH_2)$–CH_2–C(NH)(NH_2)–CH=CH–COOH	Imidazole (–NH–CH=N–)
Isoleucine	Ile	I	CH_3–CH_2–$CH(CH_3)$–$CH(NH_2)$–COOH	Amino (–NH_2), carboxylic acid (–COOH)
Leucine	Leu	L	CH_3–CH_2–CH_2–$CH(CH_3)$–$CH(NH_2)$–COOH	
Lysine	Lys	K	NH_2–$(CH_2)_4$–$CH(NH_2)$–COOH	
Methionine	Met	M	CH_3–S–CH_2–CH_2–$CH(NH_2)$–COOH	Thioether (–S–), amino (–NH_2), carboxylic acid (–COOH)

(continued)

Table 14.1 (continued)

Amino acid	3-letter abbreviation	1-letter abbreviation	Chemical structure	Functional group
Phenylalanine	Phe	F	C_6H_5–CH_2–CH(NH_2)–COOH	Phenyl group, amino (–NH_2), carboxylic acid (–COOH)
Proline	Pro	P	(CH_2)$_3$–CH(COOH)–NH	Secondary amine (–NH–), carboxylic acid (–COOH)
Serine	Ser	S	HO–CH_2–CH(NH_2)–COOH	Hydroxyl (–OH), amine (–NH_2), carboxylic acid (–COOH)
Threonine	Thr	T	CH_3–CH(OH)–CH(NH_2)–COOH	
Tryptophan	Trp	W	(CH_2)$_2$–C_6H_4–NH–CH(COOH)–CH_3	Indole, amino (–NH_2), carboxylic acid (–COOH)
Tyrosine	Tyr	Y	HO–C_6H_4–CH_2–CH(NH_2)–COOH	Phenol (–OH), amino (–NH_2), carboxylic acid (–COOH)
Valine	Val	V	CH_3–CH(CH_3)–CH(NH_2)-COOH	Amino (–NH_2), carboxylic acid (–COOH)

Peptide can be synthesized using several approaches including liquid-phase peptide synthesis (LPPS), solid-phase peptide synthesis (SPPS), native chemical ligation (NCL), expressed protein ligation (EPL), chemical ligation, and native peptide ligation (Table 14.2). SPPS enables the development of a specific sequence or a library of peptides targeting a particular entity. It can be also used for mass production. The summary for common synthesis procedures is given in Table 14.2.

Nanomaterials or nanoparticles are objects with a size of 1–200 nm. Peptides can be conjugated or modified with nanoparticles [11]. There are several synthesis methods for peptide-based nanomaterials (Table 14.3). Metallic nanoparticles (NPs), e.g., silver (Ag) nanoparticles and gold (Au) nanoparticles can form monolayers of the peptide on their external surface. The presence of a thiol functional group on the peptide enabled the self-assembly monolayer (SAM) of the peptide on metallic nanoparticles, e.g., Au NPs. Metal oxide nanoparticles such as magnetic nanoparticles (Fe_3O_4, MNPs) or iron oxide nanoparticles (IONPs) can be modified with peptides via covalent and non-covalent interactions. The presence of functional

Table 14.2 Common synthesis procedure for peptide synthesis

Method	Principle	Advantages	Disadvantages
Liquid-phase peptide synthesis (LPPS)	Stepwise addition of amino acids to a growing peptide chain using coupling reagents	Simple, flexible, can be used to synthesize peptides of various lengths and sequences	Low yields, time-consuming, and require purification
Solid-phase peptide synthesis (SPPS)	Stepwise addition of amino acids to a solid support using coupling reagents	High yields, easy purification, can be used to synthesize large peptides	Requires specialized equipment and reagents, potential for side reactions
Native chemical ligation (NCL)	Chemoselective reaction between a peptide thioester and a peptide N-terminal cysteine	High yields, can be used to synthesize long peptides with complex structures	Requires specific amino acid sequences, which can be challenging to optimize
Expressed protein ligation (EPL)	Enzymatic ligation of a synthetic peptide with a recombinant protein	High yields, can be used to synthesize large proteins with complex structures	Requires recombinant protein expression, can be expensive
Chemical ligation	Chemoselective reactions between peptides or amino acids	High yields, can be used to synthesize peptides with complex structures	Requires specific amino acid sequences, which can be challenging to optimize
Native peptide ligation	Enzymatic ligation of peptides	High yields, can be used to synthesize peptides with complex structures	Requires specific amino acid sequences, which can be challenging to optimize

groups such as hydroxyl (OH) enabled the chemical bond formation with groups such as silane. Metal chalcogenides and oxides such as zinc oxide/sulfide (ZnO/ZnS), and cadmium sulfide (CdS) quantum dots (QDs) were also reported. Carbon nanomaterials were reported inducing one-dimensional (e.g., carbon nanotube (CNTs), single-wall CNTs, multi-wall CNTs), two-dimensional (2D) carbon nanomaterials (e.g., graphene oxide (GO), reduced GO (rGO), graphene (G), and graphitic carbon nitride (g-C_3N_4)). In general, nanoparticles can be conjugated with peptides via in situ and ex situ synthesis procedures (Table 14.3). Nanoparticles can be synthesized in the presence of peptides, i.e., in situ or one-pot procedure. This procedure enables interaction between the nanoparticles and the peptides' functional groups. The peptide is a template for the synthesized nanoparticles controlling the nanoparticle's properties such as size and morphology. On the other side, the ex situ procedure involved the synthesis of both peptides and nanoparticles and then conjugate them in one composite.

Advanced porous polymeric materials, e.g., hybrid polymers (metal–organic frameworks (MOFs) [12]) and organic polymers (e.g., covalent organic frameworks (COFs)) [13] were also conjugated with peptides (Table 14.3) [11]. A new class was also reported including peptide-derived MOFs (PMOFs) [14]. The integration of peptides imparts conformational flexibility, inherent chirality, and improve the molecular recognition capabilities to the frameworks. Thus, the materials can significantly enhance the utility in enantiomeric separation, and affinity interactions with target analytes. The functional groups of COF materials can be functionalized with peptide's functional groups via covalent bonds.

Two-dimensional (2D) inorganic metal carbides, nitrides, and carbonitrides, denoted as MXene were conjugated with peptides. They exhibit layer morphology that ensures the encapsulation of peptides into the interplanar distance materials [15]. Peptides can undergo self-assembly into the layers of MXene [15].

Peptide-based nanomaterials can be characterized using a wide number of analytical methods. Particle size can be characterized using nanoparticle tracking analysis (NTA), dynamic light scattering (DLS) for a solution, and electron microscopy (transmission or scanning electron microscopy (TEM or SEM)) in a solid state. The morphology can be evaluated using optical microscopy, TEM, or SEM images. The surface charge before and after modification can be determined using zeta potential measurements. The formation of a monolayer of peptides on a substrate such as a gold (Au) electrode can be characterized using CV, EIS, and atomic force microscopy (AFM) [16].

Conjugation of peptides with nanoparticles is important to fulfill the requirements for materials to be used for sensor or biosensor fabrication. Nanoparticles play an important role in peptide-based nanomaterials. They can be the source of the signal, e.g., fluorescence emission. For example, the near-infrared emission of nanoparticles such as single-wall carbon nanotubes (SWCNTs) enabled the detection in biological fluids. On the other side, peptides ensure the selectivity toward certain analytes. Chemosensors based on peptides can be selective toward metal ions such as Cu^{2+} using diglycyl-l-histidine that mimics the specific transport site of human albumin [17].

Table 14.3 List of nanoparticles used for the formation of peptide-based nanomaterials, size, type of interaction and their synthesis conditions

Nanoparticle	Size (nm)	Type of interaction	Synthesis conditions
Au NPs	1—100	Electrostatic, hydrophobic, covalent	Reduction of Au ions in the presence of peptides, self-assembly
Ag NPs	1–100	Electrostatic, hydrophobic, covalent	Reduction of Ag ions in the presence of peptides, self-assembly
IONPs	1–100	Electrostatic, magnetic	Co-precipitation of iron oxide nanoparticles with peptides, self-assembly
Semiconductor nanoparticles (e.g., CdSe, ZnS)	2–10	Electrostatic, hydrophobic, covalent	Synthesis of nanoparticles in the presence of peptides, self-assembly
CNTs	1–100	π–π interactions, hydrophobic interactions	Non-covalent functionalization of CNTs with peptides, covalent modification
GO	1–10 (thickness)	Electrostatic, π–π interactions, hydrophobic interactions	Functionalization of GO with peptides, self-assembly
QDs	2–10	Electrostatic, hydrophobic, covalent	Synthesis of QDs in the presence of peptides, self-assembly
MOFs	25–200	Electrostatic, coordination bonds	Synthesis of MOFs in the presence of peptides, post-synthetic modification
COFs	10–200	Covalent bonds	Synthesis of COFs in the presence of peptides, post-synthetic modification
MXene	1–10 (thickness)	Electrostatic, hydrophobic,	Post-synthetic procedures

14.3 The Role of Peptides-Based Nanomaterials in (Bio)Sensors

Peptides can be an important part of sensors and biosensors (Table 14.4) [18]. They are utilized as recognition elements in biosensing (Table 14.4). They are frequently employed as probes in biosensors. Edited peptides can maintain their strong affinity for the target analyte. A particular peptide sequence, with unique enzyme substrates, is essential in tests for enzymatic activity and the screening of enzymatic inhibitors.

Various approaches have been employed to construct peptide-based biosensors by the conjugation of peptides with signal markers. Peptides can be used for analyte adsorption such as glucose [19]. In summary, the common roles of peptides in biosensors are recognition elements, immobilization agents, signal transduction elements, and amplification elements (Table 14.4).

Peptide functionalization or editing is also required to induce certain properties. For instance, a peptide-based chemosensor was designed using a synthetic peptidyl template and a covalently linked fluorescent reporter that responds to metal-induced conformational alterations in the polypeptide structure [20]. Environmentally sensitive fluorophores, including the derivatives of benzamide, naphthalene-sulfonamide, and coumarin were reported for peptide modification [20]. The ligand sphere was manipulated by removing one thiolate ligand to improve the oxidative stability of

Table 14.4 Common techniques used for peptide-based nanomaterials

Technique	Principle	Advantages	Disadvantages
Electrochemical	Electrochemical signal changes (current, potential) due to peptide-analyte interactions	High sensitivity, low cost, rapid response	Susceptible to interference from other ions or molecules
Optical	Changes in fluorescence intensity, color, or polarization due to peptide-analyte interactions	High sensitivity, non-destructive, label-free	Requires specialized equipment, potential for photobleaching
Field-effect transistors (FETs)	Changes in electrical conductivity due to peptide-analyte interactions	High sensitivity, label-free, real-time monitoring	Susceptible to noise
Transducers	Changes in physical properties (e.g., temperature, pressure) due to peptide-analyte interactions	High sensitivity, label-free, can be used in vivo	Requires specialized equipment, high experience
Quartz crystal microbalances (QCM)	Changes in resonance frequency due to mass changes caused by peptide-analyte interactions	High sensitivity, label-free, real-time monitoring	Requires specialized equipment, sensitive to environmental noise
Surface plasmon resonance (SPR)	Changes in refractive index due to peptide-analyte interactions	Label-free, real-time monitoring, can be used with live cells	Requires specialized equipment, sensitive to environmental factors
Raman spectroscopy	Changes in Raman spectra due to peptide-analyte interactions	Label-free, can provide structural information	Requires specialized equipment, sensitive to environmental factors

the chemosensor [20]. A fluorescent peptidyl chemosensor via fluorenylmethoxy-carbonyl (Fmoc) solid-phase peptide synthesis for heavy metal ions was reported based on the fluorescence resonance energy transfer (FRET) mechanism [21]. It was designed using fluorophores tryptophan (donor) and dansyl chloride (acceptor), and comprises the amino acids glycine and aspartic acid. It can be applied for Cu^{2+} ions analysis via a metal chelating unit that showed fluorescence quenching of both the donor and acceptor fluorophores.

Most of the peptide synthesis is based on trial-and-error procedure. The metal-binding peptides are engineered to mimic the natural peptides. In some cases, an extra amino acid is included to enhance the stability. Thompson and Lia reported the synthesis of DKDGDGYITAAE for the detection of uranyl ion (U^{6+}), derived from the calcium-binding protein calmodulin (Fig. 14.1) [22]. The synthetic peptide was reported for an electrochemical peptide-based (E-PB) sensor. It contains a 12-amino acid with a sequence based on calmodulin (CaM) that was reported as a Ca^{2+}-binding protein. However, the authors included a phosphothreonine in the synthesized peptides to enhance its affinity for U^{6+} over Ca^{2+} (Fig. 14.1) [22]. The peptide was modified with thiol and methylene blue (MB) dye to ensure modification of the gold electrode and improve the electrochemical mechanism of the metal detection (Fig. 14.1).

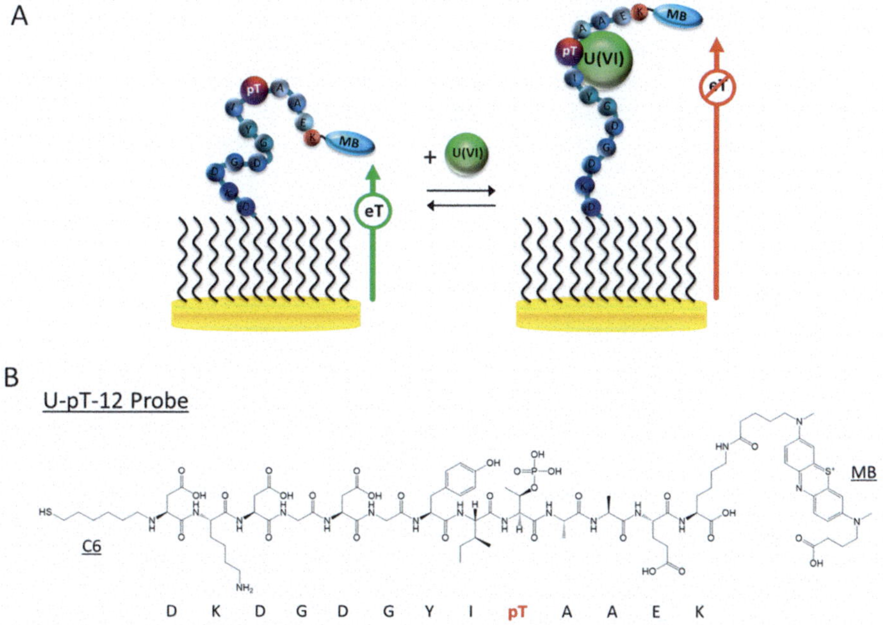

Fig. 14.1 **a** Scheme for the sensing mechanism of the E-PB for U^{6+} sensor and **b** structure of peptide probe. Figure reprinted from Ref. [22].

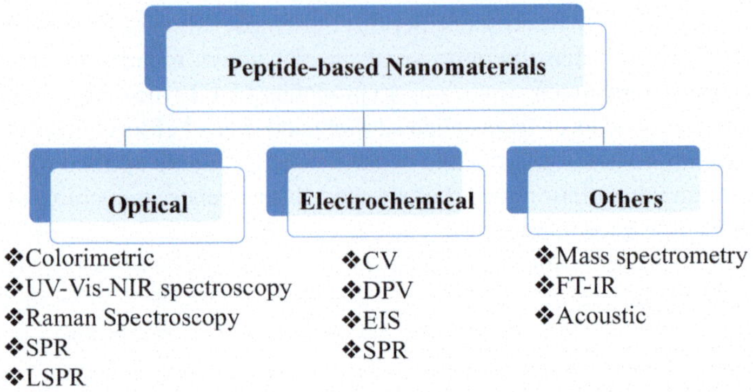

Fig. 14.2 Common methods used for peptide-based nanomaterials

Peptides were applied for various techniques for the analysis of a wide number of analytes. They were reported for both fluorescence [23], transducer [8], and electrochemical biosensors (Table 14.4) [24]. Advanced and combined techniques such as electrochemiluminscence were also reported [25]. They can be used for heavy metal analysis (e.g., zinc (Zn^{2+}) [20, 26, 27], copper (Cu^{2+}) [21], cadmium (Cd^{2+}) [28], silver (Ag^{+}) [29], mercury (Hg^{2+}) [30], uranium (U^{6+}) [22], etc.), cancer biomarker [31], and glucose [16]. Peptide-based biosensors offer a low limit of detection (LOD). For instance, it showed an LOD of 32 $\mu g\ L^{-1}$ for Cu^{2+} ions [21].

Peptide-based biosensors can be categorized based on signal transduction into optical, electrochemical, and mechanical measurements (Fig. 14.2). Optical and electrochemical biosensors are widely applied due to their simplicity and compatibility with emerging technology.

14.4 Peptide-Based Optical Sensors

Peptide-based optical sensors or biosensors are widely used for several analytes and biomarkers. They offer several techniques including colorimetric, UV-Vis-NIR spectroscopy, SPR, LSPR, and Raman spectroscopy (Fig. 14.2). SPR method is a label-free optical detecting technique. They have high sensitivity, a superior signal-to-noise ratio, and resistance to signal interference. Similar to SPR, LSPR produces amplified plasmon resonances on the surfaces of nanostructures, which are absent on the unstructured metallic surface. LSPR sensors exhibit enhanced sensitivity owing to the pronounced UV absorbance band produced by these nanostructures.

Colorimetric biosensors are a direct detection method that depends on the binding of a peptide to its target, causing changes in the structure that lead to spectrum shifts. These alterations can be converted into observable color changes, facilitating the qualitative or quantitative assessment of target compounds. This procedure is straightforward and intuitive, observable without the necessity for costly or intricate instruments. Cyclic peptide-based colorimetric sensing technique was reported for glucose [19]. The analysis technique is based on the catalytic properties of gold nanoparticles (Au NPs). The chemosensor includes a synthesized cyclic peptide that selectively interacts with glucose. The peptides showed adsorption on the Au NPs surface preventing the adsorption of 4-nitrophenol that can undergo reduction using Au NPs as catalyst. The reduction of 4-nitrophenol to 4-aminophenol using $NaBH_4$ showed a color change. The selective interaction of glucose into peptide hinders the adsorption of 4-nitrophenol. Consequently, the solution's color exhibited different shades of yellow according to the concentration of glucose. The approach exhibited a rapid response time of 10 min and showed a linear response across glucose concentrations ranging from 0.1 to 20 mM, with a LOD of 0.04 mM. The approach was effectively utilized for detecting glucose in spiked food samples and spiked rabbit blood, with recovery rates of 88.04–103.28% and 94.27–101.53%, respectively [19]. A colorimetric sensor utilizing peptides/AuNPs/magnetic NPs to engage with *Escherichia coli* (*E. coli*) O157:H7 proteases can precisely quantify *E. coli* levels in food [32]. The intensity of the golden color on the sensor surface was reported as a function of the *E. coli* cell numbers, thereby markedly diminishing possible health hazards associated with food contamination [32]. Peptide-based colorimetric methods are a low-cost method, rendering them appropriate for extensive applications across diverse detection domains. They can offer high sensitivity compared to well-established techniques such as ELISA. For example, a colorimetric sensor of AuNPs/polymorphic peptides and ELISA offered a LOD of 0.481 and 0.979 for *Toxoplasma gondii*, respectively [33]. Peptide/AuNPs/magnetic nanoparticles (MNPs) were also reported for colorimetric sensors. It can quantify the proteolytic activity of certain protein enzymes from copper-green Pseudomonas that turned back when obscured by MNPs [34]. The MNPs can be separated and moved via an external magnet.

The colorimetric method was also reported using porous materials such as MOFs. Peptide-functionalized MOFs were used as signal-amplifying tags for the detection of secreted proteins acidic and rich in cysteine (SPARC) [12]. Enzyme-MOF composites were synthesized via a co-precipitation method. Enzyme of horseradish peroxidase (HRP) and MOF, e.g., ZIF-90 and ZIF-90. MOF materials are effective protective matrix for HRP immobilization. HRP@ZIF-90 was applied as a colorimetric sensor application. The colorimetric method is based on the release of the encapsulated HRP under acidic conditions. The method offered a LOD of 30 fg/mL [12].

UV-Vis-NIR spectroscopy can be used for the analysis of several analytes using peptide-based biosensors. It is based on measuring light absorption at a certain wavelength. Chemical conjugation of the peptide with GO can be used for a UV absorbance response toward an analyte such as 2,4,6-trinitrotoluene (TNT) [35]. The analysis method offers a LOD of 4.40×10^{-9} mM. It can be selective toward TNT over similar analytes, e.g., 2,6-Dinitrotoluene (DNT). UV-Vis-NIR spectroscopy using peptide-based nanomaterials exhibits high sensitivity and good selectivity, facilitating a practical approach for the miniaturization and integrated detection of optical sensors.

Electromagnetic waves such as near-infrared (NIR) waves can be also used for biosensing applications. A NIR biosensor was reported utilizing peptide/SWCNT hybrids for the detection of TNT explosives [36]. The binding of TNT with peptide was immediately affixed to the sidewall of the SWCNTs by π–π interactions between the aromatic amino acids and SWCNTs. The interaction leads to the formation of peptide-SWCNT hybrids for NIR absorption spectral analysis. Peptide-SWCNT hybrids-based NIR optical biosensors showed sensitivity and excellent selectivity for the detection of TNT explosives [36].

Upon the chemical conjugation of the particular peptide with GO, a UV absorbance response can be elicited by TNT concentrations as low as 4.40×10^{-9} mM. 2,6-dinitrotoluene (DNT), the chemical counterpart of TNT, exhibits insensitivity during detection [90]. Consequently, this type of biosensor exhibits good sensitivity and selectivity, facilitating a practical approach for the miniaturization and integrated detection of optical sensors.

Peptide-based fluorescence sensors or biosensors are intensively used for the detection of several analytes [37]. The fluorescence emission of peptides can be due to (1) intrinsic amino acids in the peptide structure including tyrosine, tryptophan, or phenylalanine; (2) attaching fluorescent dyes, e.g., fluorescein, rhodamine, or Cy5; and (3) incorporating nanoparticles with fluorescence emission properties, e.g., quantum dots, graphene oxide, etc.

The spectral properties of peptide-based fluorophores are significantly influenced by their surrounding environment (Fig. 14.3). Consequently, they can be used for sensing or biosensing of target analytes. The detection process can be due to several mechanisms including excimer, fluorescence resonance energy transfer (FRET), photo-induced electron transfer (PET), intramolecular charge transfer (ICT), and probe–quencher 'Turn-OFF' or probe-enhancement 'Turn-ON' (Fig. 14.3). The production of an excimer, a transitory dimeric or heterodimeric molecule with two species, allows detection. Species like excimer emit longer wavelengths. The wavelength change in emission from the excimer-monomer transition can be used to generate biosensor signal output. Energy transfer between two nearby chromophores between 1 and 10 nm is called FRET. Non-radiative dipole–dipole coupling transfers energy. The donor emission spectrum and acceptor absorption spectrum must overlap for the FRET effect. FRET-induced fluorescence quenching or amplification can be used to construct biosensor signal transducers. Probe–quencher pair methods use static quenching instead of FRET because a ground state non-fluorescent complex forms (Fig. 14.3).

An excimer-type peptide beacon was reported for the detection of anti-HIV antibodies (Fig. 14.4a) [38]. The detection process is based on the affixation of two pyrene units into the terminal ends of the peptide. The interaction within the peptides caused folding and generated a pyrene excimer with an emission at 475 nm due to hydrophobic interactions among the pyrene units (Fig. 14.4a). The anti-HIV antibodies-antibody interaction with the peptide caused conformation change in the pyrene excimer, leading to a shift in the emission band from 475 to 375 nm and a reduction in the intensity of the excimer emission band [38]. The excimer mechanism is widely applied for several peptides for the analysis of various analytes.

A molecular biosensor was reported utilizing a probe–quencher pair mechanism for the detection of HIV, i.e., anti-HIV antibodies (Fig. 14.4b) [39]. It comprised

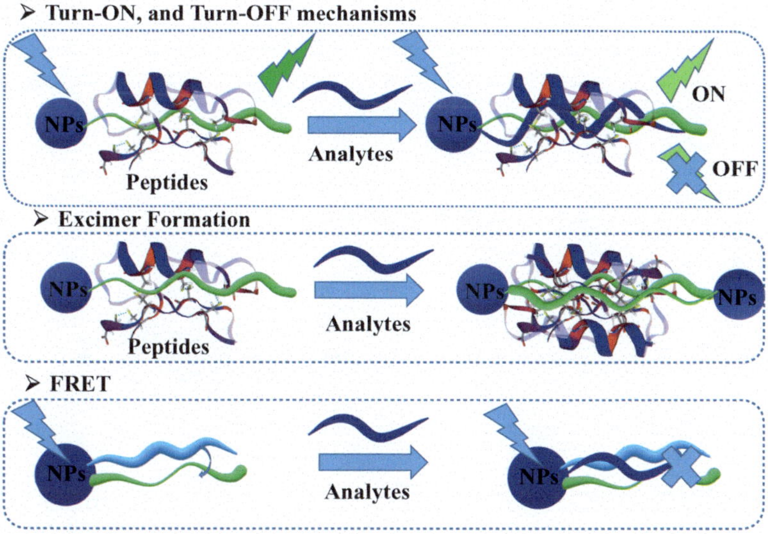

Fig. 14.3 Common mechanisms for fluorescence spectroscopy using peptide-based nanomaterials

a ruthenium (II) bisbipyridine-phenanthroline chelate as a fluorophore 'probe' and methyl viologen serving as an electron-accepting quencher. The probe and quencher were affixed at opposing ends of the peptide beacon. The interaction between the HIV antibody and the peptide obstructed electron transfer (ET) between the probe and the quencher, leading to an increase in the fluorescence emission of the probe (Fig. 14.4b). The HIV molecular biosensor might attain a six-fold enhancement in fluorescence emission intensity upon binding to HIV antibodies. The detection threshold for anti-HIV antibodies with this method may reach as low as 250 pM [39]. The pair-quenching mechanism is widely used for the analysis of analytes and biomarkers, e.g., human carbonic anhydrase II (HCA II) [40, 41], antibody anti-octreotide (AOC) [42].

A fluorescent biosensor was reported for analyzing peptide-receptor interactions utilizing GO and FITC-labeled octreotide (FOC) [42]. FOC peptide demonstrated a strong adsorption affinity for GO, leading to effective quenching of FITC. The specific interaction of the anti-octreotide antibody (AOC) with FOC competitively displaces FOC from the GO surface, resulting in the restoration of fluorescence. The GO-based fluorescent platform enabled the detection of AOC with an LOD of 2 ng/mL. It was applied to probe somatostatin receptor subtype 2 overexpressed in AR42J tumor cells, indicating significant potential for molecular imaging in cancer detection [42].

Several nanoparticles were reported with peptides to analyze several analytes using fluorescence spectroscopy. A peptide-functionalized MoS_2 nanosheets (NSs) was reported for the detection of cysteine protease-3 with an LOD of 0.33 ng/mL [43]. A peptide/Fe_3O_4@SiO_2–Au nanocomposite was reported as a fluorescence biosensor

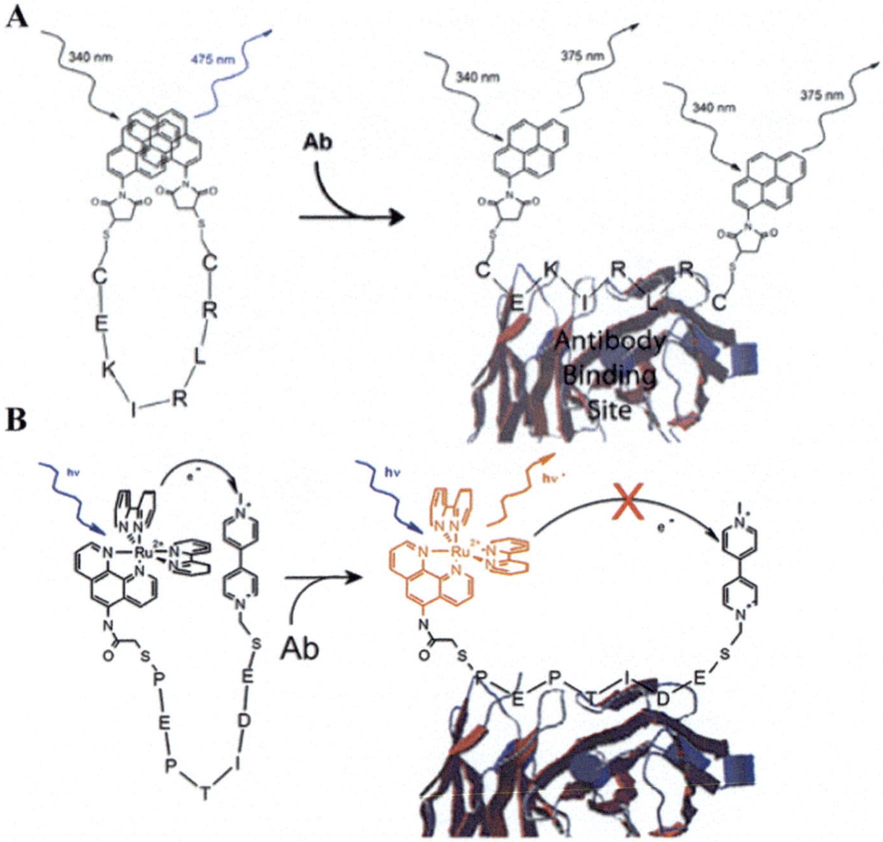

Fig. 14.4 **a** Excimer mechanism for peptide-modified pyrene for antibody (Ab) analysis. Figure reprinted with permission from Ref. [38]. Copyright © 2006 American Chemical Society; **b** probe-quencher pair mechanism. Figure reprinted with permission from Ref. [39]. Copyright © 2007 American Chemical Society

for the detection of prostate-specific antigen (PSA) [44]. It exhibited highly selective and good sensitivity (LOD of 0.3 pg/mL). A fluorescent sensor array utilizing two-dimensional MOFs (2D MOFs) linked with functional peptides offered precise detection of diverse pathogens with a high accuracy of 100% [45]. It can effectively distinguish mixed pathogen samples without requiring additional procedures [45].

Biosensors based on Raman Spectroscopy characterize the electrostatic interactions between the peptide and a target analyte. They offered structural information on the target via spectroscopy. They offer benefits including the prevention of peak overlap in intricate systems and the absence of photobleaching or self-quenching problems. It was once regarded as a highly promising optical technology, although it was not extensively utilized due to its weak signal. Nonetheless, sensitivity-enhancing methods like SERS were employed to rectify these deficiencies. The

SERS method was reported for the analysis of cadmium ions using a composite of Au/Starch/COF [46].

Other optical methods using peptides were also reported. An LSPR sensor was reported for bioanalytes such as cholera toxin subunit B (CTX-B) with an LOD of 1.89 ng/mL [47], and explosive materials [48]. The plasmonic biosensor of the AuNPs-coated glass chip was constructed by immobilizing affinity peptides, referred to as norovirus binding peptides, onto the localized LSPR layer for the detection of noroviral capsid proteins and human norovirus [49]. The biosensors offered a minimum detection value for noroviral capsid protein and human norovirus of 0.1 ng/mL and 0.9 copies/mL, respectively. LSPR using peptide-based nanomaterials offered label-free biosensors of analyte [50]. It can be optimized to ensure high sensitivity and better selectivity [51].

14.5 Peptide-Based Electrochemical Biosensors

Peptide immobilization on the electrode surface can occur via non-covalent or covalent interactions. The approaches encompass (1) electrostatic interactions, (2) direct interactions such as thiolated peptide and gold electrode interactions, (3) secondary interactions such as biotin-avidin interactions, (4) covalent bonds e.g., coupling reactions, or (5) encapsulation techniques. The SAM utilizing thiol linkages is frequently employed in peptide-based sensors. Peptides can self-assemble into nanostructures via non-covalent intermolecular interactions, including electrostatic forces, hydrogen bonding, van der Waals forces, and hydrophobic interactions. Peptide-modified electrodes can be then used in electrochemical biosensors. The common electrode is a thiol-containing peptide binding to a gold substrate. The electrode contains also a spacer sequence to preserve the flexibility or accessibility of enzymatic activity and recognition ligand. The signal production and amplification in electrochemical biosensors are contingent upon their operational modes, including amperometry, voltammetry, or potentiometry.

Peptide-based electrochemical biosensors generally comprise a particular peptide sequence, tagged with a redox reporter and modified with a thiol-containing group to facilitate immobilization on an Au electrode surface. They generally comprise a short peptide sequence, tagged with a redox reporter and modified with a thiol group to facilitate immobilization on a gold electrode surface. A spacer is frequently included between the thiol group and the peptide to enhance enzyme accessibility and provide flexibility to the probe. A study reported the impact of polyethylene glycol (PEG) spacer length on biosensor performance to enhance understanding and optimize a peptide-based electrochemical sensor. A peptide with multiple PEG spacers (e.g., PEG-4, PEG-6, PEG-8, and PEG-12) was synthesized and assessed by including a target enzyme (trypsin) and bovine serum albumin (BSA) to evaluate their analytical efficacy. An alkyl-based spacer was evaluated to compare the impacts of its composition. All proposed probes facilitated effective protease detection; however,

PEG-6 exhibited superior antifouling properties, underscoring the critical importance of the spacer in the design of peptide-based probes [52].

An electrochemical sensor was reported using zwitterionic peptides-modified a conducting polymer of citrate-doped poly(3,4-ethylenedioxythiophene) (PEDOT) through nickel cation coordination (Fig. 14.5) [53]. The sensor fabrication involved the synthesis of a layer of the conducting polymer PEDOT via direct electrodeposited onto a glassy carbon electrode (GCE) (Fig. 14.5). In the deposition process, citrate ions containing three negatively charged carboxyl groups were integrated into the synthesized PEDOT conducting polymer as a dopant, with certain carboxyl groups on the surface utilized for subsequent peptide immobilization. Nickel cations were introduced to connect carboxyl groups from the PEDOT film surface to peptides via Ni−O coordination bonds. Peptides affixed to the PEDOT film were subsequently employed for the immobilization of the BRCA1 recognition probe (terminated with a carboxyl group) via the creation of amide bonds, therefore successfully fabricating the DNA sensor. The resultant peptide-modified PEDOT exhibited exceptional antifouling properties and satisfactory conductivity, subsequently facilitating the immobilization of a DNA probe to develop an electrochemical biosensor for the breast cancer biomarker BRCA1. The DNA biosensor exhibited high sensitivity with a limit of detection of 0.03 fM and high selectivity. It can used for detecting BRCA1 in 5% (v/v) human plasma with commendable accuracy and minimal fouling. The integration of antifouling and biocompatible peptides with conducting polymers offered effective electrochemical biosensors that can detect targets in complicated biological environments with high sensitivity and resistance to biofouling (Fig. 14.5) [53].

A synthetic cyclic peptide (CP), (cyclo-[−CNDNHCRDNDC−]), was reported as a recognition receptor for glucose analysis and gold nanoparticles supported graphitic carbon nitride (AuNPs/g-C_3N_4) [25]. The electrode was synthesized based

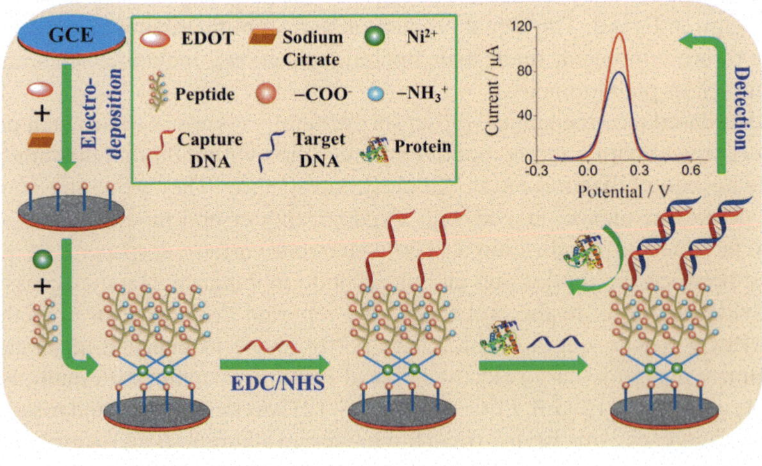

Fig. 14.5 The fabrication process of the DNA sensor. Figure reprinted with permission from Ref. [53] Copyright © 2024 Elsevier B.V

on AuNPs/g-C_3N_4 for the enhancement of ECL [25]. The synthetic CP receptor with a natural active center of Glu binding protein can replicate the interactions between Glu and Glu binding protein to selectively bind Glu. The AuNPs were decreased on g-C_3N_4, resulting in a nanohybrid suitable for use as an ECL emitter. The AuNPs/g-C_3N_4 significantly enhanced the ECL response of unmodified g-C_3N_4. Glu quantification was performed by measuring ECL shifts resulting from the interaction of Glu with CPs. The linear detection range of the constructed CPs-based ECL sensor was in the range of 1–100 mmol L^{-1}, with an LOD of 0.57 nmol L^{-1} ($S/N = 3$). The CP-based ECL sensor offered the advantages of high specificity, good repeatability, high stability, and favorable recoveries in sample analysis [25].

Peptide-based nanomaterial electrochemical biosensors offered several advantages such as high sensitivity, excellent selectivity, and in some cases specificity. They have advanced clinical diagnoses [54]. They can be used for miniaturization sensing devices. They can be applied for the integration with smartphones. Peptide-based nanomaterials using electrochemical methods are widely used for several analytes. Ac-ACSSWLPRGCGGGS-NH_2 peptide was reported for the analysis of juvenile idiopathic arthritis (JIA) IgG using DPV and EIS [55]. Modified methylene blue HS–(CH2)11-EAAEWDRVHP-K-MB peptide sequence was reported for the detection of anti-p24 antibodies (HIV detection) [56]. QHIMHLPHINTL was reported to detect human norovirus with a detection limit of 7.8 copies/mL [57]. A phage peptide ligand, i.e., EHDRMHAYYLTR offered low LOD for dengue fever biomarker NS1 using the impedance and SWV-based detection with a LOD of 1.5–25 μg/mL and in the range of 10^5 to 10^9 copy/mL [58]. Electrochemical-based methods can be also used for measuring the enzymatic activity. For example, a trypsin biosensor was reported utilizing an Fc-labeled peptide modified with a flexible PEG linker [59]. The technique is based on the trypsin cleavage of the peptide sequence causing a drop in the current measured using the CV test [59].

MXene exhibits good electrochemical performance making them promising for peptide-based biosensors. Chimeric peptide-functionalized titanium carbide MXenes (Pep-Ti_3C_2) were reported for the activity testing of post-translational modification (PTM) enzymes through integration with carboxypeptidase Y (CPY)-mediated peptide cleavage [15]. A peptide-on-AuNPs/MXene sensor was developed for the selective detection of cathepsin B [60]. It demonstrated 2–fourfold lower values for clinical samples compared to the other sensors, indicating high recovery (86–102%) with relative standard deviation (%RSD, < 11%), and the ability to differentiate various phases of Crohn's disease [60]. The peptide aptamer/Au@BP@Ti_3C_2-MXene and peptide aptamer/magnetic Au@$ZnFe_2O_4$@COF nanocomposites were examined for the detection of the foodborne pathogen Norovirus (NoV) [61]. The response currents of the electrochemical biosensor exhibited a direct correlation with NoV concentrations spanning from 0.01 to 10^5 copies/mL, with a LOD of 0.003 copies/mL [61].

14.6 Peptides for Field-Effect Transistors (FET)

Peptides can be used for ion-selective sensors of heavy metal ions. A tripeptide of Gly-Gly-His selectively interacts with copper, while its isomer Gly-His-Gly cross-reacts with both copper and zinc [62]. The peptide can be modified for selective detection of heavy metal ions. A copper ion sensor with high sensitivity and specificity was reported employing single crystalline silicon nanowires (SiNWs) configured as FETs [62]. The surface of SiNWs is modified with a histidine (His)-containing tripeptide that serves as a copper-sensitive layer. The complexation of copper ions with histidine-containing tripeptides improves the conductance of silicon nanowires, with the location of the histidine residue in the tripeptide markedly affecting copper ion binding. Data analysis showed that peptide (Gly-Gly-His)/SiNWs was a selective biosensor for Cu^{2+} ions, even in the presence of Zn^{2+} ions [62].

The copper complexation mechanism by the peptide was directly investigated on the surface of a gold microcantilever (Fig. 14.6) [63]. The authors investigated various conditions including copper concentration and duration of contact. The results reveal the presence of two separate copper–peptide complexes: a monodentate, low-affinity complex is swiftly established by histidine, succeeded by a gradual shift to a high-affinity tetradentate complex, which provokes a varied response from the cantilever due to the change in the volume of the peptide monolayer (Fig. 14.6).

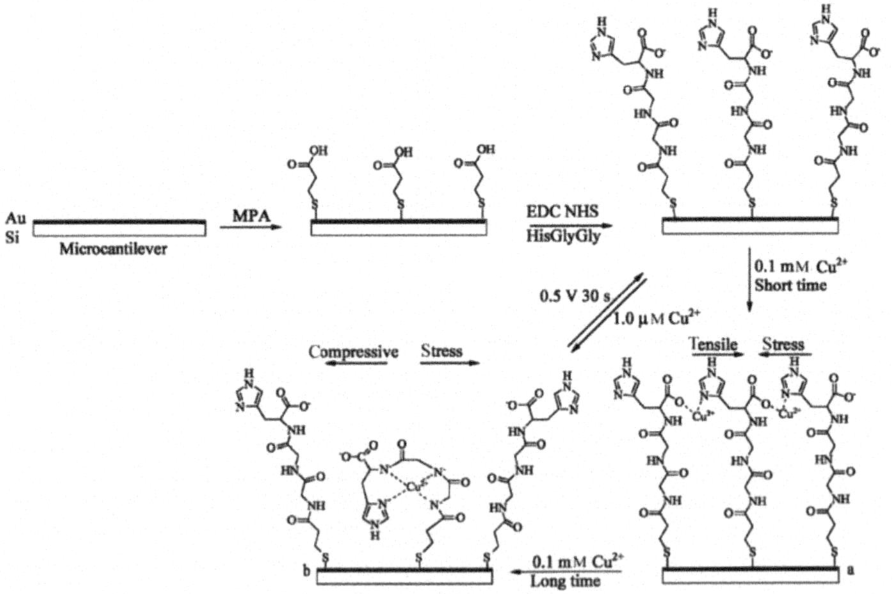

Fig. 14.6 Process of interactions between peptide-modified microcantilever and Cu^{2+} ions. Figure reprinted with permission from Ref. [63]. Copyright © 2009 Changchun Institute of Applied Chemistry, Chinese Academy of Sciences. Published by Elsevier B.V

14.7 Peptide-Based Quartz Crystal Microbalance (QCM)

A set of cyclic peptides (CPs), with a sequence of cyclo[–**CNDNHCRDND**C–], glucose receptors was engineered to replicate the binding sites of glucose binding protein (GBP) [16]. The characteristics of these CPs employed as a glucose receptor or alternative to the GBP were examined with a QCM approach. The binding affinity of the CPs to saccharides (e.g., galactose, fructose, lactose, sucrose, and maltose) was examined, revealing that the CPs' sensitivity and selectivity for glucose were contingent upon their configuration, namely, the amino acid sequence of the CPs. The cyclic unit featuring a cyclo[–CNDNHCRDNDC–] sequence exhibited the greatest selectivity and sensitivity for glucose detection. This study proposes that a synthetic peptide with a specific functional sequence may be utilized for the development of a new generation of glucose receptors, with significant applications in biological, life sciences, and clinical diagnostics [16].

14.8 Important Aspects of Peptide-Based Sensors and Biosensors

Peptide design is a challenge and in some cases is time-consuming. Thus, in silico techniques were additionally employed to investigate peptide–peptide or peptide–small molecule interactions relevant to biosensor applications. In silico, methods can be employed to investigate the kinetics and thermodynamics of the interactions between a peptide and a target analyte. The peptide design can be enhanced by various in silico techniques to get increased sensitivity toward its target analyte. The peptides may be formulated as an affinity-based biosensor or as a substrate for enzymatic activity. The last stage involves the formulation of the sensing matrix utilizing the peptide as BRE. An appropriate electrode is initially selected according to the sample type or the operational electrochemical potential. A layer of nanomaterials is often coated or transferred over the electrode surface to enhance the surface area or catalytic properties of the sensor. Nanomaterials may consist of metal nanoparticles or polymers. The peptide is subsequently affixed to the sensor surface by covalent or ligand-receptor interactions (e.g., –SH linkage), followed by the application of a blocking layer (e.g., ethanolamine) to mitigate the antifouling effects of the matrix. Subsequently, the sensor interacts with the target analyte, and a signal amplification approach is implemented to obtain the appropriate sensor signal. The selection of a suitable electrochemical approach is required for the detection of the target analyte, contingent upon the nature of the interaction and the materials utilized.

The suitable selection of nanoparticles and peptides is important for the composite stability and analytical methods. For example, metallic nanoparticles, e.g., Ag and Au can be used for colorimetric and optical sensors. However, AgNPs possess a molar extinction coefficient that is 100 times larger than that of AuNPs. Thus, it offered high sensitivity and visibility of sensors. Thus, AgNPs/peptides nanomaterials offered lower LOD of the cancer biomarker Mdm2 protein compared to AuNPs/peptides [64]. It can enhance the limit of detection (LOD) by almost an order of magnitude. In the malathion detection experiment, AgNPs served as amplification agents, achieving a LOD of 0.5 pM. Consequently, peptide-AgNPs may possess greater promise than peptide-AuNPs for colorimetric sensors aimed at detecting trace analytes in intricate environmental samples. However, silver nanoparticles exhibit higher toxicity and low photostability compared to AuNPs.

Peptides with different morphology (e.g., fibrils, spheres, and tubes) can be employed for the immobilization of a greater quantity of biomarkers in biosensing applications. For example, dipeptide diphenylalanine, cyclic peptides, and octapeptides are frequently used because of their capability to generate a highly ordered structure under basic assembly circumstances. Peptide nanotubes can be precisely positioned at specific locations using molecular recognition or dielectrophoresis and can be readily changed with various recognition components for multiplexing. The peptide architectures exhibit biocompatibility and can also stabilize antibodies and enzymes within biosensors. This indicates that peptides can function as a BRE molecule and can also be utilized in the building of a sensing matrix.

Several experimental key parameters should be considered in optical-based biosensors. The signal of optical techniques such as LSPR can be influenced by several operational aspects, including the morphology and dimensions of nanoparticles, interparticle spacing, ambient conditions, and the state of particles in solution (i.e., monodispersed or aggregated). Consequently, it is essential to meticulously plan experiments and sensor assemblies to achieve precise results. On the other side, electrochemical-based methods suffer from voltammetric baseline fluctuations and capacitance current drifts [65]. However, new methods based on the electrode and peptide modification can circumvent these drawbacks [65]. The selection of the suitable analytical method and the conditions of the experiments play an important role in the successful analysis.

14.9 Conclusion

Peptides offer several roles in sensors or biosensors. They can be used as recognition parts, selective probes, and for separation purposes. They offered several advantages such as high stability against denaturation, ease of acquisition, increased specificity, cost-effectiveness, high accessibility, ease of modification, versatility, and capability for chemical combination. Peptide-based nanomaterials for sensors or biosensor devices demonstrate exceptional characteristics, enabling mass production and prolonged preservation. They played a crucial role in the advancement of current biosensors in various domains, including clinical diagnostics, biosecurity, chemical/biological analysis, environmental monitoring, and health care. The interaction between peptide-based nanomaterials and the analyte can be measured using many sensor platforms, including optical-based biosensors (e.g., colorimetric, UV-Vis-NIR spectroscopy, SPR), electrochemical-based methods (e.g., voltammetry, amperometry, cyclic voltammetry, etc.), FET, quartz crystal microbalances, and surface plasmon resonance. However, significant challenges remain, including the detection of bacteria in real samples, enhanced sensitivity, and increased selectivity.

References

1. J. Wu, H. Liu, W. Chen, B. Ma, H. Ju, Device integration of electrochemical biosensors. Nat. Rev. Bioeng. **1**(5), 346–360 (2023). https://doi.org/10.1038/s44222-023-00032-w
2. H.N. Abdelhamid, M. Dowaidar, M. Hällbrink, Ü. Langel, Gene delivery using cell penetrating peptides-Zeolitic Imidazolate frameworks. Microporous Mesoporous Mater. **300**, 110173 (2020). https://doi.org/10.1016/j.micromeso.2020.110173
3. H.N. Abdelhamid, M. Dowaidar, Ü. Langel, Carbonized chitosan encapsulated hierarchical porous Zeolitic Imidazolate frameworks nanoparticles for gene delivery. Microporous Mesoporous Mater. **302**, 110200 (2020). https://doi.org/10.1016/j.micromeso.2020.110200
4. M. Dowaidar, H.N. Abdelhamid, M. Hällbrink, X. Zou, Ü. Langel, Graphene oxide nanosheets in complex with cell penetrating peptides for oligonucleotides delivery. Biochim. Biophys. Acta—Gen. Subj. **1861**(9), 2334–2341 (2017). https://doi.org/10.1016/j.bbagen.2017.07.002
5. M. Dowaidar, H. Nasser Abdelhamid, M. Hällbrink, Ü. Langel, X. Zou, Chitosan enhances gene delivery of oligonucleotide complexes with magnetic nanoparticles–cell-penetrating peptide. J. Biomater. Appl. **33**(3), 392–401 (2018). https://doi.org/10.1177/0885328218796623
6. M. Dowaidar, H.N. Abdelhamid, M. Hällbrink, K. Kurrikoff, K. Freimann, X. Zou, Ü. Langel, K. Kurrikoff, X. Zou, Ü. Langel, Magnetic nanoparticle assisted self-assembly of cell penetrating peptides-oligonucleotides complexes for gene delivery. Sci. Rep. **7**(1), 9159 (2017). https://doi.org/10.1038/s41598-017-09803-z
7. A. Lakshmanan, S. Zhang, C.A.E. Hauser, Short self-assembling peptides as building blocks for modern nanodevices. Trends Biotechnol. **30**(3), 155–165 (2012). https://doi.org/10.1016/j.tibtech.2011.11.001
8. S. Pavan, F. Berti, Short peptides as biosensor transducers. Anal. Bioanal. Chem. **402**(10), 3055–3070 (2012). https://doi.org/10.1007/s00216-011-5589-8
9. P. Colas, B. Cohen, T. Jessen, I. Grishina, J. McCoy, R. Brent, Genetic selection of peptide aptamers that recognize and inhibit cyclin-dependent kinase 2. Nature **380**(6574), 548–550 (1996). https://doi.org/10.1038/380548a0
10. B. Sequeira-Antunes, H.A. Ferreira, Nucleic acid aptamer-based biosensors: a review. Biomedicines **11**(12), 3201 (2023). https://doi.org/10.3390/biomedicines11123201
11. D. Zhu, H. Kong, Z. Sun, Y. Xu, P. Han, Y. Xi, G. Wei, Recent advance in tailoring the structure and functions of self-assembled peptide nanomaterials for biomedical applications. Coord. Chem. Rev. **494**, 215374 (2023). https://doi.org/10.1016/j.ccr.2023.215374
12. S. Wu, Z. Sun, Y. Peng, Y. Han, J. Li, S. Zhu, Y. Yin, G. Li, Peptide-functionalized metal-organic framework nanocomposite for ultrasensitive detection of secreted protein acidic and rich in cysteine with practical application. Biosens. Bioelectron. **169**, 112613 (2020). https://doi.org/10.1016/j.bios.2020.112613
13. L. Wang, H. Xie, Y. Lin, M. Wang, L. Sha, X. Yu, J. Yang, J. Zhao, G. Li, Covalent organic frameworks (COFs)-based biosensors for the assay of disease biomarkers with clinical applications. Biosens. Bioelectron. **217**, 114668 (2022). https://doi.org/10.1016/j.bios.2022.114668
14. Y. Li, H. Gao, Y. Jin, R. Zhao, Y. Huang, Peptide-derived coordination frameworks for biomimetic and selective separation. Anal. Bioanal. Chem. **415**(18), 4079–4092 (2023). https://doi.org/10.1007/s00216-023-04761-0

15. S. Wang, P. Zeng, X. Zhu, C. Lei, Y. Huang, Z. Nie, Chimeric peptides self-assembling on titanium carbide MXenes as biosensing interfaces for activity assay of post-translational modification enzymes. Anal. Chem. **92**(13), 8819–8826 (2020). https://doi.org/10.1021/acs.analchem.0c00243
16. C. Li, X. Chen, F. Zhang, X. He, G. Fang, J. Liu, S. Wang, Design of cyclic peptide based glucose receptors and their application in glucose sensing. Anal. Chem. **89**(19), 10431–10438 (2017). https://doi.org/10.1021/acs.analchem.7b02430
17. S.-J. Lau, T.P.A. Kruck, B. Sarkar, A peptide molecule mimicking the Copper(II) transport site of human serum albumin. J. Biol. Chem. **249**(18), 5878–5884 (1974). https://doi.org/10.1016/S0021-9258(20)79900-X
18. Q. Liu, J. Wang, B.J. Boyd, Peptide-based biosensors. Talanta **136**, 114–127 (2015). https://doi.org/10.1016/j.talanta.2014.12.020
19. X. Wu, J. Yin, J. Liu, Y. Gu, S. Wang, J. Wang, Colorimetric detection of glucose based on the binding specificity of a synthetic cyclic peptide. Analyst **145**(22), 7234–7241 (2020). https://doi.org/10.1039/D0AN00211A
20. G.K. Walkup, B. Imperiali, Fluorescent chemosensors for divalent zinc based on zinc finger domains. Enhanced oxidative stability, metal binding affinity, and structural and functional characterization. J. Am. Chem. Soc. **119**(15), 3443–3450 (1997). https://doi.org/10.1021/ja9642121
21. B. White, J. Holcombe, Fluorescent peptide sensor for the selective detection of Cu^{2+}. Talanta **71**(5), 2015–2020 (2007). https://doi.org/10.1016/j.talanta.2006.09.009
22. C.C. Thompson, R.Y. Lai, Threonine phosphorylation of an electrochemical peptide-based sensor to achieve improved uranyl ion binding affinity. Biosensors **12**(11), 961 (2022). https://doi.org/10.3390/bios12110961
23. H. Barzinmehr, S. Ramezanpour, P. Shiri, E. Meghrazi Ahadi, S. Mohammadi, F. Yazdian, P. Tavatoni, A review of fluorescent peptide-based chemosensors with selectivity for metal ions. Coord. Chem. Rev. **518**, 216055 (2024). https://doi.org/10.1016/j.ccr.2024.216055
24. V. Vanova, K. Mitrevska, V. Milosavljevic, D. Hynek, L. Richtera, V. Adam, Peptide-based electrochemical biosensors utilized for protein detection. Biosens. Bioelectron. **180**, 113087 (2021). https://doi.org/10.1016/j.bios.2021.113087
25. Y. Gu, Y. Hu, F. Zhang, L. Yi, Y. Shang, D. Ren, Z. Ge, Electrochemiluminescence sensor based on cyclic peptides-recognition and Au nanoparticles assisted graphitic carbon nitride for glucose determination. Microchim. Acta **188**(5), 151 (2021). https://doi.org/10.1007/s00604-021-04797-y
26. A.R. Reddi, M. Pawlowska, B.R. Gibney, Evaluation of the intrinsic Zn(II) affinity of a Cys 3 His 1 site in the absence of protein folding effects. Inorg. Chem. **54**(12), 5942–5948 (2015). https://doi.org/10.1021/acs.inorgchem.5b00718
27. K.L. Chan, I. Bakman, A.R. Marts, Y. Batir, T.L. Dowd, D.L. Tierney, B.R. Gibney, Characterization of the Zn(II) binding properties of the human Wilms' tumor suppressor protein C-terminal zinc finger peptide. Inorg. Chem. **53**(12), 6309–6320 (2014). https://doi.org/10.1021/ic500862b
28. Y. Li, L. Li, X. Pu, G. Ma, E. Wang, J. Kong, Z. Liu, Y. Liu, Synthesis of a ratiometric fluorescent peptide sensor for the highly selective detection of Cd^{2+}. Bioorg. Med. Chem. Lett. **22**(12), 4014–4017 (2012). https://doi.org/10.1016/j.bmcl.2012.04.088
29. J.-M. Kim, C.R. Lohani, L.N. Neupane, Y. Choi, K.-H. Lee, Highly sensitive turn-on detection of $Ag+$ in aqueous solution and live cells with a symmetric fluorescent peptide. Chem. Commun. **48**(24), 3012 (2012). https://doi.org/10.1039/c2cc16953c
30. B.R. White, H.M. Liljestrand, J.A. Holcombe, A 'Turn-on' FRET peptide sensor based on the mercury Bindingprotein MerP. Analyst **133**(1), 65–70 (2008). https://doi.org/10.1039/B711777A
31. Y. Cao, L. Zhou, Z. Fang, Z. Zou, J. Zhao, X. Zuo, G. Li, Application of functional peptides in the electrochemical and optical biosensing of cancer biomarkers. Chem. Commun. **59**(23), 3383–3398 (2023). https://doi.org/10.1039/D2CC06824A

32. G.A.R.Y. Suaifan, S. Alhogail, M. Zourob, Paper-based magnetic nanoparticle-peptide probe for rapid and quantitative colorimetric detection of Escherichia coli O157:H7. Biosens. Bioelectron. **92**, 702–708 (2017). https://doi.org/10.1016/j.bios.2016.10.023
33. S. Sousa, A. Castro, J.M. Correia da Costa, E. Pereira, Biosensor based immunoassay: a new approach for serotyping of Toxoplasma gondii. Nanomaterials **11**(8), 2065 (2021). https://doi.org/10.3390/nano11082065
34. S. Alhogail, G.A.R.Y. Suaifan, F.J. Bikker, W.E. Kaman, K. Weber, D. Cialla-May, J. Popp, M.M. Zourob, Rapid colorimetric detection of Pseudomonas aeruginosa in clinical isolates using a magnetic nanoparticle biosensor. ACS Omega **4**(26), 21684–21688 (2019). https://doi.org/10.1021/acsomega.9b02080
35. Q. Zhang, D. Zhang, Y. Lu, Y. Yao, S. Li, Q. Liu, Graphene oxide-based optical biosensor functionalized with peptides for explosive detection. Biosens. Bioelectron. **68**, 494–499 (2015). https://doi.org/10.1016/j.bios.2015.01.040
36. J. Wang, Near infrared optical biosensor based on peptide functionalized single-walled carbon nanotubes hybrids for 2,4,6-trinitrotoluene (TNT) explosive detection. Anal. Biochem. **550**, 49–53 (2018). https://doi.org/10.1016/j.ab.2018.04.011
37. A. Karimzadeh, M. Hasanzadeh, N. Shadjou, M. de la Guardia, Peptide based biosensors. TrAC Trends Anal. Chem. **107**, 1–20 (2018). https://doi.org/10.1016/j.trac.2018.07.018
38. K.J. Oh, K.J. Cash, K.W. Plaxco, Excimer-based peptide beacons: a convenient experimental approach for monitoring polypeptide−protein and polypeptide−oligonucleotide interactions. J. Am. Chem. Soc. **128**(43), 14018–14019 (2006). https://doi.org/10.1021/ja0651310
39. K.J. Oh, K.J. Cash, V. Hugenberg, K.W. Plaxco, Peptide beacons: a new design for polypeptide-based optical biosensors. Bioconjug. Chem. **18**(3), 607–609 (2007). https://doi.org/10.1021/bc060319u
40. K. Enander, G.T. Dolphin, B. Liedberg, I. Lundström, L. Baltzer, A versatile polypeptide platform for integrated recognition and reporting: affinity arrays for protein–ligand interaction analysis. Chem. A Eur. J. **10**(10), 2375–2385 (2004). https://doi.org/10.1002/chem.200305391
41. K. Enander, G.T. Dolphin, L. Baltzer, Designed, functionalized helix−loop−helix motifs that bind human carbonic anhydrase II: a new class of synthetic receptor molecules. J. Am. Chem. Soc. **126**(14), 4464–4465 (2004). https://doi.org/10.1021/ja038799c
42. B. Feng, L. Guo, L. Wang, F. Li, J. Lu, J. Gao, C. Fan, Q. Huang, A graphene oxide-based fluorescent biosensor for the analysis of peptide-receptor interactions and imaging in somatostatin receptor subtype 2 overexpressed tumor cells. Anal. Chem. **85**(16), 7732–7737 (2013). https://doi.org/10.1021/ac4009463
43. X. Li, Y. Li, Q. Qiu, Q. Wen, Q. Zhang, W. Yang, L. Yuwen, L. Weng, L. Wang, Efficient biofunctionalization of MoS_2 nanosheets with peptides as intracellular fluorescent biosensor for sensitive detection of caspase-3 activity. J. Coll. Interface Sci. **543**, 96–105 (2019). https://doi.org/10.1016/j.jcis.2019.02.011
44. L. Yang, N. Li, K. Wang, X. Hai, J. Liu, F. Dang, A novel peptide/Fe_3O_4@SiO_2-Au nanocomposite-based fluorescence biosensor for the highly selective and sensitive detection of prostate-specific antigen. Talanta **179**, 531–537 (2018). https://doi.org/10.1016/j.talanta.2017.11.033
45. Z. Sun, S. Wu, Y. Peng, M. Wang, M. Jalalah, M.S. Al-Assiri, F.A. Harraz, J. Yang, G. Li, Sensor array for rapid pathogens identification fabricated with peptide-conjugated 2D metal-organic framework nanosheets. Chem. Eng. J. **405**, 126707 (2021). https://doi.org/10.1016/j.cej.2020.126707
46. J. Li, Y. Shu, C. Li, Z. Jiang, Highly catalytic nanoenzyme of covalent organic framework loaded starch-surface-enhanced Raman scattering/absorption bi-mode peptide as biosensor for ultratrace determination of cadmium. Front. Nutr. **9** (2023). https://doi.org/10.3389/fnut.2022.1075296.
47. J.M. Lim, N.S. Heo, S.Y. Oh, M.Y. Ryu, J.H. Seo, T.J. Park, Y.S. Huh, J.P. Park, Selection of affinity peptides for interference-free detection of cholera toxin. Biosens. Bioelectron. **99**, 289–295 (2018). https://doi.org/10.1016/j.bios.2017.07.075

48. D. Zhang, Q. Zhang, Y. Lu, Y. Yao, S. Li, J. Jiang, G.L. Liu, Q. Liu, Peptide functionalized nanoplasmonic sensor for explosive detection. Nano-Micro Lett. **8**(1), 36–43 (2016). https://doi.org/10.1007/s40820-015-0059-z
49. N.S. Heo, S.Y. Oh, M.Y. Ryu, S.H. Baek, T.J. Park, C. Choi, Y.S. Huh, J.P. Park, Affinity peptide-guided plasmonic biosensor for detection of noroviral protein and human norovirus. Biotechnol. Bioprocess Eng. **24**(2), 318–325 (2019). https://doi.org/10.1007/s12257-018-0410-6
50. T. Endo, K. Kerman, N. Nagatani, Y. Takamura, E. Tamiya, Label-free detection of peptide nucleic acid−DNA hybridization using localized surface plasmon resonance based optical biosensor. Anal. Chem. **77**(21), 6976–6984 (2005). https://doi.org/10.1021/ac0513459
51. A.D. Chowdhury, F. Nasrin, R. Gangopadhyay, A.B. Ganganboina, K. Takemura, I. Kozaki, H. Honda, T. Hara, F. Abe, S. Park, T. Suzuki, E.Y. Park, Controlling distance, size and concentration of nanoconjugates for optimized LSPR based biosensors. Biosens. Bioelectron. **170**, 112657 (2020). https://doi.org/10.1016/j.bios.2020.112657
52. E. González-Fernández, M. Staderini, N. Avlonitis, A.F. Murray, A.R. Mount, M. Bradley, Effect of spacer length on the performance of peptide-based electrochemical biosensors for protease detection. Sens. Act. B Chem. **255**, 3040–3046 (2018). https://doi.org/10.1016/j.snb.2017.09.128
53. G. Wang, R. Han, X. Su, Y. Li, G. Xu, X. Luo, Zwitterionic peptide anchored to conducting polymer PEDOT for the development of antifouling and ultrasensitive electrochemical DNA sensor. Biosens. Bioelectron. **92**, 396–401 (2017). https://doi.org/10.1016/j.bios.2016.10.088
54. B. Purohit, W. E. Svendsen, Peptide-based electrochemical nanobiosensors for clinical diagnosis, in *Surface Engineering and Functional Nanomaterials for Point-of-Care Analytical Devices* (Springer Nature Singapore, Singapore, 2023), pp. 199–222. https://doi.org/10.1007/978-981-99-3025-8_9.
55. V.R. Rodovalho, G.R. Araujo, E.R. Vaz, C. Ueira-Vieira, L.R. Goulart, J.M. Madurro, A.G. Brito-Madurro, Peptide-based electrochemical biosensor for juvenile idiopathic arthritis detection. Biosens. Bioelectron. **100**, 577–582 (2018). https://doi.org/10.1016/j.bios.2017.10.012
56. A.J. Zaitouna, A.J. Maben, R.Y. Lai, Incorporation of extra amino acids in peptide recognition probe to improve specificity and selectivity of an electrochemical peptide-based sensor. Anal. Chim. Acta **886**, 157–164 (2015). https://doi.org/10.1016/j.aca.2015.05.037
57. H.J. Hwang, M.Y. Ryu, C.Y. Park, J. Ahn, H.G. Park, C. Choi, S.-D. Ha, T.J. Park, J.P. Park, High sensitive and selective electrochemical biosensor: label-free detection of human norovirus using affinity peptide as molecular binder. Biosens. Bioelectron. **87**, 164–170 (2017). https://doi.org/10.1016/j.bios.2016.08.031
58. J.M. Lim, J.H. Kim, M.Y. Ryu, C.H. Cho, T.J. Park, J.P. Park, An electrochemical peptide sensor for detection of dengue fever biomarker NS1. Anal. Chim. Acta **1026**, 109–116 (2018). https://doi.org/10.1016/j.aca.2018.04.005
59. J. Adjémian, A. Anne, G. Cauet, C. Demaille, Cleavage-sensing redox peptide monolayers for the rapid measurement of the proteolytic activity of trypsin and α-thrombin enzymes. Langmuir **26**(12), 10347–10356 (2010). https://doi.org/10.1021/la100397g
60. J.H. Kim, J.H. Shin, B. Park, C.H. Cho, Y.S. Huh, C.-H. Choi, J.P. Park, Harnessing protein sensing ability of electrochemical biosensors via a controlled peptide receptor-electrode interface. J. Nanobiotechnol. **21**(1), 100 (2023). https://doi.org/10.1186/s12951-023-01843-0
61. H. Liu, S. Ma, G. Ning, R. Zhang, H. Liang, F. Liu, L. Xiao, L. Guo, Y. Zhang, C.-P. Li, H. Zhao, A "peptide-target-aptamer" electrochemical biosensor for norovirus detection using a black phosphorous nanosheet@Ti_3C_2-Mxene nanohybrid and magnetic covalent organic framework. Talanta **258**, 124433 (2023). https://doi.org/10.1016/j.talanta.2023.124433
62. X. Bi, A. Agarwal, N. Balasubramanian, K.-L. Yang, Tripeptide-modified silicon nanowire based field-effect transistors as real-time copper ion sensors. Electrochem. Commun. **10**(12), 1868–1871 (2008). https://doi.org/10.1016/j.elecom.2008.09.027
63. Y.-M. Xu, H.-Q. Pan, S.-H. Wu, B.-L. Zhang, Interaction between tripeptide Gly-Gly-His and Cu^{2+} probed by microcantilevers. Chin. J. Anal. Chem. **37**(6), 783–787 (2009). https://doi.org/10.1016/S1872-2040(08)60106-1

64. M. Retout, B. Gosselin, A. Mattiuzzi, I. Ternad, I. Jabin, G. Bruylants, Peptide-conjugated silver nanoparticles for the colorimetric detection of the oncoprotein Mdm2 in human serum. Chempluschem. **87**(4) (2022). https://doi.org/10.1002/cplu.202100450
65. M. Puiu, A. Idili, D. Moscone, F. Ricci, C. Bala, A modular electrochemical peptide-based sensor for antibody detection. Chem. Commun. **50**(64), 8962 (2014). https://doi.org/10.1039/C4CC02858A

The manufacturer's authorised representative in the EU is Springer Nature Customer Service Centre GmbH, Europaplatz 3, 69115 Heidelberg, Germany. If you have any concerns regarding our products, please contact ProductSafety@springernature.com

Printed and bound by CPI Group (UK) Ltd, Croydon, CR0 4YY
01/10/2026
02225926-0007